GENOMES OF
PLANTS AND ANIMALS

STADLER GENETICS SYMPOSIA SERIES

Library of Congress Cataloging-in-Publication Data

Stadler Genetics Symposium (21st : 1995 : University of Missouri-
-Columbia)
     Genomes of plants and animals : 21st Stadler Genetics Symposium /
edited by J. Perry Gustafson and R.B. Flavell.
        p.   cm. -- (Stadler genetics symposia series)
     "Proceedings of the Twenty-First Genetics Symposium, Genomes of
plants and animals, held May 22-24, 1995, at the University of
Missouri, Columbia, Missouri"--T.p. verso.
     Includes bibliographical references and index.
     ISBN 0-306-45372-X
     1. Genomes--Congresses. 2. Plant genetics--Congresses. 3. Gene
mapping--Congresses.   I. Gustafson, J. P. II. Flavell, R. B.
(Richard B.) III. Title. IV. Series.
     QH447.S695 1995
     574.87'3282--dc20   96-21922
                                                              CIP

Proceedings of the Twenty-first Stadler Genetics Symposium, Genomes of Plants and Animals, held May 22–24, 1995, at the University of Missouri, Columbia, Missouri

ISBN 0-306-45372-X

© 1996 Plenum Press, New York
A Division of Plenum Publishing Corporation
233 Spring Street, New York, N. Y. 10013

10 9 8 7 6 5 4 3 2 1

GENOMES OF PLANTS AND ANIMALS

21st Stadler Genetics Symposium

Edited by

J. Perry Gustafson

USDA–ARS
University of Missouri
Columbia, Missouri

and

R. B. Flavell

John Innes Centre
Norwich, England

PLENUM PRESS • NEW YORK AND LONDON

ACKNOWLEDGMENT

The editors gratefully acknowledge the generous support of the following contributors: College of Agriculture, Food, and Natural Resources, Department of Agronomy, Plant Science Unit, Division of Biological Sciences, Plant Interdisciplinary Group, Molecular Biology Program, Genetics Area Program, Graduate School, School of Veterinary Medicine, and School of Medicine, University of Missouri–Columbia; United States Department of Agriculture–Agricultural Research Service; Anheuser–Busch Companies, Inc.; Northrup King Company; and Monsanto Company, who made the 21st Stadler Genetics Symposium a success.

The speakers, who spent a tremendous amount of time preparing their manuscripts and lectures, are gratefully acknowledged. Without their expertise and dedication, the symposium could not have taken place.

We wish to thank the local chairpersons for their efforts to see that everyone in the respective sessions was well taken care of during the symposium.

The behind-the-scene and on-site preparations were excellently handled by Joy Williams from Conferences and Specialized Courses, University of Missouri, who tirelessly handled all of our peculiar requirements and made sure everything was well organized.

A special thanks goes to Judy Spillman, USDA–ARS, and Jeni Fox, John Innes Centre, for their excellent secretarial help in handling all of the correspondence as well as keeping both of us organized.

J.P. Gustafson
Columbia, Missouri

R.B. Flavell
Norwich, England

CONTENTS

ASSESSING GENETIC DIVERSITY IN PLANTS WITH SYNTHETIC TANDEM REPETITIVE DNA PROBES

Steven H. Rogstad

Department of Biological Sciences ML6
University of Cincinnati
Cincinnati, OH 45221-0006
USA

INTRODUCTION

The palette of molecular techniques available to population biologists is extensive (Avise, 1994), and is growing richer at a rapid pace. Although all of these techniques offer various perspectives in population biology investigations, no one of them is optimally applicable in all situations, and two different molecular techniques employed to examine the same question may produce different answers (e.g., allozymes versus DNA data; see Avise, 1994; Mitton, 1994). Such differences in informativeness among different molecular approaches most likely arise from the different evolutionary rates of, and forces upon, different portions of genomes. Discrepancies should not necessarily prompt the acceptance of one technique over another, but rather, potentially offer new insights into evolutionary processes. In fact, given that such discrepancies have commonly been found, it seems reasonable that investigations of population biology issues would profit from being examined from at least two independent, relevant approaches to seek compatible interpretations.

Among techniques that utilize more rapidly evolving segments of genomes are those that explore variation at variable number tandem repeat (VNTR) loci (Nakamura et al., 1987). VNTR loci as currently observed are usually delimited either by restriction endonuclease recognition sites or polymerase chain reaction (PCR) priming sequences that flank a set of tandem repeats of a "core" sequence. Core sequences commonly investigated range in size from

2-20 nucleotides, and alleles at a locus differ primarily in the number of repeats of the core sequence. It is often (but not always) the case that loci with tandem repetitive sequences have several different alleles available among individuals in populations, resulting in high heterozygosities for such loci.

Alleles at VNTR loci can thus often be included among the most variable genetic markers yet known, implying that mutation rates should be relatively high at such loci. While an extremely variable locus in humans has been shown to have an extremely high mutation rate (Jeffreys et al., 1988), and surely similar loci will be found in other organisms, in general, transmission of VNTR markers has been shown to primarily follow Mendelian inheritance (e.g., Dallas, 1988; Prodohl et al., 1994; Rogstad, 1994a; Arens et al., 1995), with a level of meiotic mutation that is often detectable and high relative to other types of genetic markers (e.g., allozymes, most single-copy RFLP probes). For example, three of 595 VNTR bands (0.5%) transmitted to offspring were non-parental in *Brassica rapa* L. (turnip; Brassicaceae; Rogstad 1994a), while mutation frequencies of 0.001 for human CAC VNTR marker transmission (Nurnberg et al., 1991) and less than 0.0003 for GATA VNTR marker transmission to the F_2 population from a *Lycopersicum esculentum* Mill. X *L. pennellii* Mill. (tomato; Solanaceae) cross (Arens et al., 1995) have been detected. Although VNTR loci mutation rates may be relatively high, they usually are not so high that many types of analyses (e.g., see below) would be rendered meaningless. In fact, it is this relatively elevated rate of mutation that provides the high degree of variation that makes VNTR markers so useful for linkage mapping, population variation and inbreeding studies, clonal differentiation analysis, chromosome identification via *in situ* VNTR probe hybridization, pedigree determination, phylogeny reconstruction, and cell, organ, and cultivar identification (see the extensive references in Amos and Pemberton, 1993; Avise, 1994; Weising et al., 1995). In the next sections, I will both review some recent developments in the application of synthetic probes for VNTR loci, and also demonstrate the use of these probes to examine aspects of breeding systems in plants.

THE USE OF SYNTHETIC PROBES FOR VNTR LOCI IN PLANTS

Numerous kinds of core sequences for VNTR loci have been identified. Core sequences range in size from two to greater than 100 basepairs (bp; see references in Rogstad, 1994a), and also have a wide range of composition in terms of relative proportions and placement of the nucleotides. Although some studies have demonstrated that different core sequences are represented to different degrees in some genomes (e.g., Benson and Waterman, 1994; Borstnik et al., 1994; Rothuizen et al., 1994), general patterns informative about which core sequences are most universally applicable in detecting variation have not yet emerged (and may never).

In fact, several studies have demonstrated that repeats of two to 14 randomly chosen nucleotides (e.g., Ali et al., 1986; Litt and Luty, 1989; Vergnaud 1989; Weber and May, 1989; Rogstad, 1994a) can be used to reveal variation at VNTR loci. One method that has been developed by which numerous probes can be produced to seek variation at different VNTR loci is to use the polymerase chain reaction to create synthesized tandem repeats (STR's). To summarize this method, first a short oligomer (the 20-30 bp template) of tandem repeats of a selected core sequence is created with a DNA synthesizer. A second sequence (the 20-30 bp complement) composed of repeats of the complementary core sequence is also synthesized. Equimolar mixes of the template and complement sequences are then subjected to several, more or less standard, PCR reaction cycles. During the annealing step of each cycle, some of the template and complement strands will anneal out of register in a manner (with 5' overhangs) permitting extension at internal 3' ends. This extension produces new longer strands composed of core repeats. Over several cycles, longer and longer strands are created (this can be visualized on ethidium bromide gels by comparing reactions with varying numbers of cycles). For each core sequence (template-complement combination), a series of reactions using different cycling conditions (Mg concentrations; annealing temperatures) are explored to determine the most stringent conditions under which extension will take place. These optimal conditions for a particular template-complement combination need only be determined once, after which they can be used in all subsequent reactions.

To generate a PCR-STR probe, the optimal conditions are used in a reaction that evenly produces strands over a broad size range, and strands of 300-700 bp (determined with known size standards in adjacent lanes) are then excised from an agarose gel, the DNA being recovered after digestion of the agarose with agarase. A fraction of the recovered DNA is then used in a one-sided PCR reaction along with the proper oligomer that will give the greatest incorporation of a radioactive nucleotide (the corresponding non-radioactive nucleotide is not included in the PCR nucleotide mix). The resultant labelled PCR-STR probe is used in more or less standard hybridization protocols with Southern blots. Using longer repetitive DNA probes may yield more stable, clearer results than obtained under similar conditions with end-labelled, shorter oligomer repeats (e.g., Arens et al., 1995; personal observations). More details of the creation and use of PCR-STR probes are given in Rogstad (1993, 1994a).

For each new species, an array of PCR-STR probes can be used with test blots of digested genomic DNA from at least two individuals to determine which probes reveal the clearest markers with the desired degree of variation for the biological question to be investigated. For example, in population differentiation studies, markers with a high degree of intra- or interpopulation variation might be required, whereas investigations of the occurrence of hybrid offspring might profit from VNTR markers that are fixed in, but differ between, the putative parental species. In tests with over 20 species of plants spanning a wide taxonomic range, we have found intraspecific variation in all with at least one PCR-STR probe. Some cores seem more widely applicable than others. For example, GATA and GACA (Ali et al., 1986; Epplen, 1988) PCR-STR probes often produce clear, variable bands. Surveying several species, core sequences based on the stop codon (TAA, TAAA, and TAAG; Brown et al., 1990) and core sequences that are self-hybridizing (e.g., GATC, GTAC, etc.) have performed poorly in almost all species surveyed. Other core sequences (listed in Table 1 of Rogstad 1994a) appear to be more variable in the degree to which VNTR markers are revealed across species.

An example of the above noted discrepancies that become apparent when applying two different molecular approaches to examining population biology characteristics is the difference in genetic variation detected with allozymes compared to VNTR markers. In summaries of allozyme diversity in plant species, Hamrick and Murawski (1991) report a mean H (heterozygosity) for 468 plant taxa of 0.113, and a mean H for long-lived woody perennials of 0.149. Hamrick and Godt (1990) calculate a mean total genetic diversity (H_T) for 406 plant taxa of 0.310. One of the highest heterozygosities detected with allozymes for any plant species is H=0.481 reported for *Opuntia basilaris* Engelm. & Big. (Cactaceae; Table 5 in Hamrick et al., 1979).

Algorithms for estimating H from VNTR marker variation in populations have been introduced (Stephens et al., 1992; Jin and Chakraborty, 1993), and these have been incorporated into a computer program, GELSTATS (manuscript in prep.), that calculates these estimates, as well as other summary statistics for population analyses with VNTR loci. It is difficult to directly compare VNTR and allozyme loci investigations. For example, sampling design may not be comparable and different loci are probably examined, but such problems often exist also for comparison of allozyme studies. Still, the following comparisons serve to demonstrate differences in the two approaches. In all of the following reports of population heterozygosities (H) based on VNTR population band frequencies, GELSTATS was used to provide an estimated H (Jin and Chakraborty, 1993) using data sets in which redundant examples of possibly completely linked bands (reported by GELSTATS) have been removed.

Jelinski and Cheliak (1992) used allozymes to detect an H_T=0.31 for six aspen (*Populus tremuloides* Michaux; Salicaceae) populations, with H ranging from 0.28 to 0.35. An analysis of 21 population VNTR markers revealed with the M13 probe (data from Rogstad et al., 1991) from individuals sampled from eight locations separated to ensure that a clone was not re-sampled yields an estimated H=0.760. Allozyme variation yielded estimated H values of 0.26 for Ohio buckeye and 0.23 for yellow buckeye (*Aesculus glabra* Willd. and *A. flava* Solander, respectively; Hippocastanceae; DePamphilis, 1988). Three PCR-STR probes (cores GACA, GATC, GCTGGTGG) were used to probe 22 individuals sampled from each of three populations of *A. glabra* and one population of 44 individuals of *A. flava*. Average estimated heterozygosities of H=0.463 (s.d.=0.057 for 42 probings of resampling gels) for *A. glabra* and H=0.497 (s.d.=0.041 for 12 probings of resampling gels) for *A. flava* were obtained. The estimated heterozygosity average for four populations (22 individuals sampled per population) of spicebush (*Lindera benzoin* (L.) Blume; Lauraceae) investigated with the GATA PCR-STR probe was H=0.456 (s.d.=0.098 for 18 probings of resampling gels). Similar studies on four populations (22 individuals each) of *Rubus moluccanus* L. (a Philippine bramble; Rosaceae) probed with the GATA and GACA PCR-STR probes yielded an average estimated H=0.627 (s.d.=0.106 for 48 probings of resampling gels). The heterozygosity estimates from VNTR analyses provided above are data summaries that form parts of larger collaborative studies (see acknowledgements; details to be submitted elsewhere), and are presented here merely to support what has been previously advocated, namely, that variation measured across VNTR loci in plants exceeds that usually found with allozymes. A similar finding has been found for a wild yam species (*Dioscorea tokoro* Makino.; Dioscoreaceae) when surveyed at six microsatellite loci (Terauchi and Konuma, 1994) yielding an observed H=0.54 compared to H=0.23 with allozymes. These findings are also consistent with what has been found for animal populations (e.g., see Scribner et al., 1994).

VNTR MARKERS TO DETECT LEVELS OF OUTCROSSING IN PLANTS

The considerable variation found at many VNTR loci provides a new tool for investigating rates of inbreeding versus outcrossing in plants. By comparing the VNTR markers in offspring with markers present in their pistillate ("maternal") parent, the number of those offspring produced via outcrossing should often be discernible as follows.

First, all offspring that are the products of selfing by the pistillate parent should have some subset (including the complete set) of markers found in the pistillate parent. Second, any marker that is present in an offspring that is not present in its pistillate parent must either have been contributed from a staminate ("paternal") parent (a paternal allele not found in the pistillate plant), or be the product of mutation. As noted above, mutation rates across all loci are usually low enough that new, mutant markers in the pistillate parent would occur at low frequencies (e.g., approximately 1 band in 200 transmitted in the turnip study noted above (Rogstad, 1994a)). The rate of occurrence of mutant bands, if known or estimated, could be mathematically accounted for in analyses of outbreeding.

Consider a VNTR marker that occurs in a staminate plant that does not occur in a pistillate plant that is, along with its offspring, under consideration on a Southern blot. If that marker is homozygous in the staminate plant, then it will be transmitted to all of its offspring, and thus, for the given Southern blot, any offspring sired by that staminate individual will possess a marker not found in the pistillate individual. Detection of outcrossing for such crosses will be 100%, and only one marker of this type is required.

However, if a marker heterozygous in the staminate individual distinguishes it from that pistillate individual, then there is a 50% probability of transmission of that marker to offspring, and thus a 50% chance of detecting outcrossing between these two parents for that one marker. As the staminate individual differs at more and more markers from the pistillate individual and is heterozygous for those markers, the probability (P) of detecting outcrossing (at least one non-pistillate, distinguishing marker transmitted) increases according to the equation, $P = 1 - 0.5^n$, where n is equal to the number of such unlinked heterozygous markers in the staminate parent (modified Bernoulli trial; Batschelet, 1979). For example, in a cross between a staminate parent that is distinguished by only two heterozygous markers from the pistillate parent, the probability of detecting outcrossing among the offspring is 0.75. If just seven differentiating heterozygous markers occur in the staminate individual, P exceeds 99%.

For most plant species examined thus far, at least 30-40 markers per individual are detectable with three to six different PCR-STR probes. Given the high variation in banding patterns among individuals of most populations investigated, it should not be difficult to detect 8 or more bands by which potential staminate individuals differ from a given pistillate individual. Typically, we have found that the average dissimilarity ($D = 1 - ((2 * B_{XY})/(B_X + B_Y))$, where B_{XY} is the number of bands shared between individuals X and Y, and B_X and B_Y are the total number of bands in individuals X and Y, respectively) between random individuals is in the neighborhood of 0.4 to 0.6. Choosing the more conservative numbers, if typically, 40% of the 20 markers per individual differ from another individual, then eight markers should distinguish individuals, and the probability of detecting outcrossing between most individuals is approximately 100%. Again, if any one of those distinguishing markers is homozygous in the staminate parent, then the probability of detecting outcrossing is automatically 100%.

Considering the nine PCR-STR probes that revealed VNTR markers in an investigation of transmission of VNTR markers to offspring of a controlled cross in turnip, 17 markers heterozygous in and distinguishing the staminate individual, and 24 markers heterozygous in and distinguishing the pistillate individual were found (Rogstad 1994a). More importantly for the issue at hand, four cases were found in each parent of homozygous VNTR markers that distinguished them from the other parent. In species or populations where markers differentiate individuals to this degree, detection of rates of outcrossing will usually be possible by comparing VNTR markers in offspring with their pistillate parent.

DETECTING OUTCROSSING: AN EXAMPLE FROM THE TROPICS

Early theorists studying tropical lowland rain forest (tlrf) tree populations predicted that levels of inbreeding should be high based on two independent lines of reasoning (see references in Rogstad, 1994b): 1) a correlate of the high species diversity found in tlrf's is that most species have very low population densities, and thus pollen transfer between distant, isolated individuals must often be difficult; and 2) individuals of a population may often flower out of synchrony with one another due to the relative climatic constancy of the tropics which provides few cues of sufficient intensity and unambiguity to induce synchronized population flowering across microhabitats. However, experimental observations and allozyme investigations indicate that the majority of tlrf tree species studied are predominantly outbreeding (see references in Rogstad, 1994b).

Observations of flowering in *Polyalthia lateriflora* (Bl.) King (Annonaceae), a main canopy tree at Pasoh Forest Reserve, Negeri Sembilan, W. Malaysia, suggest that, unlike some other species in the genus (Rogstad, 1994b), geitonogamy may be possible due to overlap in timing of staminate and carpellate function among flowers on one tree (personal observations). The flowers of *P. lateriflora* have an odor similar to rotting fruit, and are visited by a several species of beetles and flies. One mature seed produced from each of 30 flowers was collected from one pistillate tree of *P. lateriflora*. Eleven seed were destroyed, apparently by boring insects as determined from the presence of exit holes. The remaining 19 seeds germinated, and tissue from the seedlings, along with tissue from the pistillate tree and from the nearest mature conspecific individual (7 m distant) was preserved and exracted for DNA according to Rogstad (1992). These separate DNA samples (5 μg each) were digested with *Hae*III, electrophoresed, Southern blotted, and probed with PCR-STR probes according to the protocols in Rogstad (1993, 1994a).

Fragment profiles revealed with a GACA PCR-STR probe for the pistillate parent and its offspring, and the adjacent mature individual are shown in Figure 1. Note that all of the offspring have markers not present in the pistillate plant, and several have markers not attributable to either the pistillate plant or the adjacent mature individual, indicating that all seeds are products of outcrossing, often with trees at least 100 m away (the nearest trees other than the adjacent individual). Further support for a conclusion of 100% outcrossing for the 19 offspring of *P. lateriflora* comes from employing 7 other PCR-STR probes in re-probings of the same filter (cores = CAC, CACTCC, GATA, GATGTGGG; GCAC, TTCCA, and the M13 probe; Rogstad 1994a), in which bands not present in the pistillate parent were revealed in all offspring with at least three, and up to six, probes.

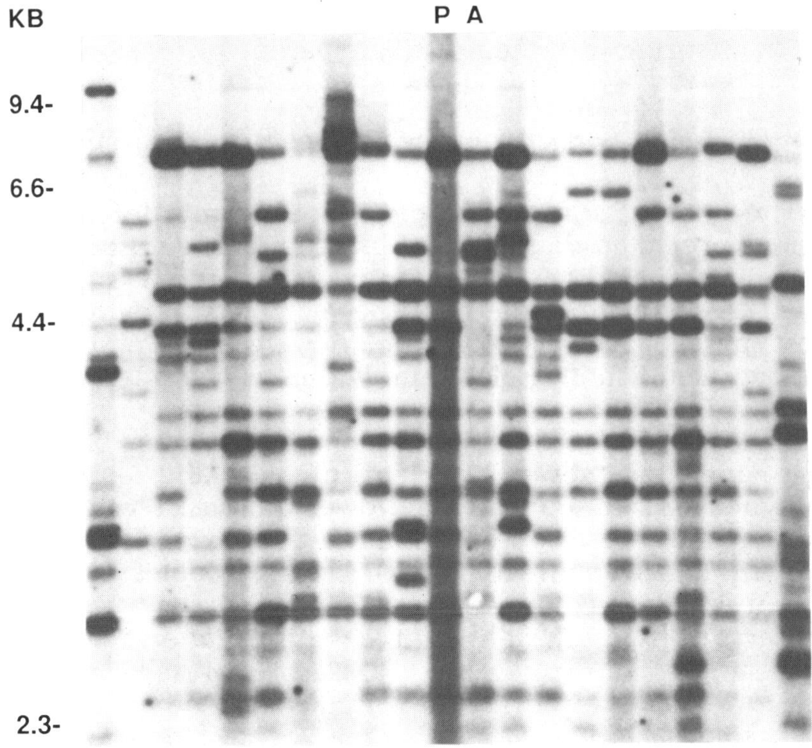

Figure 1. 100% outcrossing in *Polyalthia lateriflora* revealed with the GACA PCR-STR probe. P=pistillate individual; A=the only adjacent mature individual (7 m distant; other mature individuals were at least 100 m distant). All other lanes are offspring of the pistillate individual. Note that all offspring have at least one marker not found in the pistillate individual. KB=kilobase pairs.

Although it has been suggested that agamospermy may be widely represented in tlrf trees (Kaur et al., 1978), no evidence for this in *P. laterifolia* was found (Figure 1; and see next section).

DISTINGUISHING BETWEEN SELFING AND APOMIXIS

It is not unusual for plants to reproduce apomictically (asexually) in a variety of ways, the main two means being either through vegetative reproduction or through agamospermy (asexual production of embryos and seeds). Vegetative apomixis is common, and it has been demonstrated that VNTR markers can be used to reveal the clonal nature of plant populations (e.g., Rogstad et al., 1991; Neuhaus et al., 1993).

All of the forms of agamospermy (e.g., see Richards, 1986) produce offspring that are genetically identical to the parent. Note, however, that with selfing, it is possible in some cases for an offspring to also appear to be genetically identical to the pistillate plant when examined with VNTR markers. For example, if all of the markers revealed with a particular probe in the pistillate plant are homozygous, then all of those markers will be transmitted to all of the offspring. Thus, distinguishing offspring produced via agamospermy versus via selfing depends on the occurrence of at least some loci that are heterozygous in the pistillate plant.

But how many heterozygous loci are needed in the "maternal" plant to "ensure" that selfing can be distinguished from agamospermy? In this case, the probability (P) that all bands heterozygous in a maternal plant will be detectable in offspring produced by selfing is P = 0.75^n (both homozygous and heterozygous cases detected), where n is the number of those heterozygous alleles. If eleven heterozygous bands are present in the pistillate plant, the probability that all eleven will be detectably transmitted to an offspring are 0.042, and thus the chance that at least one band will not be transmitted (agamospermy ruled out and selfing detected) is 0.958 (1-P). From another perspective, considering one band that is heterozygous in the maternal plant, the probability of detectable transmission to n offspring produced via selfing is 0.75^n, while the probability of detectable transmission via agamospermy (barring mutation) is 100%. Attempts to distinguish between selfing and agamospermy via maternal parent-offspring marker comparisons will thus benefit from large offspring arrays.

DEMONSTRATION OF AGAMOSPERMY IN TARAXACUM

The common dandelion, *Taraxacum officinale* Wiggers (sensu lato; Asteraceae), typically produces offspring via agamospermy (parthenogenesis involving unreduced eggs (Nogler, 1984)), and thus, unless mutation or autosegregation occur, all offspring should be genetically identical to their pistillate, "maternal" plant. While normally triploid, a diploid sexual form of *T. officinale* has been described (Grant, 1981), and in such cases, offspring may not be genetically identical with their maternal parent.

As part of an investigation of genetic variation in *T. officinale*, the use of VNTR marker transmission to offspring was examined as a means of detecting apomixis. Seeds were collected from individual plants along with a leaf tissue sample of the maternal, seed-producing individual. Seeds were germinated and grown in sterilized soil in an enclosed greenhouse until harvesting for DNA extraction. Separate DNA samples (5 ug each) were digested with *Taq*I, electrophoresed, Southern blotted, and probed with PCR-STR probes according to the protocols in Rogstad (1993, 1994a). Figure 2 shows the results of one trial in which VNTR marker profiles are revealed with the PCR-STR GATGTGGG probe in a dandelion maternal parent and 21 of its offspring. At least 18 bands can be seen to be faithfully transmitted to all offspring in this autoradiograph.

In examining VNTR variation in this species, we have found extensive variation among individuals (in accordance with what has been found with morphology and allozymes (Richards, 1986)), and also

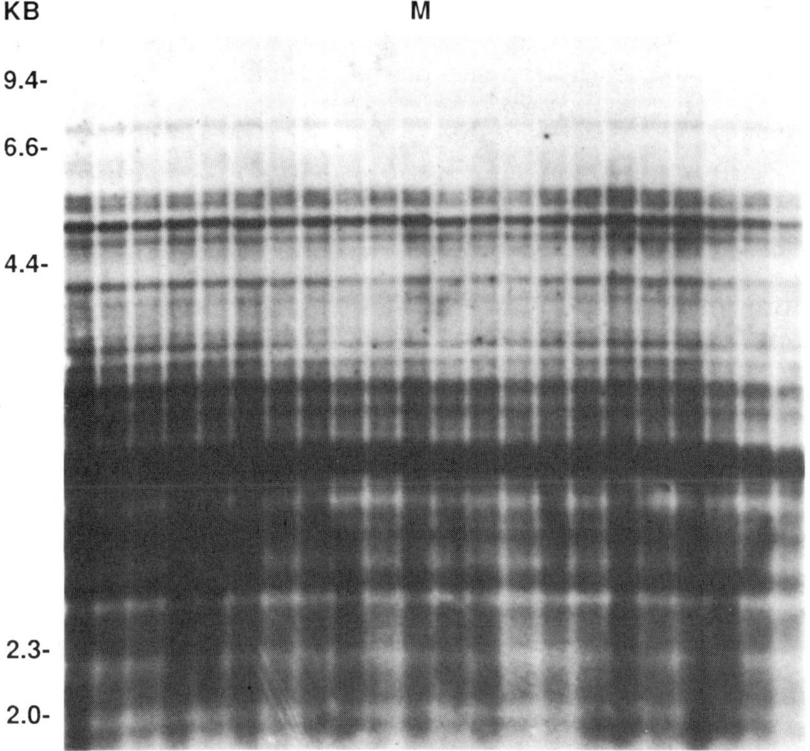

Figure 2. Agamospermy detected in *Taraxacum officinale* with the GATGTGGG PCR-STR probe. M="maternal" or seed-producing individual; all other 21 lanes are offspring of this maternal individual. KB=kilobase pairs.

have detected several instances of mutation (unpublished data), which should have the effect of creating heterozygous alleles. Both of these findings suggest the possibility that *T. officinale* individuals may be heterozygous at some VNTR loci. If the parent in Figure 2 was heterozygous for any one of the VNTR markers surveyed, then the probability of detectable transmission of that marker via selfing (or crossing with an identical clone) to all 21 offspring would be $0.75^{21} = 2.38 * 10^{-3}$. If the parent was heterozygous for two detectable VNTR markers, the probability of detectable transmission via selfing (or crossing with an identical clone) would be $0.75^{42} = 5.66 * 10^{-6}$. Any cross with non-clonal individuals should produce offspring markers not found in the pistillate plant since all individuals in the local population surveyed differed in numerous markers from one another, excepting hypothesized clonal individuals that were identical with one another (unpublished data). Considering the above, the most consistent conclusion is that this PCR-STR probe demonstrates that the offspring of the *T. officinale* parent were produced via agamospermy.

FUTURE APPLICATIONS

VNTR DNA markers have been utilized to investigate a wide range of genetic phenomena (Amos and Pemberton, 1993), and applications with plants are increasing (Weising et al., 1995). One limitation to their exploitation has been the availability of VNTR probes. PCR-STR probes (Rogstad 1993, 1994a) offer one approach by which a virtually limitless array of VNTR probes can be produced. Surveying a particular species with a series of PCR-STR probes to investigate genetic variation is relatively rapid since cloning, clone screening, sequencing, and other steps involved in alternate methods are not involved.

Improved computer programs for analyses of VNTR loci variation revealed with multilocus probes are needed. One new program, (GELSTATS; manuscript in prep.) provides summary statistics (e.g., bands per individual; frequency of population bands; similarity (band sharing); estimates of number of alleles/locus and heterozygosity) for data sets in which the user can define the groups. The program also conducts permutation trials (Good, 1994) to statistically test whether groups differ in selected values. Due to the high variation levels of many VNTR loci, cases where offspring produced via outcrossing versus selfing (e.g., the *P. lateriflora* example above) or via agamospermy (e.g., in species where inter-individual variation is high, such as has been found for *T. officinale*) may frequently be obvious. However, computer programs for analyzing VNTR marker patterns with regard to estimating outcrossing rates from data sets including large parent-offspring arrays probed with several PCR-STR probes are needed. For plant species that lack extensive variation, or for populations in which a history of inbreeding exists, statistical approaches to estimating rates of outcrossing (perhaps involving comparative estimation of variation among adults) analagous to those developed for allozymes (e.g., Shaw et al., 1981; Ritland, 1990) will be necessary. Obligate and facultative apomixis are widely distributed in plants (Richards, 1986), and their relative importance in some communities (e.g., tropical lowland rain forests (Kaur et al., 1978)) remains to be determined. VNTR markers offer a new means of investigating these modes of reproduction.

Yet another potential application involves the recent determination that several independent human genetic diseases are associated with expansions of unstable trinucleotide repeats at disease-related, protein coding loci (Davies, 1993). For example, in fragile X syndrome (Feng et al. 1995), it has been demonstrated that the disease-related FMR1 gene, with a mode of 30 CGG repeats in the normal population, is translationally inhibited when the number of repeats exceeds approximately 200. It seems plausible that related genetic circumstances occur in other organisms. In conclusion, PCR-

STR probes offer a new means to investigate this and other population biology phenomena.

ACKNOWLEDGMENTS

I am extremely grateful to the following for permission to briefly note in this review of the wide applicability of PCR-STR probes findings that are part of larger collaborative efforts, including: John Beresh (dandelion data); Dan Busemeyer (Philippine bramble data); Susan Frede (spicebush data); Hae Lim (buckeye data); and Stephan Pelikan (development of GELSTATS). Full details about all of these projects will hopefully appear in the near future. In addition, my deepest thanks to Chester Burton, Colleen Daley, Shane Heschel, Dr. Anil Kumar, Encik Abd. Samad bin Latif, Sheila Mehra, Ioanna Popescu, and Preeti Rout, as well as to the Director and Staff at the Forest Research Institute Malaysia. This research was in part supported by funds from the National Science Foundation (BSR 9096317); the National Geographic Society; and from the University of Cincinnati, for which I am also thankful.

REFERENCES

Ali, S., Muller, C.R., and Epplen, J.T., 1986, DNA fingerprinting by oligonucleotide probes specific for simple repeats. *Human Genetics* 74:239.

Amos, B., and Pemberton, J., 1993, A bibliography of DNA fingerprinting studies, *Fingerprint News* 5:2.

Arens, P., Odinot, P., van Heusden, A.W., Lindhout, P., and Vosman, B., 1995, GATA- and GACA-repeats are not evenly distributed throughout the tomato genome, *Genome* 38:84.

Avise, J.C., 1994, "Molecular Markers, Natural History, and Evolution," Chapman & Hall, N.Y.

Batschelet, E., 1979, "Introduction to Mathematics for Life Sciences," 3rd ed., Springer-Verlag, Berlin.

Benson, G., and Waterman, M.S., 1994, A method for fast database search for all k-nucleotide repeats, *Nuc. Acids Res.* 22:4828.

Borstnik, B., Pumpernik, D., Lukman, D., Ugarkovic, D., and Plohl, M., 1994, Tandemly repeated pentanucleotides in DNA sequences of eucaryotes, *Nuc. Acids Res.* 22:3412.

Brown, C.M., Stockwell, P.A., Trotman, C.N.A., and Tate, W.P., 1990, Sequence analysis suggests that tetra-nucleotides signal the termination of protein synthesis in eukaryotes, *Nuc. Acids Res.* 18:6339.

Dallas, J.F., 1988, Detection of DNA "fingerprints" of cultivated rice by hybridization with a human minisatellite DNA probe, *Proc. Natl. Acad. Sci. USA*, 85:6831.

DePamphilis, C.W., 1988, "Hybridization in Woody Plants: Population Genetics and Reproductive Biology in Southeastern Buckeyes," Ph.D. Diss., Univ. Georgia, Athens.

Epplen, J.T., 1988, On simple repeated GATA-GACA sequences in animal genomes: a critical reappraisal, *J. Heredity* 79:409.

Feng, Y., Zhang, F., Lokey, L.K., Chastain, J.L., Lakkis, L., Eberhart, D., and Warren, S.T., 1995, Translational suppression by trinucleotide repeat expansion at FMR1, *Science* 268:731.

Good, P., 1994, "Permutation Tests," Springer-Verlag, Berlin.

Grant, V., 1981, "Plant Speciation," 2nd ed., Columbia Univ. Press, N.Y.

Hamrick, J.L., and Godt, M.J.W., 1990, Allozyme diversity in plant species, *in*: "Plant Population Genetics, Breeding, and Genetic Resources," A.H.D. Brown, M.T. Clegg, A.L. Kahler, and B.S. Weir, eds., Sinauer Associates Inc., Sunderland.

Hamrick, J.L., Linhart, Y.B., and Mitton, J.B., 1979, Relationships between life history characteristics and electrophoretically detectable genetic variation in plants, *Ann. Rev. Ecol. Syst.* 10:173.

Hamrick, J.L., and Murawski, D.A., 1991, Levels of allozyme diversity in populations of uncommon neotropical tree species, *J. Trop. Ecol.* 7:395.

Jeffreys, A.J., Royle, N.J., Wilson, V., and Wong, Z., 1988, Spontaneous mutation rates to new length alleles at tandem-repetitive hypervariable loci in human DNA, *Nature* 332:278.

Jelinski, D.E., and Cheliak, W.M., 1992, Genetic diversity and spatial subdivision of *Populus tremuloides* (Salicaceae) in a heterogeneous landscape, *Amer. J. Bot.* 79:728.

Jin, L., and Chakraborty, R., 1993, A bias-corrected estimate of heterozygosity for single-probe multilocus DNA fingerprints, *Mol. Biol. Evol.* 10:1112.

Kaur, A., Ha, C.O., Jong, K., Sands, V.E., Chan, H.T., Soepadmo, E., and Ashton, P.S., 1978, Apomixis may be widespread among trees of the climax rain forest, *Nature* 271:440.

Litt M., and Luty, J.A., 1989, A hypervariable microsatellite revealed by in vitro amplification of a dinucleotide repeat within the cardiac muscle actin gene, Amer. *J. Human Genetics* 44:397.

Mitton, J.B., 1994, Molecular approaches to population biology, *Ann. Rev. Ecol. Syst.* 25:45.

Nakamura, Y., Leppert, M., O'Connel, P., Wolff, R. Holm, T., Culver, M., Martin, C., Fujimoto, E., Hoff, M., Kumlin, E., and White, R., 1987, Variable number of tandem-repeat (VNTR) markers for human gene mapping, *Science* 235:1616.

Neuhaus, D., Kuhl, H., Kohl, J.G., Dorfel, P., and Borner, T., 1993, Investigation on the genetic diversity of *Phragmites* stands using genomic fingerprinting, *Aquatic Bot.* 45:357.

Nogler, G.A., 1984, Gametophytic apomixis, *in*: "Embryology of Angiosperms," B.M. Johri, ed., Springer-Verlag, Berlin.

Nurnberg, P., Barth, I., Fuhrmann, E., Lenzner, C., Lozanova, T., Peters, C., Poche, H., and Thiel, G., 1991, Monitoring genomic alterations with a panel of nucleotide probes specific for various simple repeat motifs, *Electrophoresis* 12:186.

Prodohl, P.A., Taggart, J.B., and Ferguson, A., 1994, Single locus inheritance and joint segregation analysis of minisatellite (VNTR) DNA loci in brown trout (*Salmo trutta* L.), *Heredity* 73:556.

Richards, A.J., 1986, "Plant Breeding Systems," George Allen & Unwin, London.

Ritland, K., 1990, A series of FORTRAN computer programs for estimating plant mating systems, *J. Heredity* 81:235.

Rogstad, S.H., 1992, Saturated NaCl-CTAB solution as a means of field preservation of leaves for DNA analyses, *Taxon* 41:701.

Rogstad, S.H., 1993, Surveying plant genomes for variable number of tandem repeats loci, Meth. *Enzymology* 224:278.

Rogstad, S.H., 1994a, Inheritance in turnip of variable number tandem repeat genetic markers revealed with synthetic repetitive DNA probes, *Theor. Appl. Genet.* 89:824.

Rogstad, S.H., 1994b, The biosystematics and evolution of the *Polyalthia hypoleuca* species complex (Annonaceae) of Malesia. III. Floral ontogeny and breeding systems, *Amer. J. Bot.* 81:145.

Rogstad, S.H., Nybom, H., and Schaal, B.A., 1991, The tetrapod "DNA fingerprinting" M13 repeat probe reveals genetic diversity and clonal growth in quaking aspen (*Populus tremuloides*), *Plant Syst. Evol.* 175:115.

Rothuizen, J., Wolfswinkel, J., Lenstra, J.A., and Frants, R.R., 1994, The incidence of mini- and micro-satellite repetitive DNA in the canine genome, *Theor. Appl. Genet.* 89:403.

Scribner, K.T., Arntzen, J.W., and Burke, T., 1994, Comparative analysis of intra- and interpopulation genetic diversity in   *Bufo bufo*, using allozyme, single-locus microsatellite, minisatellite, and multilocus minisatellite data, *Mol. Biol. Evol.* 11:737.

Shaw, D.V., Kahler, A.L., and Allard, R.W., 1981, A multilocus estimator of mating system parameters in plant populations, *Proc. Natl. Acad. Sci.* USA 78:1298.

Stephens, J.C., Gilbert, D.A., Yuhki, N., and O'Brien, S.J., 1992, Estimation of heterozygosity for single-probe multilocus DNA fingerprints, *Mol. Biol. Evol.* 9:729.

Terauchi, R., and Konuma, A., 1994, Microsatellite polymorphism in *Dioscorea tokoro*, a wild yam species, *Genome* 37:794.

Vergnaud, G., 1989, Polymers of random short oligonucleotides detect polymorphic loci in the human genome, *Nuc. Acids Res.* 17:7623.

Weber, J.L., and May, P.E., 1989, abundant class of human DNA polymorphisms which can be typed using the polymerase chain reaction, Amer. *J. Human Genet.* 44:388.

Weising, K., Nybom, H., Wolff, K., and Wieland, M., 1995, "DNA Fingerprinting in Plants and Fungi," CRC Press, Boca Raton.

MICROSATELLITE HETEROZYGOSITY: FUNCTIONAL CONSTRAINTS ON THE DEVELOPMENT AND USE OF COMPREHENSIVE GENOMIC MAPS FOR LIVESTOCK

Craig W. Beattie

USDA, ARS
U.S. Meat Animal Research Center
P.O. Box 166
Clay Center, NE 68933

Microsatellites (ms) are abundant, multi-allelic, repetitive elements uniformly distributed throughout the genome of numerous species, including man (Litt and Luty, 1989; Weber and May, 1989). As ms are inherited codominantly, they have the potential to replace serum protein and red blood cell antigen polymorphisms for parental identification in livestock (Kappes et al., 1994; Glowatzki-Mullis et al., 1995; Usha et al., 1995). They currently provide the markers required to rapidly produce the linkage maps essential to identifying and assigning segregating loci of interest in cattle (Bishop et al., 1994), swine (Rohrer et al., 1994) and sheep (Crawford et al., 1995) including blood group antigens (Kappes et al., 1994).

In contrast, DNA fingerprinting or profiling with minisatellites, large tandem repetitive elements (VNTR's), or restriction fragment length polymorphisms (RFLP's) has become the strategy of choice in paternity verification and forensic pathology in humans (Jeffreys et al., 1985). Fingerprints are generated when genomic DNA is digested with a restriction endonuclease and subjected to Southern analysis with either a minisatellite, microsatellite or random oligonucleotide probe. Marker choice for each profile is dictated, in part, by the level of heterozygosity of an individual type of marker. In genetic mapping, polymorphisms that distinguish two parents or parental species segregate in a mapping population. RFLP's suffer from low heterozygosity, while minisatellites possess extreme genetic variability making profile interpretation difficult (Jeffreys et al., 1991). Although minisatellite probes can reveal a large number of hybridization bands, in cattle, the actual number of polymorphic loci can be low (Georges et al., 1988) possibly due to band sharing. More importantly, the degree of polymorphism between breeds is notably different (Buitkamp et al., 1991). In spite of the difficulty inherent in genotyping minisatellites in livestock, they exhibit pronounced differences between unrelated individuals (Haberfield et al., 1993), segregate according to Mendelian laws and have been used successfully to determine identity and paternity in cattle (Buitkamp et al., 1991; Trommelen et al., 1993) and detect leucochimerism in bovine twins (Plante et al., 1992), a difficult assignment for a generally biallelic system such as an RFLP. However, minisatellites, as multilocus probes, in contrast to earlier promise (Georges et al., 1988, 1990) currently do not appear to have the potential that single locus probes, such as ms, have for linkage mapping of economically important loci in livestock.

Microsatellite heterozygosity is generally higher than RFLP's, and unlike minisatellites, they arise from defined loci which makes interpretation of a genotypic profile easier. The level of heterozygosity is generally sufficient to obviate the use of endonuclease restricted DNA. The degree of polymorphism at a marker locus influences the probability of detection to an index locus (rare dominant allele which segregates). Informativeness (polymorphism) represented by the probability that a given offspring of a parent carrying the rare allele at the index locus allows deduction of the parental genotype at the marker locus. Each single marker locus can be evaluated for its polymorphism information content (PIC) by summing the mating frequencies multiplied by the probability that an offspring will be informative (Botstein et al., 1980). The suggested value of PIC for cost-effective linkage (or paternal) studies is greater than >.70 (Hearne et al., 1992). In livestock, polymorphic ms markers with PIC > .70 enable over 50% of families to be fully informative. Initial studies in cattle suggest that a panel of ms standardized for relatively high within and between breed allelic variation, ease of co-amplification by the polymerase chain reaction (PCR) and electrophoretic multiplexing offer the opportunity to improve parental identification over conventional blood typing (Glowatzki-Mullis et al., 1995; Usha et al., 1995). Exclusion efficiencies and probabilities, while variable for an individual marker within breed, increase significantly to greater than 98% with a ms panel selected across an entire genome (Glowatzki-Mullis et al., 1995). This should not suggest that any panel containing an economically viable number of ms primers can be standardized for use in parental analysis in all breeds irrespective of species. Initial observations (Beattie et al., unpublished) suggest that as breeds without any information on ms heterozygosity are incorporated into a DNA fingerprinting strategy, typing panels will have to be modified accordingly. A conservative, yet realistic strategy, might also include at least temporary retention of decades of blood antigen records and concurrent exclusion analysis within and between breeds to firmly establish the concordance between the two approaches.

Perhaps the most significant potential use of ms in livestock parental identification is in closely related species such as the ruminant *artiodactyla*. Here, the significant incidence of cross amplification and ms informativeness associated with primer pairs developed during the construction of the bovine genetic linkage map in water buffalo, sheep, goats, and deer promises to extend the use of ms beyond that of parental identification into population dynamics, genetic diversity, comparative linkage mapping and characterization of economically important loci (Beattie, 1994).

The initial observation of Moore et al. (1991) that conservation of DNA sequence flanking ms in cattle and sheep yielded primer pairs able to reciprocally amplify the genome of either species, led to additional studies demonstrating the potential informativeness of *bovid* microsatellites across closely related species (Kemp et al., 1993,

Table 1. Marker heterozygosity in sire and dam breeds comprising map reference populations of cattle and swine.

Sex	Breed	Heterozygosity (%)
Cattle sires	Gelbvieh—Simmental	59.8
	Gelbvieh—Simmental	60.1
	Brahman[a]—Angus	73.7
	Brahman—Hereford	76.8
Cattle dams	Piedmontese—Angus	57.4
	Piedmontese—Hereford	61.9
	Longhorn—Angus	63.1
	Longhorn—Hereford	60.5
	Nelore[a]—Hereford	76.7
	Angus	46.3
	Hereford	45.0
	All western breeds	49.5
	Across indicus—taurus	75.7
Swine sires	WC[b]	56.8
Swine dams	WC—Duroc[c]	64.5
	WC—Fengjing	78.2
	WC—Minzhu	80.3
	WC—Meishan	82.1

[a] Indicus breed.
[b] WC (White Composite) is 1/4 Chester White, 1/4 Large White, 1/4 Landrace, 1/4 Yorkshire.
[c] Western breed.

1995; Vaiman et al., 1994; Bishop et al., 1994; Crawford et al., 1995; Moore et al., 1995) and attempts to use bovine ms as markers for parentage control in goats (Pepin et al., 1995). Individual marker heterozygosity was low, but the cumulative exclusion probability was ~90%, suggesting that ms could have a significant role in establishing registered pedigrees, a first step in marker assisted selective breeding (Pepin et al., 1995). It is the lower level of ms heterozygosity within a species breed and closely related breeds (Fredholm et al., 1993; Rohrer et al.,1994; Bishop et al., 1994; Table 1), a property shared with minisatellites, that can limit the use of individual ms in marker-assisted selection (MAS) within intrabreed, "purebred", crosses. This has not limited their use in determining evolutionary relationships or estimating genetic diversity where ms have largely replaced protein and mitochondrial DNA (mtDNA) polymorphisms and minisatellites as a strategy of choice.

Microsatellites, minisatellites and RFLP's are not the only means to establish genomic or individual identity, however. Differences in the fingerprints between two genomic DNA templates amplified with arbitrary (sequence) oligonucleotides are generally due to mutations that create or destroy primer binding sites or by insertion or deletion events that change the interval between primer-binding sites. These polymorphisms are often referred to as random amplified polymorphic DNA's (RAPD's) (Williams et al., 1990) and can be used to determine whether two organisms are genetically distinct (McClelland et al., 1995). In population or phylogenetic analysis, each polymorphic marker can be treated as a character (McClelland et al., 1995). The relatively widespread use of RAPD's in genetic mapping in plants has not been translated into mammalian species as RAPD's are not co-dominant markers, do not segregate as such in outbred populations, generally are not transferable to populations outside the one used to develop the polymorphisms and are difficult to reproduce technically. However, in specific instances, RAPD's have been used to fingerprint pooled DNA samples representing distinct cattle populations and revealed population-specific polymorphisms (Gwakisa et al., 1994; Kemp and Teale, 1994). When cloned, and used as a hybridization probe, *Bos indicus* specific polymorphisms can discriminate *Bos taurus* and provide a means to detect introgression of Zebu genes into *B. taurus* cattle populations (Teale et al., 1995).

MICROSATELLITES IN POPULATION ANALYSIS

In contrast to the numerous attempts to use genetic variation at minisatellite loci in the context of parental identification, animal breeding, and pedigree analysis, variation at minisatellite loci has only recently been adapted to assessing genetic diversity or conservation within and between natural populations. Here, polymorphisms in mitochondrial DNA have long been exploited to address problems in population structure, phylogenetics and resource management not resolved by variation at the protein level. Although the limits to which multilocus probes can be applied to estimate kinship and analyze the structure of relatedness in natural populations have been stressed (Lynch, 1988; Lewin, 1989; Capy and Brookfield, 1991), allelic frequency at minisatellite loci has recently been used to establish that human ethnic groups can be differentiated (Balazs et al., 1992). The use of VNTR to survey genetic variability in diverse natural populations has also demonstrated that low levels of polymorphism at VNTR loci in some sub-species of lion may reflect poor reproductive capacity as a result of a population bottleneck and significant inbreeding. A similar case could not be made for the Cheetah, *Acinonyx jubatus*, (O'Brien, 1994), or European beaver, *Castor fiber*, where low levels of polymorphism at hypervariable loci apparently reflect bottlenecks within individual populations of this species, but do not necessarily reflect biologic viability (Ellegren et al., 1993). Allelic frequency(s) at minisatellite loci also provide evidence for population clustering by geographic region as well as major genetic distinction between geographic regions in species as diverse as salmonid fish (Taylor, 1995) and the house sparrow (Wetton et al., 1987). However, when used to survey genetic variation in diverse natural populations, multilocus probes estimate only the average degree of kinship between individuals and determines whether they are related or not, since all distributions of identity can broadly overlap even when a large number of loci are typed (Capy and Brookfield, 1991).

In contrast, the rapid advent of monolocus probes such as ms to determine kinship, intra- and interpopulation differences, social structure and evolutionary relationships is demonstrated by the recent wealth of information on both conservation and relative diversity within and between a myriad of species including livestock. These range from studies using ms to assess population structure within species as well as transpecies comparisons.

Transpecies conservation of microsatellite sequence is apparently not limited to mammalian species diverging 15-25 *(Artiodactyla)* (Moore et al., 1991) and 35-40 *(Cetacea)* million years ago (Schlotterer et al., 1991). Microsatellite sequences can also apparently persist across families of marine and fresh water turtles that span approximately 300 million years of divergent evolution, although levels of heterozygosity are higher in the species from which primer pairs were designed limiting direct cross-species comparisons of variability (FitzSimmons et al., 1995). Within species, ms variation between divergent populations is not only consistent with earlier results with protein or mtDNA polymorphisms, but as highly polymorphic, monolocus, codominantly inherited probes are able to distinguish interpopulation differences as well as parenteral inheritance in such diverse species including *Camponotus* ants (Gertsch et al., 1995), red deer (*Cervus elaphus*) (Pemberton et al., 1995), cattle (*Bos taurus*) (Glowatzki-Mullis et al., 1995), and sheep (Paszek et al.,

1996a) particularly if a panel of sufficiently heterozygous ms are used. Panels of ms have also been useful in conservation genetics to assess interpopulation differences in genetically depaupurate mammalian species where initial studies employing alloenzyme variation were inconclusive due to a combination of low levels of polymorphism, low population density and small effective population sizes. Analytic DNA profiling with such panels has shown that social groups or pods of pilot whales, *(Globicephalia melas,* Delphinidae) are strongly matrifocal with both sexes remaining within the pod, yet males do not mate within their natal pods. A behavior pattern unusual for mammals (Amos et al., 1993). A panel of ms developed to analyze genetic variance and population structure in the North American black bear (*Ursus americanus*) has proven extremely useful for similar analysis in closely related species of bears (Paetkau and Strobeck, 1994; Paetkau et al., 1995). Analysis of ms loci in grizzly bear (*Ursus arctos)* family groups has demonstrated that individual cubs in a litter can be sired independently, although only approximately half of the males in a healthy population reproduce (Craighead et al., 1995). In geographically isolated populations of the North American black bear, ms have been used to demonstrate that, in spite of levels of heterozygosity lower than that of the endangered Cheetah (O'Brien, 1994), biologic viability remains high (Paetkau and Strobeck, 1994). This again suggests that current estimates of genetic variation may not entirely reflect population vulnerability. The use of ms to define populations is not limited to non-primates. Microsatellites have been combined with minisatellites as probes to define human racial populations (Iwasaki et al., 1992; Matsutani et al., 1992).

PHYLOGENETIC RELATIONSHIPS

Mini-and microsatellites, as well as random oligonucleotides, all uncover DNA fingerprint patterns following Southern analysis of genomic DNA digested by restriction endonucleases. The polymorphic nature of the resultant fragments has been attributed to variation in the number of minisatellite sequences tandemly repeated within the fragment. Kashi et al. (1994) suggested the basis for ms and random oligonucleotides also uncovering this (minisatellite) polymorphism is the presence of a large class of interspersed tracts consisting of at least two different minisatellites separated by stretches of unique DNA. Restriction sites are apparently eliminated in these tracts creating a large genomic fragment that permits hybridization to a variety of probe sequences (Kashi et al., 1994). Interspersed repetitive elements such as just described and others, e.g., short interspersed elements (SINES) and Alu or Alu-like sequences, are ubiquitous in mammalian genomes and are apparently ancient within the mammalian radiation (Jurka et al., 1995). The presence of interspersed repetitive elements in and near coding sequences have been documented to alter gene expression (Brini et al., 1993; Onda et al., 1993; Vidal et al., 1993) and may have been a major force in the evolution of mammals (Jurka et al., 1995). Instability of tandemly repeated ms associated with subsequent genetic disorder(s) is well documented (Richards and Sutherland, 1992). The significant number of ms located within or near such diverse elements provide a genomic basis for selecting ms as a strategy of choice in assessing phylogenetic relationships as well as evolutionary diversity. Genomic fingerprinting by microsatellite-primed PCR amplifies genomic DNA samples from a variety of plant species and allows intra-and interspecific distinction among plant taxa (Weising et al., 1995). Closer to home evolutionarily, ms have been employed to a great effect in improving the resolution of relationships within individual livestock species. Buchanan et al. (1994) have resolved the earlier dating of the divergence of the British Poll Dorset and Merino breeds of sheep using blood protein loci from 69,700 yrs to 1,094 yrs with a panel of ms, a more realistic figure since sheep have been domesticated for not more than 11,000 yrs. Moreover, the divergence of Australian from New Zealand Merinos was calculated at 227 yrs, a figure quite in line with importation dates. Recent work (Cepica et al., 1995) with blood protein loci in pigs clustered Large White, Landrace and Duroc as a branch separate from Hampshire. Estimates of standard genetic distances placed equal values on Landrace and Duroc, but no estimates of time of divergence were presented. Paszek et al. (1996b), using a panel of ms restricted to one chromosome, estimated genetic distance between similar breeds and the Chinese Meishan and found western breeds clustered in a similar manner (Figure 1). However, calculated genetic distances were 2-5 times greater than that of Cepica et al. (1995) with little distance between western breeds (Table 2). Phylogenetic results suggest the

Table 2. Genetic distances and estimated time of breed diversion for swine breeds.

Swine Breeds	Yorkshire	Hampshire	Duroc	Landrace	Meishan
Yorkshire	—	0.2354[a]	0.5338[a]	0.3828[a]	1.2214[a]
Hampshire	261[b]	—	0.4568[a]	0.4296[a]	1.3369[a]
Duroc	593[b]	507[b]	—	0.5337[a]	1.3056[a]
Landrace	425[b]	477[b]	593[b]	—	1.1868[a]
Meishan	1357[b]	1485[b]	1451[b]	1318[b]	—

[a] Genetic distances.
[b] Time (in years) of breed divergence.

time of divergence of Yorkshire and Hampshire breeds as well as western and Chinese breeds correlates well with the historical record (Porter, 1993) and the cluster tree provided by Li and Enfield (1989) based on analysis of phenotypic data. Genetic absolute dating based on ms also fixes the deepest split in the human phylogeny occurred about 156,000 yrs ago (Goldstein et al., 1995). Microsatellite polymorphisms also reveal phylogenetic relationships between primate species including man, with at least the potential for a correlation between repeat array sizes and the phylogenetic distances of primates from man (Meyer et al., 1995). Smaller arrays, hence

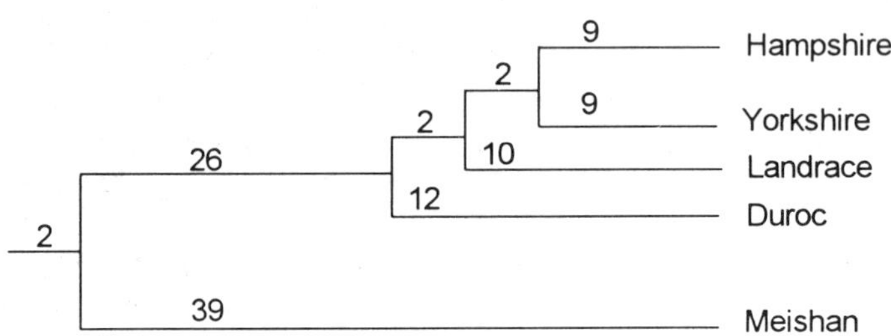

^a Calculated based on genetic distances using KITSCH algorithm of the PHYLIP package.

Figure 1. Phenogram^a showing genetic similarities among five swine breeds. Numbers at the nodes represent the percentage of a group occurrence in 250 bootstrap replications.

less polymorphism and a decrease in allele size and frequency, are putatively associated with an ancestral state in microsatellite (Edwards et al., 1991), and minisatellite (Gray and Jeffreys, 1991) development. A case for the directionality and variation in rate of microsatellite evolution between species has been made in primates, with longer microsatellites in humans (Rubinsztein et al., 1995a,b), but questioned on the basis of an ascertainment bias in the selection of loci analyzed; ms chosen on the basis of being highly polymorphic in one species will generally tend to harbor shorter repeats in a related species (Ellegren et al., 1995; FitzSimmons et al., 1995; Forbes et al., 1995). Although ascertainment bias may be unlikely due to the significant incidence of polymorphic loci in chimpanzees (Rubinsztein et al., 1995b), it remains to be seen whether reciprocal amplification of ms from closely related species yields similar degrees of polymorphism. A cautionary note for the assumptions that all ms alleles are equally likely to expand and contract, that the probability of mutation does not vary with repeat number, and mutation rates are similar in diverse species argues that the unqualified use of ms as "biological clocks" to calculate divergence times between populations and species may not be warranted (Rubinsztein et al., 1995b; Forbes et al., 1995).

FUNCTIONAL CONSTRAINTS ON DEVELOPMENT AND USE OF COMPREHENSIVE GENETIC MAPS

Development of genetic linkage maps has taken advantage of the increase in heterozygosity observed when either coding sequences (RFLP's) (Copeland et al., 1993) or ms (Bishop et al., 1994; Rohrer et al., 1994; Ellegren et al., 1994; Crawford et al., 1995) are screened across interspecific or interbreed crosses (Beattie, 1994). In livestock, the heterozygosity of ms markers is significantly reduced within breed and crosses irrespective of species, and also lower in those interbreed crosses that incorporate more recently diverged breeds (Beattie, 1994; Paszek et al., 1996a). This appears true whether ms loci are selected at random from across the entire genome (Table 1) or are representative of a smaller region of the genome (Paszek et al., 1996a). The tacit assumption that ms variability appears to correlate with repeat number (Weber, 1990) and by inference the number and size of alleles, does not appear to effect the level of polymorphism across breeds when (dinucleotide) ms greater than 11-12 repeats in length are cloned and genotyped (Paszek et al., 1996a; Beattie et al., unpublished observations; Table 3). This extends to species more divergent than those "western" breeds commonly employed in commercial breeding and selection programs e.g., Chinese breeds of swine (Paszek et al., 1996a). As noted above, reciprocal amplification and genotyping of ms (at least 11-12 repeats in length) selected on the basis of polymorphism for an individual breed in a closely related species should answer this question, particularly if those ms which prove monomorphic in the breed or species from which they are developed prove informative in a closely related species. It should be noted that the ~60% reduction in the number of ms that amplify and are informative in closely related species, such as ruminants (Crawford et al.,1995; deGortari and Beattie, unpublished observations), leaves a significant proportion of ms available for construction of comparative linkage maps, and in conjunction with comparative mapping of coding sequences, allows the reconstruction of evolutionary events within and among relatively diverse genera and species (Beattie, 1994).

The lack of heterozygosity in outcrossed commercial species becomes extremely important when ms are used to first identify and then refine the genomic location of genes accounting for significant portions of the genetic variation of an economically important phenotype(s) (Table 3). The problem is confounded when ms used to mark an interval are employed in marker-assisted selection, particularly within breeds in which one or more loci was not originally identified. A cluster of highly polymorphic markers linked to a locus of interest would facilitate identification of appropriate

Table 3. Means and standard deviations of heterozygosities for nine chromosome 6 framework (map) microsatellites.

	Swine breed				
Marker	Yorkshire	Hampshire	Duroc	Landrace	Meishan
S0035	.22±.42	.09±.29	0.0	.10±.31	NT
Sw1057	.76±.32	.30±.38	.68±.35	.64±.43	.44±.46
S0087	.54±.50	.50±.51	.46±.48	.50±.51	.92±.28
Sw1067	.44±.50	.65±.48	.53±.51	.52±.51	.84±.37
Sw1129	.80±.41	.61±.49	.75±.44	.69±.47	.67±.48
Sw316	.53±.50	.51±.51	.21±.42	.38±.49	.48±.51
S0031	.35±.48	.51±.51	.38±.49	.77±.44	.28±.46
Sw353	.45±.50	.47±.50	.60±.50	.57±.50	.43±.51
Sw824	.40±.49	.26±.44	.19±.40	.63±.49	.54±.51
Mean[a]	.38±.25	.47±.13	.47±.20	.58±.12	.57±.21

[a] Without microsatellite S0035.

individuals to initiate introgression of favorable alleles into a breeding population. Such clusters will probably be needed before embarking on intrabreed matings, as the number and heterozygosity of markers affects the efficiency of identification of economic trait loci (ETL's) and MAS. The efficiency of either process increases with the number of polymorphic markers up to ~500 markers; beyond this number, efficiency remains fairly constant. Maximum efficiency of selection is achieved when at least 500 polymorphic markers are available for each parent (J. Keele, unpublished observations; Beattie, 1994). As the number of markers likely to prove useful in identifying an economically relevant locus depends on marker number and level of marker (ms) heterozygosity (Table 4), a systematic approach to identify economically important trait loci (ETL's) in livestock might be to screen unrelated animals within the breeds to be crossed using a panel of ms that are uniformly spread across the genome before large-scale genotyping of a resource population of several hundred animals is initiated.

When taken to the logical conclusion of using ms in an extended selection program, it might be well to consider the impact of generation interval and selection intensity on the relative

Table 4. Number of microsatellite markers needed to insure 500
polymorphic markers per parent.

Breed composition needed	Heterozygosity	Number of markers
Cattle		
Bos indicus _ *Bos indicus* (F$_1$)	73.1	684
Among *Bos taurus*	58.4	856
Angus (purebred)	44.8	1,116
Hereford (purebred)	43.2	1,157
Swine		
Chinese _ WC[a] (white composite)	80.8	619
Duroc _ WC	64.5	775
WC	56.8	880

[a] WC = 1/4 Chester White, 1/4 Large White, 1/4 Landrace, 1/4 Yorkshire

heterozygosity of those ms flanking a genomic locus in animals
selected to introgress the associated positive phenotype. A short
(poultry) or relatively short (swine) interval and intensive selection
pressure may act to lower the heterozygosity or allelic frequency of
the ms used in the selection process. Here again, clusters of ms
flanking a locus may be of some advantage. As desirable alleles in
question become fixed and marker disequilibrium lost during selection
as the process achieves its goals, we will need to move from selection
on those marker alleles identifying the locus and develop more
strategies (and markers) to account for the additional genetic variation
within each trait.

REFERENCES

Amos, B., Schlotterer, C., and Tautz, D., 1993, Social structure of pilot whales
revealed by analytical DNA profiling, *Science* 260:670.

Balazs, I., Neuweiler, J., Gunn, P., Kidd, K.K., Kuhl, J., and Mingjun, L., 1992,
Human population genetic studies using hypervariable loci. I. Analysis
of Assamese, Australian, Cambodian, Caucasian, Chinese and Melanesian
populations, *Genetics* 131:191.

Beattie, C.W., 1994, Livestock genome maps, *T.I.G.* 10:334.

Bishop, M.D., Kappes, S.M., Keele, J.W., Stone, R.T., Sunden, S.L.F., Hawkins, G.A.,
Solinas Tolda, S., Fries, R., Grosz, M.D., Yoo, J-Y., and Beattie, C.W., 1994, A
genetic linkage map for cattle, *Genetics* 136:619.

Botstein, D., White, R.L., Skolnick, M., and Davies, R.W., 1980, Construction of a
genetic linkage map in man using restriction fragment length
polymorphisms, *Am. J. Hum. Genet.* 32:314.

Brini, A.T., Lee, G.M., and Kinet, J-P., 1993, Involvement of Alu sequences in the
cell-specific regulation of transcription of the gamma chain of the Fc and
T-cell receptors, *J. Biol. Chem.* 268:1355.

Buchanan, F.C., Adams, L.J., Littlejohn, R.P., Maddox, J.F., and Crawford, A.M., 1994, Determination of evolutionary relationships among sheep breeds using microsatellites, *Genomics* 22:397.

Buitkamp, J., Zischler, H., Epplen, J.T., and Geldermann, H., 1991, DNA fingerprinting in cattle using oligonucleotide probes, *Anim. Genet.* 22:137.

Capy, P., and Brookfield, J.F.Y., 1991, Estimation of relatedness in natural populations using highly polymorphic genetic markers, *Genet. Sel. Evol.* 23:391.

Cepica, S., Wolf. J., Hojny, J., Vackova, I., and Schroffel, J. Jr., 1995, Relations between genetic distance of parental pig breeds and heterozygosity of their F_1 crosses measured by genetic markers, *Anim. Genet.* 26:135.

Copeland, N.G., Jenkins, N.A., Gilbert, D.J., Eppig, J.T., Maltais, L.J., Miller, J.C., Dietrich, W.F., Weaver, A., Lincoln, S.E., Steen, R.G., Stein, L.D., Nadeau, J.H., and Lander, E.S., 1993, A genetic linkage map of the mouse: Current applications and future prospects, *Science* 262:57.

Craighead, L., Paetkau, D., Reynolds, H.V., Vyse, E.R., and Strobeck, C., 1995, Microsatellite analysis of paternity and reproduction in arctic grizzly bears, *J. Hered.* 86:255.

Crawford, A.M., Dodds, K.G., Ede, A.J., Pierson, C.A., Montgomery, G.W., Garmonsway, H.G., Beattie, A.E., Davies, K., Maddox, J.F., Kappes, S.W., Stone, R.T., Nguyen, T.C., Penty, J.M., Lord, A.E., Broom, J.E., Buitkamp, J., Schwaiger, W., Epplen, J.T., Matthew, P., Matthews, M.E., Hulme, D.J., Beh, K.J., McGraw, R.A., and Beattie, C.W., 1995, An autosomal genetic linkage map of the sheep genome, *Genetics* 140:703.

Edwards, A., Civitello, A., Hammond, H.A., and Caskey, C.T., 1991, DNA typing and genetic mapping with trimeric and tetrameric tandem repeats, *Am. J. Hum. Genet.* 49:746.

Ellegren, H., Chowdhary, B.P., Johansson, M., Marklund, L., Fedholm, M., Gustavsson, I., Andersson, L., 1994, A primary linkage map of the porcine genome reveals a low rate of genetic recombination, *Genetics* 137:1089.

Ellegren, H., Hartman, G., Johansson, M., and Andersson, L., 1993, Major histocompatibility complex monomorphism and low levels of DNA fingerprinting variability in a reintroduced and rapidly expanding population of beavers, *Proc. Natl. Acad. Sci. USA* 90:8150.

Ellegren, H., Primmer, C.R., and Sheldon, B.C., 1995, Microsatellite 'evolution': directionality or bias, *Nature Genet.* 11:360.

FitzSimmons, N.N., Moritz, C., and Moore, S.S., 1995, conservation and dynamics of microsatellite loci over 300 million years of marine turtle evolution, *Mol. Biol. Evol.* 12:432.

Forbes, S.H., Hogg, J.T., Buchanan, F.C., Crawford, A.M., and Allendorf, F.W., 1995, Microsatellite evolution in congeneric mammals: Domestic and Bighorn sheep, *Mol. Biol. Evol.* 12:1106.

Fredholm, M., Winterø, A.K., Christensen, K., Kristensen, B., Nielsen, P.B., Davies, W., and Archibald, A., 1993, Characterization of 24 porcine $(dA-dC)_n$ - $(dT-dG)_n$ microsatellites: Genotyping of unrelated animals from four breeds and linkage studies, *Mamm. Genome* 4:187.

Georges, M., Lequarre, A.S., Castelli, M., Hanset, R., Vassert, G.G., 1988, DNA fingerprinting in domestic animals using four different minisatellite probes, *Cytogenet. Cell Genet.* 47:127.

Georges, M., Lathrop, M., Hilbert, P., Marcotte, A., Schwers, A., Swillens, S., Vassart, G., and Hanset, R., 1990, On the use of DNA fingerprints for linkage studies in cattle, *Genomics* 6:461.

Gertsch, P., Pamilo, P., and Varvio, S.-L., 1995, Microsatellites reveal high genetic diversity within colonies of *Camponotus* ants, *Mol. Ecol.* 4:257.

Glowatzki-Mullis, M-L., Gaillard, C., Wigger, G., and Fries, R., 1995, Microsatellite-based parentage control in cattle, *Anim. Genet.* 26:7.

Goldstein, D.B., Ruiz Linares, A., Cavalli-Sforza, L.L., and Feldman, M.W., 1995, Genetic absolute dating based on microsatellites and the origin of modern humans, *Proc. Natl. Acad. Sci. USA* 92:6723.

Gray, J.C., and Jeffreys, A.J., 1991, evolutionary transience of hypervariable minisatellites in man and primates, *Proc. R. Soc. Lond. [Biol]* 243:241.

Gwakisa, P.S., Kemp, S.J., and Teale, A.J., 1994, Characterization of Zebu cattle breeds in Tanzania using random amplified polymorphic DNA markers, *Anim. Genet.* 25:89.

Haberfield, A., Kalay, D., Weisberger, P., Gal, O., and Hilliel, J., 1993, Application of multilocus molecular markers in cattle breeding. 1. Minisatallites and microsatellites, *J. Dairy Sci.* 76:645.

Hearne, C.M., Gosh, S., and Todd, J.A., 1992, Microsatellites for linkage analysis of genetic traits, *T.I.G.* 8:288.

Iwasaki, H., Stewart, P.W., Dilley, W.G., Holt, M.S., Steinbrueck, T.D., Wells, S.A., and Donis-Keller, H., 1992, A minisatellite and a microsatellite polymorphism within 1.5 kb at the human muscle glycogen phosphorylase (PYGM) locus can be amplified by PCR and can have combined informativeness of PIC 0.95, *Genomics* 13:7.

Jeffreys, A.J., Macleod, A., Tamaki, K., Neil, D.L., and Mossekton, D.G., 1991, Minisatellite repeat coding as a digital approach to DNA typing, *Nature* 354:204.

Jeffreys, A.J., Wilson, V., and Thein, S.L., 1985, Hypervariable minisatellite regions in the human DNA, *Nature* 314:67.

Jurka, J., Zietiewicz, E., and Labuda, D., 1995, Ubiquitous mammalian-wide interspersed repeats (MIRs) are molecular fossils from the Mesozoic era, *N.A.R.* 23:170.

Kappes, S.M., Bishop, M.D., Keele, J.W., Penedo, M.C.T., Hines, H.C., Grosz, M.D., Hawkins, G.A., Stone, R.T., Sunden, S.L.F., and Beattie, C.W., 1994, Linkage of the bovine erythrocyte antigen loci B, C, L, S, Z, R' and T' and the serum protein loci post-transferrin 2(PTF2), vitamin D binding protein(GC) and albumin(ALB) to DNA microsatellite markers, *Anim. Genet.* 25:133.

Kashi, Y., Nave, A., Darvasi, A., Gruenbaum, Y., Soller, M., and Beckmann, J.S., 1994, How is it that microsatellites and random oligonucleotides uncover DNA fingerprint patterns, *Mamm. Genome* 5:525.

Kemp, S.J., Brezinsky, L., and Teale, A.J., 1993, A panel of bovine, ovine and caprine polymorphic microsatellites, *Anim. Genet.* 24:363.

Kemp, S.J., Hishida, O., Wambagu, J., Rink, A., Longeri, M.L., Ma, M.Z., Da, Y., Lewin, H.A., Barendse, W., Teale, A.J., 1995, A panel of polymorphic bovine, ovine and caprine microsatellite markers, *Anim. Genet.* 26:299.

Kemp, S.J., and Teale, A.J., 1994, Randomly primed PCR amplification of pooled DNA reveals polymorphism in a ruminant repetitive DNA sequence which differentiates *Bos indicus* and *B. Taurus, Anim. Genet.* 25:83.

Lewin, R., 1989, Limits to DNA fingerprinting, *Science* 243:1549.

Li, M.D., and Enfield, F.D., 1989, A characterization of Chinese breeds of swine using cluster analysis, *J. Anim. Breed. Genet.* 106:379.

Litt, M., and Luty, J.A., 1989, A hypervariable microsatellite revealed by *in vitro* amplification of a dinucleotide repeat within the cardiac muscle actin gene, *Am. J. Hum. Genet.* 44:397.

Lynch, M., 1988, Estimation of relatedness by DNA fingerprinting, *Mol. Biol. Evol.* 5:584.

Matsutani, A., Janssen, R., Donis-Keller, H., and Permutt, M.A., 1992, A polymorphic $(CA)_n$ repeat element maps the human glucokinase gene (GCK) to chromosome 7p, *Genomics* 12:319.

McClelland, M., Mathieu-Daude, F., and Welsh, J., 1995, RNA fingerprinting and differential display using arbitrarily primed PCR, *T.I.G.* 11:242.

Meyer, E., Wiegand, P., Rand, S.P., Kulhman, D., Brack, M., and Brinkman, B., 1995, microsatellite polymorphisms reveal phylogenetic relationships in primates, *J. Mol. Evol.* 41:10.

Moore, S.S., Evans, D., Byrne, K., Barker, J.S.F., Tan, S.G., Vankan, D., and Hetzel, D.J.S., 1995, A set of polymorphic DNA microsatellites useful in swamp and river buffalo (*Bubalus bubalis*), *Anim. Genet.* 26:355.

Moore, S.S., Sargeant, L.L., King, T.J., Mattick, J.S., Georges, M., and Hetzel, D.J.S., 1991, The conservation of dinucleotide microsatellites among mammalian genomes allows the use of heterologous PCR primer pairs in closely related specie, *Genomics* 10:654.

O'Brien, S.J., 1994, Genetic and phylogenetic analysis of endangered species, *Annu. Rev. Genet.* 28:467.

Onda, M., Kudo, S., Rearden, A., Mattei, M.G., and Fukida, M., 1993, Identification of a precursor genomic segment that provided a sequence unique to the glycophorin B and E genes, *Proc. Natl. Acad. Sci. USA* 90:7220.

Paetkau, D., Calvert, W., Stirling, I., and Strobeck, C., 1995, Microsatellite analysis of population structure in Canadian polar bears, *Mol. Ecol.* 4:347.

Paetkau, D., and Strobeck, C., 1994, Microsatellite analysis of genetic variation in black bear populations, *Mol. Ecol.* 3:489.

Paszek, A.A., Flickinger, G.H., Fontanesi, L., Beattie, C.W., Rohrer, G.A., Alexander, L.J., and Schook, L.B., 1996b, Determining evolutionary relationships of diverse swine breeds using microsatellites, *J. Mol. Evol.* (submitted).

Paszek, A.A., Flickinger, G.H., Fontanesi, L., Rohrer, G.A., Alexander, L.J., Beattie, C.W., and Schook, L.B., 1996a, Utility of framework microsatellite markers for inter- and intra-genetic selection between diverse swine breeds, *Mamm. Genome* (submitted).

Pemberton, J.M., Slate, J., Bancroft, D.R., and Barrett, J.A., 1995, Nonamplifying alleles at microsatellite loci: A caution for parentage and population studies, *Mol. Ecol.* 4:249.

Pepin, L., Amigues, Y., Lepingle, A., Bertherier, J-L., Bensaid, A., and Vaiman, D., 1995, Sequence conservation of microsatellites between *Bos taurus* (cattle), *Capra hircus* (goat) and related species. Examples of use in use in parentage testing and phylogeny analysis, *Heredity* 74:53.

Plante, Y., Schmutz, S.M., Lang, K.D.M., and Moker, J.S., 1992, Detection of leucochimerism in bovine twins by DNA fingerprinting, *Anim. Genet.* 23:295.

Porter, V., 1983, "Pigs. A Handbook to the Breeds of the World," Comstock Publishing Associates, Div. Cornell University Press, Ithaca, N.Y.

Richards, R.I., and Sutherland, G.R., 1992, Heritable unstable DNA sequences, *Nature Genet.* 1:7.

Rohrer, G.A., Alexander, L.J., Keele, J.W., Smith, T.P., and Beattie, C.W., 1994, A microsatellite linkage map of the porcine genome, *Genetics* 136:231.

Rubinsztein, D.C., Amos, W., Leggo, J., Goodburn, S., Jain, S., Li, S-H., Margolis, R.L., Ross, C.A., and Ferguson-Smith, M.A., 1995a, Microsatellite evolution-evidence for directionality in rate between species, *Nature Genet.* 10:337.

Rubinsztein, D.C., Leggo, J., and Amos, W., 1995b, Microsatellites evolve more rapidly in humans than in chimpanzees, *Genomics* 30:610.

Schlotterer, C., Amos, W., and Tautz, D., 1991, Conservation of polymorphic simple sequence loci in cetacean species, *Nature* 354:63.

Taylor, E.B., 1995, Genetic variation at minisatellite DNA loci among north pacific populations of steelhead and rainbow trout (*Oncorhynchus mykiss*), *J. Heredity* 86:354.

Teale, A.J., Wambugu, J., Gwakisa, P.S., Stranzinger, G., Bradley, D., and Kemp, S.J., 1995, A polymorphism in randomly amplified DNA that differentiates the Y chromosomes of *Bos indicus* and *Bos taurus*, *Anim. Genet.* 26:243.

Trommelen, G.J., Den Daas, J.M., Vijg, J.H.G., and Uitterlinden, A.G., 1993, DNA profiling of cattle using micro- and minisatellite core probes, *Anim. Genet.* 24:235.

Usha, A.P., Simpson, S.P., and Williams, J.L., 1995, Probability of random sire exclusion using microsatellite markers for parentage verification, *Anim. Genet.* 26:156.

Vaiman, D., Imam-Ghali, M., Moazami-Goudarzi, K., Guerin, G., Nocart, M., Grohs, C., Leveziel, H., and Saidi-Mehtar, N., 1994, Conservation of a syntenic group of microsatellite loci between cattle and sheep, *Mamm. Genome* 5:310.

Vidal, F., Mougneau, F., Gaichenhaus, N., Vaigot, P., Darmon, M., and Cuzin, F., 1993, coordinated post-transcriptional control of gene expression by molecular elements including Alu-like repetitive sequences, *Proc. Natl. Acad. Sci. USA* 90:208.

Weber, J.L., 1990, Informativeness of human (dC-dA)n.(dG-dT)n polymorphisms, *Genomics* 7:524.

Weber, J.L., and May, P.E., 1989, Abundant class of human DNA polymorphisms which can be typed using the polymerase chain reaction, *Am. J. Hum. Genet.* 44:388.

Weising, K., Atkinson, R.G., and Gardner, R.C., 1995, Genomic fingerprinting by microsatellite-primed PCR: A critical evaluation, *PCR Meth. Appl.* 4:249.

Wetton, J.H., Carter, R.E., Parkin, D.T., Walters, D., 1987, Demographic study of a wild house sparrow population by DNA fingerprinting, *Nature* 327:147.

Williams, J.G.K., Kubelik, A.R., Livak, K.J., Rafalski, J.A., and Tingey, S.V., 1990, DNA polymorphisms amplified by arbitrary primers are useful as genetic markers, *N.A.R.* 18:6531.

APPLICATIONS OF DNA FINGERPRINTS FOR THE STUDY OF GENETIC STRUCTURE OF HUMAN POPULATIONS

J. McComb[1], M.H. Crawford[1], W.R. Leonard[2], M.S. Schanfield[3], and L. Osipova[4]

[1]Laboratory of Biological Anthropology, University of Kansas, Lawrence, Kansas, 66045
[2]School of Human Biology, Guelph University, Guelph, Ontario, Canada
[3]Analytical Genetic Testing Center, Denver, Colorado
[4]Population Genetics Laboratory, Institute of Cytology and Genetics, Russian Academy of Sciences, Novosibirsk, Russia

INTRODUCTION

The highly polymorphic nature of Variable Number Tandem Repeat (VNTR) loci have made these DNA markers useful in the examination of the genetic structure of human populations (Hartmann et al., 1994; McComb et al., 1995a). A VNTR locus is characterized by a sequence of DNA bases which is repeated over a section of DNA. The number of repeats is variable, sometimes few in number, and thus exhibiting a small allele. In other cases, the repeats are numerous, and the locus has a large allele. Thus, a VNTR allele can be characterized by its length.

A VNTR locus is inherited in a Mendelian fashion. Therefore each individual receives an allele from his/her mother and another allele from his/her father. Considerable diversity exists in most VNTR loci because of the high mutation rates (Jeffreys, et al. 1988). Thus, it is common for an individual to possess two alleles of different sizes at a single locus.

A DNA fingerprint consists of several VNTR loci used to characterize an individual genome. The pattern of VNTR fragment sizes observed for that individual is relatively distinct from unrelated individuals. However, in genealogically-related individuals it is expected that some of the VNTR fragments will be identical in size because the fragments are identical-by-descent. Hence, VNTR markers are particularly useful in the characterization small isolated populations in which many of the individuals are related.

Although standard markers of the blood (blood groups, proteins and enzymes) are useful in examining the structure and evolutionary relationships between human populations, VNTRs provide a unique perspective because most noncoding VNTR loci appear to be unaffected by natural selection (Harding, 1992). Thus, VNTRs should provide information on how certain evolutionary forces etch their marks on the gene pool. In particular, because alleles may be removed from a gene pool by genetic drift, VNTR analysis is useful in examining the effects of stochastic processes on the genetic diversity of small isolated populations.

Siberian human aggregates are suited for VNTR testing because of the subdivision and cultural diversity found within the indigenous populations. Historical records and genealogical reconstructions document the relative isolation of groups such as the Evenki reindeer herders. However, other groups, such as the Altai, and the Kets, are less isolated reproductively and appear to have been more profoundly affected by the movement of Russian and Asian immigrants into Siberia. In addition to the migrations into Siberia, mitochondrial DNA evidence links the origin of the Native Americans to migrants who came out of Siberia (Torroni et al., 1993).

Evolutionary theory predicts that these historic and prehistoric events will be reflected in the gene pool and will have affected the frequency of various VNTR fragments. Like the standard blood markers, some VNTR loci may be more informative than others in reconstructing population history. However, due to the high mutation rates at VNTR loci it is possible for two VNTR fragments to possess a similar number of repeats via a mutation and not because they are identical-by-descent (Kidd et al., 1991). Although, it is improbable that such a mutation will increase to a level which affects the analysis of that locus significantly, clearly the best approach, to eliminate this bias, is achieved through the use of multiple markers in a multivariate analysis. Thus, the effects of evolutionary forces on the populations of Siberia have been examined by using multiple VNTR markers.

The Populations

The Siberian indigenous populations are represented by four samples from Central Siberia which were drawn in the 1991 - 1993 field seasons. A summary of sample sizes for each locus can be found in Table 1.

The individuals who constitute the Mountain Altai sample come from the village of Mendur - Sokkon which is home to approximately 1000 individuals. This Kizhi village is located near the Katun' river proximal to the borders of Kazakhastan, China and Mongolia. The people of the Mountain Altai are primarily pastoralists, and because of their geographic location, have historically had contact with populations of both Asia and the Europe. The language of the Altai is Turkic in origin and belongs to the Altaic language family. According to Bowles (1977), the Turkic speakers originated in the Altai region and spread eastward, into Asia, and westward, towards Europe. Ultimately, the Altai region became a crossroads between Europe and Asia. Historical records indicate that the Altai maintained trade contacts with the Chinese and the Mongolians (Popov, 1964). In addition, the Altai were also subjected to the invasions of the Mongol hordes which swept the Middle East and Europe (Popov, 1964). In 1756, the Altai allied themselves with the Russians who were expanding into Siberia.

Figure 1. Map of Central Siberia which shows the villages sampled in this study.

The 22 Ket individuals are from the village of Sulamai which contains approximately 60 Kets and 60 Russians. The village is located in near the junction of the Yenisey and Stony Tunguska rivers in the Krasnoyarsk region. This region contains only 1084 Kets who subsist primarily by hunting and fishing. The Ket language is unintelligible to the surrounding populations and is a linguistic isolate when compared to the three major linguistic families of Siberia (Altaic, Uralic and Paleoasiatic). Thus, there is much controversy about the origin of the Kets. Historically, the Kets have allied themselves with Russian settlers who were pushing into Siberia from the west (Popov and Dolgikhi, 1964). The proximity of Russian settlers has led to a mutual mixing of the of the two groups and approximately 50% of the Kets in the Krasnoyarsk region have at least one Russian parent. The Evenki samples were drawn from the two villages of Poligus and Surinda as well as 12 brigades (herding units) which are located around the Stony Tunguska river and its tributaries. Each village (including the brigades) contains approximately 600 people. The Evenki are reindeer herders who occupy a territory spanning the taiga of Central Siberia to the Amur Region of East Asia (Vailevich and Smolyak, 1964). There are approximately 30,000 Evenki making them one of the more numerous indigenous people in Siberia (Hannigan,

1991). The Evenki speak a Tungusic language which belongs to the Altaic language family. Their origins are in dispute, but based on archeological evidence, Okladnikov (1950) has postulated that the Evenki bands originated near Lake Baikal. Vasilevich (1946) notes that the Evenki were probably reindeer hunters who later domesticated reindeer to adopt the herding mode of subsistence seen historically. The wide-spread Evenki have had some historic contact with the Mongols, and formed ties with the Russians in the 17th century (Vasilevich and Smolyak, 1964). Unlike the Altai and the Kets, however, the Evenki have remained isolated on the taiga and have experienced less contact with other groups than either of the first two groups.

For comparative purposes, several groups residing in the United States were characterized genetically. These samples were drawn from paternity cases in Colorado, New Mexico and Arizona by Analytical Genetic Testing Center (Denver). Each individual reported his/her own ethnic group affiliation. These groups contain varying sample numbers as shown in Table 1.

Table 1. Sample sizes, in individuals, for each locus.

Population	D7S104	D11S129	D18S17	D20S15	D21S112
African American	695	284	253	250	553
European American	1881	461	640	387	1367
Mexican American	597	135	160	149	379
Native American	140	41	57	62	170
Mendur-Sokkon(Altai)	95	95	95	95	95
Sulamai (Ket)	22	22	22	22	22
Surinda (Evenki)	8	83	83	83	8
Poligus (Evenki)	18	18	18	18	18

DNA Analyses

In the 1991-1993 field seasons, 10 ml blood samples were collected in vacutainers from the villages of Surinda, Poligus, Sulamai, and Mendur - Sokkon. The blood was shipped to the Analytical Genetic Testing Center (Denver) and DNA was extracted from the buffy coats. Five microgoods of DNA were digested from each individual using the restriction enzyme *Pst*-I and the fragments of DNA were separated using horizontal agarose gel electrophoresis. The DNA fragments were then Southern blotted onto a nylon membranes. Several different membranes were hybridized with the biotinylated probes found in Table 2. The probes were then detected non-isotopically using the enzyme alkaline phosphatase (which was reacted with NBT and BCIP) which was linked to the probe via a biotin-streptavidin link. Fragment sizes were obtained using a DNAstar digitizer with two different people reading each membrane. (A more detailed protocol can be found in McComb et al., 1995b)

Table 2. Characteristics of the Collaborative Research Incorporated (CRI) probes used in this study.

Locus	Probe	Repeat Length (base pairs)
D7S104	CRI-PAT-pS194	50
D11S129	CRI-PAT-pR365-1	28
D18S17	CRI-PAT-pL159-1	48
D20S15	CRI-PAT-pL355-8	33 - 35
D21S112	CRI-PAT-pL427-4	26 - 30

Analytical Procedures

The digitized DNA fragment sizes were binned using +/- 2 % error bins. To verify statistical significance of the distributions the Kolmogorov - Smirnov two sample test (p = 0.05, with Bonferroni protection) was used (Sokal and Rohlf, 1981). In this paper, only the Kolmogorov - Smirnov results from locus D11S129 are presented. Discussion of other results may be found in McComb et al., 1995b.

The genetic structure within the Siberian groups was examined using a linear discriminant function. The discriminant function examines the variance and covariance of fragment sizes between the different populations and, using eigenanalysis, determines functions that can best separate the individuals into their respective groups. Using VNTR fragment sizes, this analysis examines the multivariate variation, and creates linear equations which are used to classify individuals into their population of origin.

Three VNTR loci: D11S129, D18S17 and D20S15, were chosen for this analysis because they had the largest sample size. The fragment sizes from each locus were used as variables to form the discriminant function. Because humans are diploid, each individual had two variables (fragments) assigned for each locus. The first variable was the largest fragment in that individual, and the second variable was the other (possibly equal) fragment detected at that locus. The variance-covariance matrices for the populations were sufficiently homogenous to use a pooled within-group variance-covariance matrix (p > 0.05, and accepts the null hypothesis that there is no lack of homogeneity). Because of the absence of univariate normality in the VNTR fragment distributions (as shown by n-scores), the absence of multivariate normality was assumed. Non-normal data may skew a discriminant function. Thus, the significance of the function is difficult to assess except via non-parametric means. As an estimate of significance, 500 discriminant functions were generated randomly and the proportion of individuals correctly classified was compared to the Siberian discriminant function.

In addition, the significance of the function was also assessed by randomly removing 15 Altai from the data set prior to the computation of the linear discriminant function. The village of Poligus (Evenki) was also excluded from the computation. After the Siberian function was calculated these individuals were then classified using the equation.

To examine the differences between the Siberian populations, the principal components of an R matrix were used (Harpending and Jenkins, 1973). The R matrix is a specialized variance - covariance matrix, with the elements calculated by the general equation

$$r_{ij} = \frac{1}{k} \sum_{i=1}^{k} \frac{(P_{il} - \bar{P}_l)(P_{jl} - \bar{P}_l)}{\bar{P}_l(1 - \bar{P}_l)}$$

where k is the total number of alleles. P_i and P_j are the frequency of a specific allele (l) at population i and population j. P - bar is the average allele frequency over all of the populations (Harpending and Jenkins, 1973).

Similar to the discriminant function, a principal component analysis examines the variance and covariance of the frequencies of different fragments within and between populations. It is expected that populations which share an evolutionary relationship will be genetically similar. Thus, the frequencies of different fragments within two related populations should be similar and therefore the variance will also be similar.

In a principal components analysis, the variance-covariance matrix of fragment frequencies is decomposed into eigenvectors which correspond to a proportion of the variation observed within the variance-covariance matrix. Populations which have similar variation are grouped together when this variation is observed. Essentially, this analysis examines the total multivariate variation, and select groups which are similar to one another.

The R matrix analysis was used to examine the variation in Siberian and the American reference populations. In the analysis, fragment frequencies were used from all of the five VNTR loci and the variance-covariance matrix was weighted by sample size. In addition to the R matrix plot, Nei's (1973) weighted G_{ST} is reported. It was calculated by the general equation

$$G_{ST} = D_{ST} / H_T$$

where H_T is gene diversity (heterozygosity) in the total population. D_{ST} is a measure of gene diversity between populations (Nei, 1973).

Ultimately, G_{ST} is a general measure of genetic differentiation between the populations being examined. To assess the significance of G_{ST}, 500 randomly generated G_{ST} values were calculated using the same data set used to generate the observed G_{ST}. Each G_{ST} has been corrected for random bias by subtracting the mean of the distribution of the randomly generated G_{ST} values (J. Blangero, personal communication)

To examine the effects of gene flow and isolation within the populations, the genetic distance from the centroid (r_{ii}) was plotted against the mean per locus heterozygosity (H) (Harpending and Ward, 1982). The mean per locus heterozygosity is calculated by summing the average heterozygosity expected under Hardy-Weinberg conditions at each locus and dividing by the total number of loci. The genetic distance from the centroid is a statistic which shows how genetically different a single population is from the mean of the gene frequencies over all of the populations. If a population is isolated, then it tends to lose VNTR fragment alleles (due to genetic drift) and becomes less diverse and more genetically distinct. This loss of variation increases the r_{ii} value. In contrast, populations which experience gene flow tend to gain new VNTR fragment sizes and thus have a low r_{ii} because their gene pools have similar diversity to that of the other populations.

Heterozygosity is similar in that as fragment diversity increases heterozygosity tends to become elevated. In addition, inter - populational gene flow may also increase heterozygosity. Thus, a linear relationship is established: as fragment diversity rises, so does heterozygosity. However, populations may deviate from this linear relationship if they experience substantial genetic drift (from isolation), or gene flow. Thus, a heterozygosity versus r_{ii} plot was constructed, using the five VNTR loci and the Siberian and American populations.

RESULTS

D11S129 is the most informative of all of the VNTR loci used in this research, and the population specificity of this locus is best described in simple histograms (Figure 2). The most obvious feature shown in the histograms is the presence of a 600 base-pair void that separates the distribution of VNTR fragments into two modes. From the histograms it is clear that this "great void" is present in all of the populations thus far sampled, and the few (a total of 6) alleles found within the void come from European and African Americans. More importantly, the distribution of alleles on either side of this void is variable between the different populations.

The African American and the European American samples both show a high frequency of small fragments and a low frequency of the large fragment sizes. In contrast, the Native Americans, and the Evenki display a low frequency of small fragments and a high frequency of large fragments. The Mexican Americans, the Kets and the Altai all have frequencies that appear to be intermediate between the other populations.

Statistically, a Kolmogorov - Smirnov test for the comparison of distributions (p = 0.05 with Bonferroni protection) indicates that significant differences exist between the distributions of most of the populations (Table 3). The major exception is the Kets, whose sample may be of insufficient size to detect the differences.

The high frequency of large sized fragments found in Siberia and the New World suggests a common phylogenetic relationship between the two regions. This supports the data from mtDNA and other markers (Torroni et al., 1993) which show that the Asians from Siberia were some of the original founders of the New World populations.

Based on historical reconstructions, the intermediate frequencies

Table 3. Kolmogorov - Smirnov Results at locus D11S129.

	Mexican American	African American	European American	Altai	Ket	Evenki (Surinda)	Evenki (Poligus)
Native American	0.36*	0.72*	0.66*	0.35*	0.27	0.38*	0.26
Mexican American		0.37*	0.30*	0.21*	0.12	0.59*	0.48*
African American			0.12*	0.44*	0.48*	0.83*	0.77*
European American				0.34*	0.39*	0.78*	0.72*
Altai					0.09	0.46*	0.40*
Ket						0.47*	0.36
Evenki (Surinda)							0.11

* Significant at p = 0.05 with Bonferroni Protection

found in several populations can be explained by admixture between the different groups. For example, the Indians of Mexico probably once had a distribution similar to other Native Americans. However, gene flow from both the Spanish and Africans (slaves) (Crawford et al., 1979) has molded their gene pool into the distribution shown in Figure 2. Genealogical reconstruction reveals that the Kets (Sulamai) are heavily admixed with Russians, and their gene pool also appears to be a mixture of indigenous Siberian and European alleles. Finally the mountain Altai may have a European component in their gene pool because they have been in contact with the Russians and the Turkic speakers to the west. Gene flow from Mongolia, Siberia, and Europe has apparently created the mixed distribution observed in the Mendur - Sokkon gene pool.

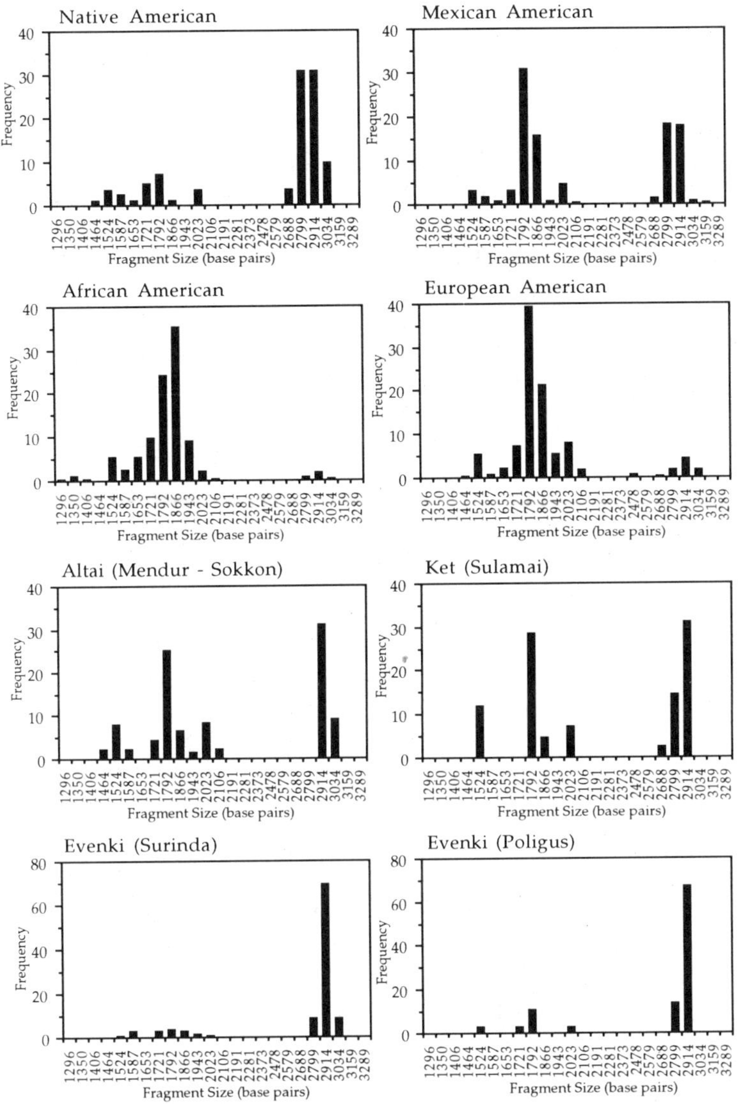

Figure 2. Histograms of fragment distributions at locus D11S129.

Although D11S129 is an informative locus, the other VNTR loci also show similar trends. By combining the different loci into a multivariate analysis the clearest picture of the genetic structure of Siberia emerges. The genetic structure of human populations can be approached on two levels: (1) the distribution of individual genotypes, within and between populations, can be examined. Second, the frequency distribution of allele types in each gene pool can be examined (without examining the specific genotypes of individuals).

By examining the distribution of different individuals with a specific genotype, it is possible to detect familial relationships. Since VNTR alleles are transmitted in a Mendelian fashion, it is expected that family members will have similarities in genotype. The populations of Siberia are particularly suited for this because they are small populations of related individuals. A linear discriminant function separates the Siberian individuals into their correct ethnic group 88.1 % of the time, using the three loci D11S129, D18S17 and D20S15 (Table 4). The coefficients of the linear discriminant function can be found in appendix 1.

500 groups were randomly generated using the Siberian data and were used to create random linear discriminant functions to compare how well the real function compared to random distributions. The results are plotted in Figure 3 which shows that the true classification is superior to the random results. Because of the nature of a discriminant function, some individuals lie on the border between groups, and are difficult to classify. The Siberian function has 5 individuals that are in such border areas and this gives a range of 87 % - 92 % correctly classified for the function. Even at the low end of the range, the Siberian function shows a trend towards correct classification.

The ultimate test of a discriminant function is to classify individuals who were not used to create the function. For the Siberian function, the Evenki from the adjoining community of Poligus and 15 randomly selected mountain Altai were excluded from the generation of the classification matrix. The Siberian linear discriminant function was then used to classify these individuals (Table 5). The equation separated the individuals 84 % of the time. In addition, it is interesting to note that the misclassified Evenki are placed into the Ket group. Poligus and the Ket village of Sulamai are reasonably close, geographically.

A principle component decomposition of an R matrix (variance-covariance matrix) can be used to achieve similar results to the linear discriminant function. However, the R matrix analysis, uses the frequency of fragment sizes to assess population affinity rather than the genotype of specific individuals. Along with the principal components, Nei's G_{ST} (1973) is often used as a measure of genetic differentiation between populations. The G_{ST} of 0.031 (corrected) for the Siberian populations is highly significant ($p < 0.002$) when compared to 500 randomly generated values using the same Siberian

Table 4. Linear Discriminant Analysis for the Siberian Groups using Loci D11S129, D18S17 and D20S15.

Put into Group	Altai	Ket	Evenki (Surinda)
		True Group	
Altai	68	0	0
Ket	0	11	9
Evenki	2	4	32
Total N 1	70	15	41
N Correct	68	11	32
Proport.	0.97	0.73	0.78

N = 126   N Correct = 111   Prop. Correct = 0.881

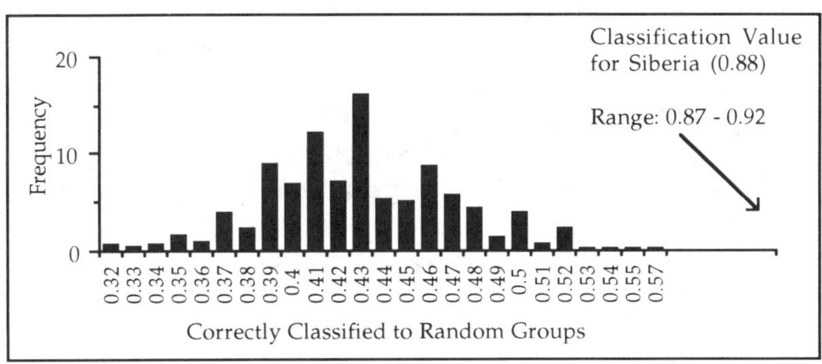

Figure 3. Results of 500 randomly generated discriminant functions.

data. Thus, there is significant genetic differentiation between the Siberian populations.

Figure 4 is a plot of the principal components of an R matrix using the five VNTR loci (D7S104, D11S129, D18S17, D20S15 and D21S112). Most of the variation is contained within the first eigenvector (67.4 %) which places the Evenki together, the Kets close to the centroid, and the Altai are distant from the other groups. This relationship is familiar as the Kets are geographically closer to the Evenki, while the Altai are geographically distant from both groups.

The second eigenvector (Y - axis) explains 24.7 percent of the variation and presents a picture that seems antithetical at first because one would expect the related Evenki groups to cluster together. The Altai and Kets are lumped together, while the two Evenki villages are separated at either end of the Y - axis. This eigenvector is indicative of gene flow into the different populations. The relationship between stochastic process versus gene flow is best approached by an examination of the mean per locus heterozygosity versus the genetic distance from the centroid (r_{ii}).

Table 5. The discriminant function was used to classify 15 Altai who were randomly removed from the sample before the Siberian function was calculated. In addition, individuals from the other Evenki village (Poligus) are also classified.

Group	Altai (Random)	Evenki (Poligus)
Altai	15	0
Ket	0	5
Evenki	0	11
Total N	15	16
N Correct	15	11
Proport.	1.00	0.69

N = 31   N Correct = 26   Prop. Correct = 0.839

Figure 5 is a plot of the mean per locus heterozygosity (H) versus r_{ii} for American populations as well as the Siberians using the five VNTR loci. A theoretical regression line has been fit to the points using the method of Harpending and Ward (1982). Evolutionary theory predicts that populations experiencing gene flow will have a high heterozygosity. In addition, populations that are intermixing with

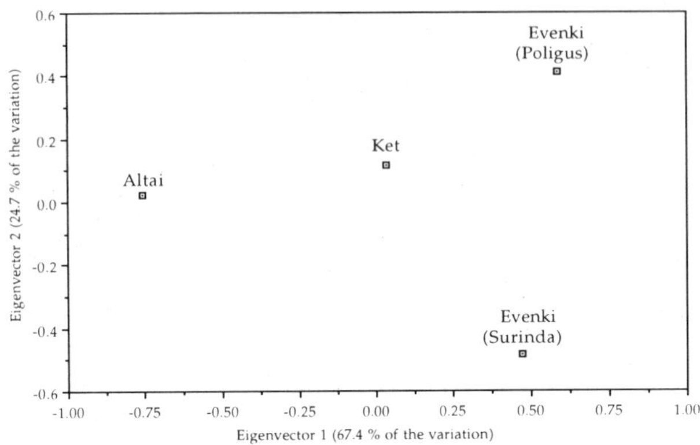

Figure 4. Principal component analysis of the Siberian R matrix using the five VNTR loci.

one another (via gene flow) will become more genetically similar, and thus have a lower genetic distance from one another. It is thus expected that isolated populations with little gene flow will have a high genetic distance and a low heterozygosity. Figure 5 shows clearly that as the genetic distance from the centroid increases, the heterozygosity decreases,

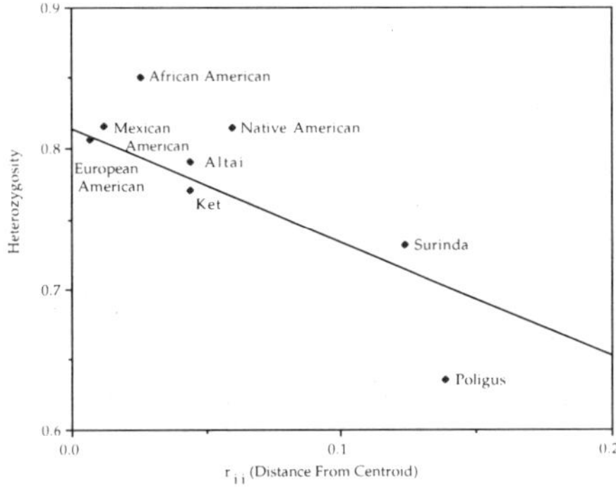

Figure 5. Plot of the genetic distance from the centroid versus the mean per locus heterozygosity (as from McComb et al., 1995b).

and therefore the populations approximate an isolation-by-distance model. The fitted regression indicates the relative flow of genes into the various populations. Populations above the line tend to be receiving gene flow, while those below the line are isolated and subject to genetic drift. Though there are no significant deviants from the regression line, the village of Poligus appears to be more isolated than the village of Surinda. In contrast, the Altai and the Kets appear to receive a similar magnitude of gene flow. This situation is reflected by the dispersal along eigenvector 2 of Figure 4: Poligus and Surinda are relatively distinct in terms of gene flow; Sulamai (Ket) and Mendur - Sokkon (Altai) have similar levels of gene flow.

The principle components of an R matrix (Figure 6) uses the frequencies of fragments from the five VNTR loci in the American and Siberian populations to examine evolutionary relationships between the populations. The two principle components account for a similar amount of variation (38.8 % and 35.5 %) and separate out the populations into three distinct clusters. The Old World populations (African Americans and the European Americans) are distinct from all the New World and Siberian groups. Even though the distribution at locus D11S129 is similar for the Old World groups, the other loci are dissimilar, making these populations unique. In addition, the Native American, Mexican American and Siberian populations cluster together. This cluster shows that the common evolutionary relationship seen in locus D11S129 is also displayed by the other loci. Also, the populations that appear to be admixed (by locus D11S129) show a closer affinity, along the first eigenvector, tothe European and African Americans than the Evenki and the Native Americans.

The corrected G_{ST} for all of the populations is 0.032 and is again significantly different ($p < 0.002$) from randomly generated G_{ST} values using the same data. Thus, the Siberian and World G_{ST} values appear to be identical. This is a direct effect of the calculation of the GST statistic: $G_{ST} = D_{ST} / H_T$ (Nei, 1973). Because the Siberian populations are isolated and small in size, they tend to have a low gene diversity, and also a low heterozygosity. Since the low heterozygosity

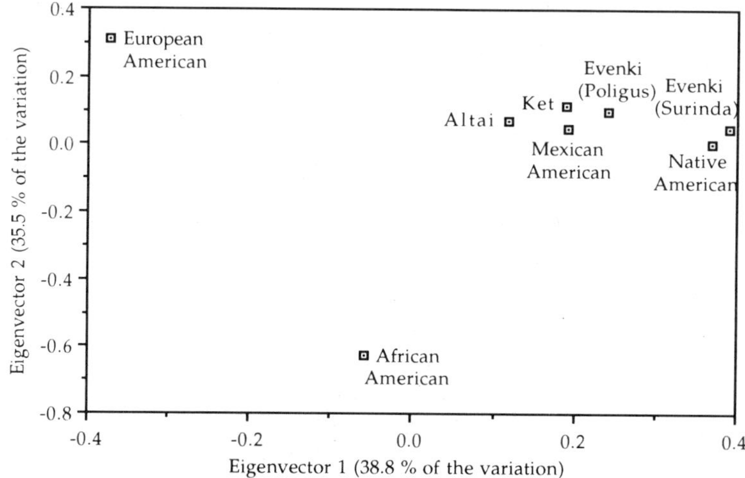

Figure 6. Principal component analysis of the Siberian and American R matrix using the five VNTR loci (as from McComb et al., 1995b).

(expressed as a percent) is in the denominator of the equation, it tends to drive the G_{ST} value up. In Siberia, this low heterozygosity makes the G_{ST} equal to the G_{ST} value observed between the populations in Figure 6.

DISCUSSION

Siberian VNTRs tell a tale of population expansion, isolation, and migration within and between the various indigenous groups. More importantly, the analyses of the Siberian VNTRs confirm the historic accounts and support previous studies of the relationship between Siberia and the New World based on mtDNA (Torroni et al., 1993). Figure 7 summarizes these relationships as reflected in population movement in Siberia.

Prior to 11,500 years ago (during the pre-Clovis period in the New World) a group of Asian Siberians crossed over the Bering Strait and into the New World. Although it is not known whether they traveled inland across Beringia or (more likely) along the coast, it does appear that the large-sized fragments of locus D11S129 were within their gene pool, and that a similar distribution is found in modern Siberian populations links the two populations.

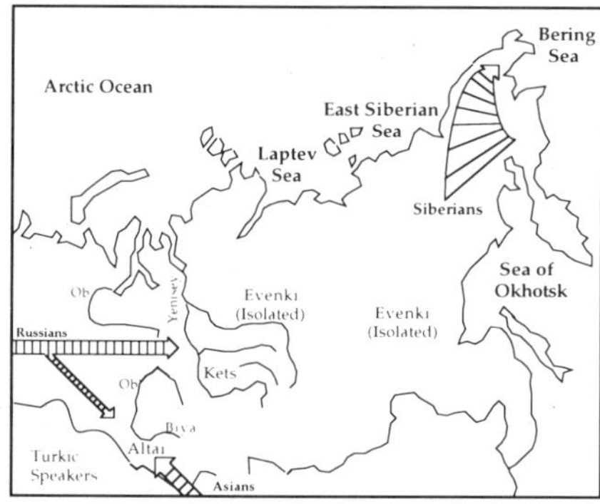

Figure 7. Map of Central Siberia which details several population movements in Siberia.

Along the taiga, the Evenki reindeer-herders expanded to encompass a region from East Asia to Central Siberia. However, the brigades remained relatively isolated, as indicated by the genetic distance from the centroid and the heterozygosity of the populations. The expansion of Russians into Siberia brought new trade prospects for the Evenki. However, the Russian expansion appears to have had little effect upon the Evenki gene pool: (1) locus D11S129 indicates few European alleles within the Evenki gene pool; (2) R matrix analysis suggests that the Evenki are distinct from other Europeans.

Although the Kets may have once been isolated, their alliance with the Russians has had an impact on their gene pool. Genealogically about 50 % of the Kets report a single Russian parent. This is supported by the distribution of fragments at locus D11S129 which appears to be a mixture of European and Siberian fragment sizes. In addition, the genetic distance from the centroid versus the mean per heterozygosity plot indicates that the Kets are less isolated then their neighbors, the Evenki.

Historically, the mountain Altai have not been reproductively and culturally isolated. Their contacts with other Turkic groups, the Asians and the Russians have molded their gene pool in various ways. Gene flow from the Europeans may have contributed the small (Old World) fragment sizes at locus D11S129. Contact with Siberians and Asians appear to have added the large fragment sizes producing the mixed distribution observed today. In addition, like the Kets, the Altai gene pool indicates less isolation than the Evenki thus supporting the historical records.

Ultimately, Variable Number Tandem Repeat loci provide useful markers in studying gene flow and the stochastic processes which operate on the gene pools of small populations. Unlike standard blood markers, many VNTRs appear to be unaffected by natural selection, which has resulted in highly polymorphic loci. When compared to classical markers, the large number of VNTR fragment sizes, provides finer resolution of the action of specific evolutionary forces in shaping a gene pool. As shown above, VNTR analysis supports historical and prehistoric events, which points to their utility in reconstructing population history. In conclusion, VNTRs provide a unique perspective in which to observe the evolutionary forces which mold a gene pool and the cultural patterns which influence these same forces.

Appendix 1: Linear Discriminant Function for the Siberian Groups

	Altai	Ket	Evenki (Surinda)
Constant	-456.10	-467.30	-482.98
D11S129 A	5.54	1.83	4.30
D11S129 B	21.33	29.01	30.07
D18S17 A	91.23	80.88	81.30
D18S17 B	43.01	51.18	50.38
D20S15 A	11.08	6.63	5.64
D20S15 B	27.68	34.02	36.29

References

Bowles, G.T., 1977, The People of Asia, Charles Scribner's Sons, New York.
Crawford, M.H., Dykes, D.D., Skradski, K., and Polesky, H.F., 1979, Gene flow and genetic microdifferentiation of a transplanted Tlaxcaltecan Indian population: Saltillo, *Am .J .Physical Anthr* .50: 401.

Hannigan, J., 1991, Statistics on the economic and cultural development of the northern aboriginal people of the USSR (for the period of 1980-1989), Bureau of Indian and Northern Affairs, Ottawa.

Harding, R.M., 1992, VNTRs in Review, *Evolutionary Anthropology* 1:62.

Harpending, H., and Jenkins, T., 1973, Genetic distances among southern African populations, *in:* "Methods and Theories of Anthropological Genetics ", M.H. Crawford and P.L. Workman (eds), University of New Mexico Press, Albuquerque.

Harpending, H., and Ward, R.H., 1982, Chemical systematics and human populations, *in:* "Biological Aspects of Evolutionary Biology ", M.H. Nitecki (ed), University of Chicago Press, Chicago.

Hartmann, J., Keister, R., Houlihan, B., Thompson, L., Baldwin, R., Buse, E., Driver, B., and Kuo, M., 1994, Diversity of ethnic and racial VNTR RFLP fixed-bin frequency distributions, *Am. J. Human Genet.* 55:1268.

Jeffreys, A.J., Royle, N.J., Wilson, V., and Wong, Z., 1988, Spontaneous mutation rates to new length alleles at tandem-repetitive hypervariable loci in human DNA, *Nature* 332:278.

Kidd, J.R., Black, F.L., Weiss, K.M., Balazs, I., and Kidd, K.K., 1991, Studies of three Amerindian populations using nuclear DNA polymorphisms, *Human Biology* 63:775.

McComb, J., Blagitko, N., Comuzzie, A.G., Schanfield, M.S., Sukernik, R.I., Leonard, W.R. and Crawford, M.H., 1995a, VNTR DNA variation in Siberian indigenous populations, *Human Biology* 67:217.

McComb, J., Crawford, M.H., Osipova, L., Karaphet, T., Posukh, O., and Schanfield M.S., 1995b, DNA inter-populational variation in Siberian indigenous populations: the mountain Altai, *Am. J. of Human Bio.*, in press.

Nei, M., 1973, Analysis of gene diversity in subdivided populations, *Proc. Nat. Acad. Sci. USA* 70 (12): 3321.

Okladnikov, A.P., 1950, The initial stages of formation of the people of Siberia: the population of the Cis-Baykal region during the Neolithic and Early Bronze ages, *Sovetskaya Etnografiya* 2.

Popov, A.A., and Dolgikh, B.O., 1964, The Kets, *in:* " The Peoples of Siberia ", M.G. Levin and L.P. Potapov (eds), University of Chicago Press, Chicago.

Popov, L.P., 1964, The Altays, *in:* " The Peoples of Siberia ", M.G. Levin and L.P. Potapov (eds), University of Chicago Press, Chicago.

Sokal, R.R., and Rohlf, F.J., 1981, Biometry, 2nd ed, Freeman, San Francisco.

Torroni, A., Sukernik, R.I., Schurr, T.G., Starikovskaya, Y.B., Cabell, M.F., Crawford, M.H., Comuzzie, A.G., and Wallace, D.C., 1993, mtDNA variation of aboriginal Siberians reveals distinct genetic affinities with Native Americans, *Am. J. Human Genet.* 53: 591.

Vasilevich, G.M., 1946, The oldest ethnonyms of Asia and the names of Evenki clans, *Sovetskaya Etnografiya* 4.

Vasilevich G.M., and Smolyak A.V., 1964, The Evenks, *in:* " The Peoples of Siberia", M.G. Levin and L.P. Potapov (eds), University of Chicago Press, Chicago.

PROGRESS TOWARD A TRANSCRIPT MAP OF THE HUMAN GENOME

Anthony R. Kerlavage and Mark D. Adams

The Institute for Genomic Research
9712 Medical Center Drive
Rockville, MD 20892 USA

INTRODUCTION

A major goal of the human genome project is the identification and mapping of the complete complement of human genes. Single-pass, partial sequencing of cDNA clones to generate Expressed Sequence Tags (ESTs) has provided a rapid method of gene discovery (Adams et al., 1991), which has been widely applied in humans and other species (Bogusky et al., 1993). In addition, the EST strategy provides a key resource for gene mapping (Polymeropoulos et al., 1993); Durkin et al., 1994; Korenberg et al., 1995). The Institute for Genomic Research (TIGR) has sequenced ESTs from over 300 human cDNA libraries made from single cells, fetal and embryonic tissues, a large number of adult organs and tissues, and cancerous tissues (Adams et al., 1995). The combination of data on gene expression and putative gene functions inferred from sequence similarity provides a powerful means of assessing the transcriptional activity of the genome in the cells and tissues of an organism.

The total number of human EST sequences from projects world-wide has surpassed 350,000. We have assembled these ESTs into contigs, reducing redundancy and allowing analysis of the tissue distribution of transcript expression. This process has resulted in identification of over 40,000 distinct, non-overlapping cDNA sequences with associated expression profiles.

TIGR has established the Human cDNA Database (HCD) to provide researchers at non-profit institutions access to cDNA/EST sequence and related data TIGR and HumanGenome Sciences (HGS). HCD currently contains over 350,000 EST sequences, including approximately 105,000 sequences obtained at TIGR along with approximately 55,000 sequences from HGS that have significant overlap with TIGR sequences or sequences in the public domain. These sequences represent between 30,000 and 35,000 unique human genes and serve as important resources for the generation of a human transcript map.

EXPRESSED SEQUENCE TAGS

The laboratory aspects of EST sequencing are quite straightforward (Adams et al., 1994), and the information management and sequence analysis aspects have been well developed (Kerlavage et al., 1993, 1995). An EST project begins with an experimental question to answer such as: 'what genes are expressed in liver' or 'what are the differences in expression between normal and transformed cells' or what are the most abundantly expressed genes in this species.' Selection of tissue samples for preparation of mRNA is driven by these questions.

Construction of a representative cDNA library is critical so that the distribution of EST sequences can be reliably assumed to reflect the steady-state mRNA levels in the tissue from which the library was made. cDNA library construction has been reviewed previously (Moreno-Planques and Fuldner, 1994). Over three hundred cDNA libraries were constructed for human EST sequencing projects at TIGR (Adams et al., 1995), the majority of which were constructed using the lambdaZAP phage system (Stratagene, La Jolla, CA). An extensive quality control regimen was developed to assess the quality of each library prior to and during large-scale sequencing (Adams et al., 1995). Over half of the libraries were rejected or re-made due to skewed representation of transcripts, unacceptably high levels of non-recombinant clones or mitochondrial transcripts, or small insert sizes. Several libraries contained large numbers of clones for a few very abundant mRNA species, reflecting high levels of expression of certain genes. While these libraries were representative, they were not suited to large-scale sequencing and were screened with total cDNA or specific clones in order to increase the diversity of clones sequenced. Nearly all libraries were directionally cloned, meaning that the polyA tail of the mRNA was cloned adjacent to a specific end of the vector. Directional cloning allows a choice of 5' or 3' ends of the cDNA for sequencing.

The primary decision point in EST sequencing is whether to sequence 5' ends, 3' ends, or both. Especially in mammalian mRNA, the 3' untranslated sequence can be quite long (e.g., - 2.4 kb in the obesity gene (Zhang et al., 1994)). Since protein-coding information is most useful for gene identification, 5' ends are preferred for gene discovery applications. However, since 3' ends of transcripts are typically unique, they provide much more robust analysis of redundancy for comparative projects since cDNAs representing transcripts from the same gene can always be grouped accurately)assuming no gross differences in polyadenylation site). Sequencing 3' ends also provides unique markers for mapping genes. Choosing to sequence both ends of clones answers both goals, but at a reduction oin throughput since two reactions are done on each clone rather than one. We have used a combination of approaches, depending upon the project. Frequently, clones are selected for sequencing from both ends based on the results of gene identification and grouping, thus minimizing the number of additional sequencing reactions which must be performed.

Plasmid templates (prepared using a procedure from AGTC, Inc., Gaithersburg, MD) were reacted using fluorescent dye-labeled sequencing primers and cycle sequencing with Taq polymerase. Reactions were run on Applied Biosystems 373 DNA Sequencers. The large volume of materials and information that are handled in a project of this scale require a careful attention to standards of quality to ensure that the resulting sequences will meet minimum standards of accuracy (Adams et al., 1994). A laboratory information management system has been developed based around the Sybase relational database management system (ESTDB, Kerlavage et al., 1993, 1995). Sofrware was written to pass information on template preparation, sequencing reactions, and sequencing gel runs to ESTDB. This system provides the basis for tracking ongoing projects and maintaining quality control. Accurate management of projects is the beginning of sequence analysis. All ESTs are subjected to an initial standard sequence analysis protocol. This includes identification of repetitive elements, classification of matches to known human genes, searches against the public protein and nucleic acid databases, comparison with other ESTs, assessment of potential to contain a protein-coding region. For an average library, between thirty and fifty percent of the ESTs can be putatively identified based on sequence similarity to a gene from human or another organism. The remainder remain unidentifiable, but substantial information can still be derived by examining the cDNA libraries in which they were found.

cDNA libraries generally reflect the mRNA abundance in the tissue from which they are made. Therefore, more ESTs are obtained from abundantly expressed transcripts than from rare transcripts. Correct association of ESTs and genes is essential to calculate redundancy and accurately assess the distribution of expression of genes in different tissues. For ESTs matching known human genes, this process is reasonably straightforward, because even non-overlapping ESTs can be assigned to the same gene unambiguously by a match to the complete sequence from GeneBank. To facilitate the association of ESTs with known human genes, we developed a canonical set of human cDNA sequences from GenBank (HT sequences, Adams et al., 1995). The HT dataset contains a non-redundant set of human genes with associated annotation such as coding regions, splice insoforms, and exon-intron boundaries. For each sequence in the HT dataset, a profile of gene expression was constructed indicating the relative level of expression of that gene in each cDNA library studied.

ESTs that do not exactly match a sequence in GenBank are more difficult to analyze in terms of redundancy and distribution. To identify the degree of redundancy and build sets of overlapping ESTs, software was developed to build assemblies of ESTs (Sutton et al., 1995). These contigs were named tentative human consensus sequences (THCs, Figure 1). The THC assembly software was designed to minimize the contribution of alterhatively spliced or chimeric ESTs.

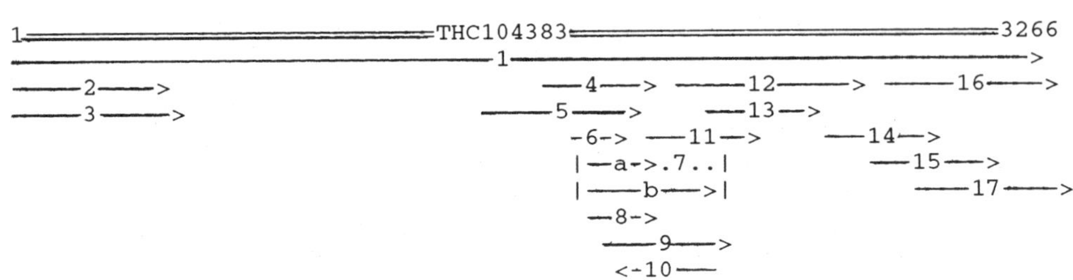

#	EST#	GB#	ATCC#	left	right	library
1	E HT1002			1	3183	
2	B EST143092			21	464	Heart
3	B EST143393			21	467	Heart
4	B EST124352			1781	2100	Brain
5	F	R12095		1862	2284	Brain
6	A EST31673			1870	2003	Embryo
7a	C EST07766		84742	1905	2078	Brain
7b	C EST06074	T08183	84742	1905	2263	Brain
8	L	F00027		1917	2089	Skeletal muscle
9	A EST56833			1978	2304	Brain
10	L	Z19202		2018	2258	Skeletal muscle
11	A EST70041			2094	2381	White blood cells
12	F	H10926		2174	2671	Brain
13	A EST77257			2504	2794	Pancreas
14	A EST55039	T28754	103497	2588	2884	Brain
15	F	R24627		2730	3038	Placenta
16	F	T90430		2807	3243	Lung
17	F	T83396		2829	3266	Liver

Sequence source codes: E= EGAD, B = HGS, F = WashU/Merck, A = TIGR, C = NIH, L = Genethon

Figure 1. A Human cDNA Database report for a THC representing leukemia virus receptor 2. The THC includes a representative HT sequence (HT1002) from TIGR's Expressed Gene Anatomy Database and 17 ESTs from this transcript (based upon sequence similarity). ESTs 7a and 7b were obtained from the same cDNA clone. The report shows the source laboratory where the sequence was obtained, GenBank and ATCC accession numbers (where available), the coordinates of each sequence in the assembly, and the tissue source of the cDNA library for each EST.

If an individual EST had a region of mismatch with the consensus sequence, it was excluded from the assembly. This resulted in construction of separate assemblies for different splice isoforms.

THCs were analyzed and annotated using the process described above for ESTs. Although many THCs can be identified by sequence similarity to known genes, the majority of THCs have no match to sequences in the public databases. Examination of the tissue distribution of THCs revealed several abundant clones that are apparently tissue-specific or widely expressed, several of which do not match any previously known gene (Table 1).

A TRANSCRIPT MAP OF THE HUMAN GENOME

The recent publication of three new genes that are mutated in inherited forms of colon cancer demonstrates the power of ESST sequencing to discover new genes involved in disease. Colon and other cancers are characterized at the molecular level by an accumulation of mutations in the DNA of transformed cells, including expansion of di- and tri-nucleotide repoeats. Because of these defects, it was proposed that DNA repair enzymes may be altered or missing in cancerous cells. One gene, called hMSH2 was cloned (Leach et al., 1993 and Fischel et al., 1993)

Table 1. Tissue-specific and widely-expressed genes. Example THCs matched by at least ten ESTs from a single tissue, but not from any other tissues are listed as potentially tissue-specific, abundant genes. Also listed are two novel TCHs containing ESTs from at least 20 of the 30 tissues from which more than 1000 ESTs were obtained are listed with the number of tissues.

THC#	Putative Identification	Number of ESTs	Tissue
81371	endothelial leukocyte adhesion molec. 1	15	endothelial cells
33408	carboxypetidase A2 isolog	16	pancreas
93542	pulmonary surfactant apoprotein	30	lung
31998	integrase-like protein FE65 isolog	32	brain
123393	regenerating protein I beta	57	pancreas
80608	nerve-terminal protein SNAP25	110	brain
31361	unknown	26	testis
9969	unknown	14	brain
9214	unknown	11	brain
11971	unknown	12	brain
16933	unknown	13	white blood cells
11412	unknown	99	many
11406	unknown	72	many

based on sequence similarity to yeast and *E. coli* DNA repair enzymes. Mutations in this gene were shown to segregate with affected members of families exhibiting hereditary non-polyposis colon cancer (HNPCC), thereby demonstrating that the DNA repair defect is likely a cause rather than an effect of transformation. Because the DNA repair pathways in bacteria and yeast are complex and involve many enzymes, it was hypothesized that additonal human repair enzyme genes might exist. Searches of TIGR's EST database with bacterial and yeast DNA repair enzyme protein sequences revealed three additional genes, each with striking similarity to the bacterial and yeast enzymes. Each now gene was mapped and determined to be in a chromosomal region associated with HNPCC. Full-length sequencing of the coding region of the genes and comparison of the sequence from normal and affected individuals resulted in identification of mutations in each gene which associeated with the disease in affected families (Papadopoulos et al., 1994 and Nicolaides et al., 1994). The ability to screen individuals who are at risk of developing HNPCC to either eliminate them as carriers of the disease genes or to identify them for more careful observation will be a dramatic improvement in pre-symptomatic treatment of HNPCC.

In the case of the HNPCC genes, the search for candidate genes was driven by a hypothesis about the involvement of DNA repair enzymes. The candidate genes were straightforward to identify based on a similarity search of the database. Additional weight was given to these genes as candidates by mapping them to regions implicated in disease. For many inherited disorders, however, no clues to the biological mechanism exist, and the only starting point is to try to identify all candidate genes within a chromosomal interal defined by the closest genetically linked markers.

As mentioned above, several hundred thousand human ESTs have been sequenced and placed in publicly accessible databases. Thus, most human genes have been tagged and are represented by multiple ESTs. The next phase of application of ESTs to genome characterization is assignment of ESTs to unique chromosomal locations. This serves to pinpoint the location of the gene matched by the EST and builds a bridge between the current physical genome maps and the genes themselve. The map of ESTs is called a 'transcript map' or 'expression map' because it combines information on gene locations and expression patterns.

An expression map should prove useful in several areas of genome research, but none more than in inherited disease identification. The current process of positional cloning is quite laborious. After demonstration of linkage of a disease in families with a particular genetic marker (and thus chromosomal location), positional cloning involves isolation of the physical DNA containing the chromosomal region of interest (e.g., a clone from the physical map) followed by attempts to identify the genes within the region. Once these candidate genes have been identified, they can be screened for mutations in patients inheriting the disease. Candidate gene identification can be approached in many ways including complete sequencing, cDNA selection (Lovett et al., 1991), and exon amplification (Buckler et al., 1991). A saturated expression map, one with all of the genes localized, would eliminate the need for all of these gene-finding techniques. A partial map, one with only some genes, will provide an immediate set of candidate genes to be evaluated for mutations while other techniques are initiated. The recent description of cloning of the breast cancer gene resulted from an extensive effort to identify all possible candidate genes (Miki et al., 1994).

The current genetic and physical maps of the human genome (Murray et al., 1994 and Chumakov et al., 1995) are based on non-gene markers such as simple sequence repeats and randomly isolated STSs. Genetic maps are, by definition, ordered because they represent recombination frequencies between polymorphic sites. Polymorphic STSs can be used as anchor points in construction of a physical map. In the case of the CEPH YAC map, YACs corresponding to 2000 polymorphic STSs were identified; these formed the skeleton on which the map was built. Other methods of identifying overlapping YACs were used to build bridges between the YACs containing polymorphic STSs. In this way, a map covering approximately 75% of the genome was constructed (Chumakov et al., 1995). There is about a 75% chance of being able to map an EST to the CEPH YAC library. The YAC location provides information on chromosome location and a physically cloned piece of DNA for further manipulation in disease gene discovery.

Another form of physical map is called a 'statistical radiation hybrid map.' Radiation hybrids are formed by irradiating human DNA to break it into small pieces of 100 kb to several megabases in size (Cox, 1992). The human DNA is transformed into a rodent cell line, where it is stably maintained. Each human-rodent somatic cell hybrid contains about twenty percent of the human genome in hundreds of small pieces. In an analogous way to construction of the YAC map, a series of markers is used to 'type' each hybrid cell line to develop a profile of the chromosomal fragments it carries. Once this profile is developed, the location of an unknown marker can be determined by comparison of its profile with those of the known markers. The map is 'statistical' in the sense that profiles will rarely be exact matches; discrepancies increase with the distance between the location of the unknown marker and the nearest anchor point. A series of well-spaced anchor markers with known order and cytogenetic location is therefore essential for the utility of radiation hybrids.

Different approaches to disease gene discovery require different materials and different methodologies, dependent upon the particular circumstances. For instance, if the location of a disease gene has been genetically defined to a small region of chromsome 4, then the location of ESTs on a radiation hybrid map is sufficient to define candidate genes. If the goal is to obtain the complete genomic sequence of a region or to use other gene-finding techniques such as cDNA selection (Lovett et al., 1991) or exon amplification (Buckler et al., 1991), the cloned DNA, such as obtained from the YAC map, is necessary.

For several (sometimes contradictory) maps to be useful, they need to be integrated (correlated to one another along their lengths) so that a position on one can be related to the others. The polymorphic

STS-defined genetic markers have proven a useful tool for map integration since they have been placed on both the YAC and radiation hybrid maps. Placement of ESTs on both radiation hybrid and YAC maps will also serve to relate the two to one another. Location of ESTs on both radiation hybrid and YAC maps will also serve to provide candidate gene identification from both entry points to the physical map. Polymorphic STSs and ESTs are likely to become the standard for integrating various types of physical maps. Also, as maps become more precise, the density of markers required increases. For instance, a 10 megabase map of the genome requires only about 400 markers, but a 100 kilobase map would require 80,000 markers. It would be tremendously useful for these thousands of markers to be genes in order to facilitate positional cloning and map construction. A variety of evidence suggests that genes are not uniformly distributed along the genome. Use of ESTs to build physical maps will result in the best map quality in regions of high gene density.

TIGR has recently embarked upon a major human transcript mapping project. A key component of this project is the development of second generation databases (see below) which allow linking of sequence, expression and mapping information for genes rather than individual EST or other sequences. We have evaluated all of the existing THCs in TIGR's HCD database to identify the best candidates for a large-scale effort to map human transcripts. We are designing STS primer pairs from the 3' ends of ~ 10,000 transcripts and demonstrating that the synthesized primers amplify human DNA. These primer pairs are being distributed to a number of collaborating labs where they are being placed on radiation hybrid, YAC and BAC maps. The primer sequences, resulting map locations, and links to THCs will be made available in the TIGR Database. The major goals of this project are to integrate these diverse maps, provide an integrated database of mapping and expression data, and to ultimately produce a transcript map of the human genome.

TIGR DATABASE

Through the TIGR Database, we have established a mechanism for making the EST data available to scientists at non-profit institutions. The Human cDNA Database (HCD) operates as a World Wide Web server and as an electronic mail server. Software allows users of the database to search by gene name (putative identification), EST or THC number from the recently published Genome Directory (Adams et al., 1995), or on the basis of sequence similarity. The database contains all of the information in Adams et al. (1995), and is updated regularly with new sequences. The goal of the database is to provide broad distribution of the data for research purposes. In additon, a curated, canonical set of human cDNA sequences from GenBank (HT set) is avaliable through the Expressed Gene Anatomy Database (EGAD). The TIGR Database also contains Sequences, SOurces, Taxa (SST), which links source, collection, and taxonomy information with DNA sequence data. All of these databases are extensively linked and provide easy accessibility to a wealth of biological information including sequences, expression data, curated cellular roles, and map positions. The TIGR Database may be accessed on the World Wide Web at the URL:http://www.tigr.org/tdb/tdb.html. More information can be obtained by sending E-mail to info@hcd.tigr.org.

REFERENCES

Adams, M.D., Kelley, J.M., Gocayne, J.D., Dubnick, M., Polymeropoulos, M.H., Xiao, H., Merril, C.R., Wu, A., Olde, B., Moreno, R.F., Kerlavage, A.R., McCombie, W.R., and Venter, J.C., 1991, Complimentary Dna sequencing: expressed sequence tags and the human genome project, *Science* 252:1651.

Adams, M.D., Kerlavage, A.R., Kelley, J.M., Gocayne, J.D., Fields, C., Fraser, C.M., and Venter, J.C., 1994, A model for high-throughput automated DNA sequencing and analysis core facilities, *Nature* 368:474.

Adams, M.D., Kerlavage, A.R., Fleischmann, R.D., Fuldner, R.A., Bult, C.J., Lee, N., Kirkness, E.F., Weinstock, K. G., Gocayne, J.D., White, O., Sutton, G., Blake, J.A., Brandon, R.C., Chiu, M.-W., Clayton, R.A., Cline, R.T., Cotton, M.D., Earle-Hughes, J., Fine, L.D., FitzGerald, L.M., FitzHugh, W.M., Fritchman, J.L., Geoghagen, N.S.M., Glodek, A., Gnehm, C.L., Hanna, M.C., Hedblom, E., Hinkle, P.S., Kelley, J.M., Klimek, K.M., Kelley, J.C., Liu, O.-I., Marmaros, S.M., Merrick, J.M., Moreno-Palanques, R.F., McDonald, L.A., Nguyen, D.T., Pellegrino, S.M., Phillips, C.A., Ryder, S.E., Scott, J.L., Saudek, D.M., Shirley, R., Small, K.V., Spriggs, T.A., Utterback, T.R., Weidman, J.F., Li, Y., Bednarik, D.P., Cao, L., Cepeda, M.A., Coleman, T.A., Collins, E.-J., Dimke, D., Feng, P., Ferrie, A., Fischer, C., Hastings, G.A., He, W.-W., Hu, J.-S., Greene, J.M., Gruber, J., Hudson, P., Kim, A., Kozak, D.L., Kunsch, C., Ji, H., Li, H., Meissner, P.S., Olsen, H., Raymond, L., Wei, Y.-F., Wing, J., Xu, C., Yu, G.-L., Ruben, S.M., Dillon, P.J., Fannon, M.R., Rosen, C.A., Hasetine, W.A., Fields, C., Fraser, C.M. and Venter, J.C., 1995, Initial assessment of human gene diversity and expression patterns based upon 52 million basepairs of cDNA sequence, *Nature* 377(Suppl):3.

Boguski, M., Lowe, T., and Tolstochev, C., 1993, dbEST - database for expressed sequence tags, *Nature Genet.* 4:332.

Buckler, A.J., Chang, D.D., Graw, S.L, Brook, J.D., Haber, D.A., Sharp, P.A., and Housman, D.E., 1991, Exon amplification: a strategy to isolate mammalian genes based on RNA splicing, *Proc Natl Acad Sci USA 88:4005.*

Chumakov, I.M., Rigault, P., LeGall, I., Bellanne-Chatelot, C., Billault, A., Guillou, S., Soularue, P., Guasconi, G., Poullier, E., Gros, I., Belova, M., Sambucy, J.-L, Susini, L., Gervy, P., Glibert, F., Beaufils, S., Bui, H., Massart, C., De Tand, M.-F., Dukasz, F., Lecouland, S., Ougen, P., Perrot, V., Saumier, M., Soravito, C., Bahouayila, R., Cohen-Akenine, A., Barillot, E., Bertrand, S., Codani, J.-J., Caterina, D., Georges, I., Lacroix, B., Lucotte, G., Sahbatou, M., Schmit, C., Sangouard, M., Tubacher, E., Dib, C., Faure, S., Fizames, C., Gyapay, G., Millasseau, P., Nguyen, S., Muselet, D., Vignal, A., Morissette, J., Menninger, J., Lieman, J., Desai, T., Banks, A., Bray-Ward, P., Ward, D., Hudson, T., Gerety, S., Foote, S., Stein. L, Page, D.C., Lander, E.S., Weissenbach, J., Le Paslier, D. and Cohen, D., 1995, A YAC contig map of the human genome, *Nature* 377(Suppl): 175.

Cox, D., 1992, Radiation hybrid mapping, *Cytogenet. Cell Genet.* 59:80.

Durkin, A.S., Nierman, W.C., Zoghbi, H., Jones, C., Kozak, C.A., and Maglott, D.R., 1994, Chromosome assignment of human brain expressed sequence tags (ESTs) by analyzing fluorescently labeled PCR products from hybrid cell panels, *Cytogenetics and Cell Genetics* 65:86.

Fishel, R., Lescoe, M.K., Rao, M.R.S., Copeland, N.G., Jenkins, N.A., Garber, J., Kane, M. and Kolodner, R., 1993, The human mutator gene homolog MSH2 and its association with hereditary nonpolyposis colon cancer, *Cell* 75:1027.

Kerlavage, A.R., Adams, M.D., Kelley, J.C., Dubnick, M., Powell, J., Shanmugam, P., Venter, J.C. and Fields, C., 1993, Analysis and management of data from high-throughput .

TOWARD THE CONSTRUCTION OF AN INTEGRATED GENETIC AND PHYSICAL MAP OF RICE

Masahiro Yano, Yoshiaki Nagamura, Nori Kurata, Takuji Sasaki, and Yuzo Minobe

Rice Genome Research Program
National Institute of Agrobiological Resources
2-1-2, Kannondai, Tsukuba, Ibaraki 305 Japan

INTRODUCTION

Rice is the major staple food for about half the population of the world. Improvement of rice varieties with improved economic values, such as yield and grain quality, is very important to keep up with the population growth of the world, especially in Asia. So far, the genetic basis of several traits with economic importance as well as those with biological importance have been clarified by many rice geneticists and breeders. However, further analyses, at the molecular level, will be necessary for further improvement of rice varieties.

Recently, rapid advances in molecular techniques resulted in a new era of genetic analysis. Genetic linkage maps with densely spaced markers have been rapidly constructed using DNA markers, such as a restriction fragment length poymorphisms (RFLPs) and randomly amplified polymorphic DNAs (RAPDs) in many plant species.

Four years ago, in 1991, the Rice Genome Research Program (RGP) was launched to analyze the molecular basis of rice genome organization and function. This program has three primary objectives, 1) construction of a genetic linkage map using DNA markers, 2) construction of a physical map of rice chromosomes, and 3) large scale sequencing and cataloguing of cDNAs for several different tissues of rice plant.

In this paper, we describe the recent progress in the construction of a high-density linkage map with DNA markers and the reconstruction of a rice chromosome map by the formation of contig of

Genomes of Plants and Animals: 21st Stadler Genetics Symposium
Edited by J. Perry Gustafson and R. B. Flavell, Plenum Press, New York, 1996

YAC clones. We also describe the genetic analysis of the traits with economic and biological values using these DNA markers and review prospects for the construction of an integrated genetic and physical map of rice.

CONSTRUCTION OF THE GENETIC LINKAGE MAP

Construction of a high density linkage map with DNA markers is an important basis for genome analysis of rice, as well as of other plant species. A high-density linkage map will be useful for the construction of a physical map of rice chromosome and for tagging genes of morphological and physiological traits. At the beginning of RGP, we planned to map more than 2000 DNA markers. We are approaching completion of this linkage mapping. Using these mapped DNA markers, we have already embarked on the tagging genes of several traits with biological and economic importance, including analyses of quantitative trait loci (QTLs).

High-Density Linkage Map with DNA Markers

The RGP has developed a high-density linkage map of rice with DNA markers (Kurata et al., 1994b). The general procedure of linkage map construction is shown in Figure 1. For the construction of the linkage map in RGP, we have been using 186 F_2 plants derived from the intra-specific cross between a *japonica* variety, Nipponbare, and an *indica* variety, Kasalath. Total DNA of each individual has been extracted from frozen leaf tissue by CTAB method (Murray and Thompson, 1980). In order to overcome a shortage of total DNA of F_2 individuals during mapping analysis, bulked F_3 seedlings derived by selfing from each F_2 individual have been used to restore F_2 DNA. In our mapping strategy, we have used mainly cDNA clones as a probes. As of March 1995, we have screened two parental lines with more than 4000 DNA clones, mainly cDNA clones from callus, root and shoot, and several genomic clones. After extensive RFLP screening, informative clones have been used to score genotypes of 186 F_2 individuals by Southern hybridization analysis.

The first high-density and high resolution rice genetic map was developed by RGP in three years (Kurata et al., 1994b). In total, 1383 DNA markers have been mapped on the linkage map. The number of markers and their categories on the 12 chromosomes are shown in Table 1. The DNA markers are distributed along 1575 cM on 12 linkage groups and consist of 883 cDNAs, 265 genomic DNAs, 147 RAPDs (random amplified polymorphic DNAs) and 88 DNAs from other sources. cDNA markers were derived from rice callus and root libraries. Genomic markers were mainly randomly selected DNA clones, with some *Not*I linking, YAC end and telomere-associated clones (Table 1). RAPD markers were used for mapping to find out whether there is a distribution bias for this type of marker and if they

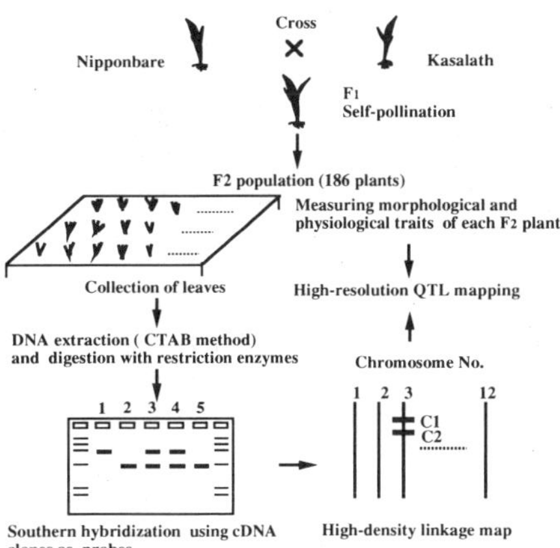

Figure 1. Schematic representation of the methods and materials used in the construction of a high-density linkage map and the QTL analysis.

can be used to fill marker-rare regions on linkage map. A bulked segregant analysis (Michelmore et al. 1991) was employed to develop markers in marker-rare regions. It is noteworthy that all the cDNA fragments and most of the genomic fragments were partially sequenced to convert them into STS (sequence tagged site) or EST (expressed sequence tag) markers. The mapped markers have been characterized in detail by Kurata et al. (1994b). So far, we have been making further efforts to construct a more high-density linkage map. Number of markers mapped on the high-density linkage map have been increasing. As of March 1995, we have mapped about 1900 DNA markers on the linkage map. cDNA libraries of green shoot and of etiolated shoot, instead of callus and root libraries, have been used as a probe sources.

Analysis of molecular markers, such as RFLP, is an effective method to reveal chromosomal rearrangements, such as translocations, deletions and duplications. Kishimoto et al. (1994) indicated, by linkage mapping of cDNA clones, that sequences between some regions of chromosome 1 and 5 have been conserved. In our linkage map of RGP, many clones showed more than one band by genomic Southern hybridization and were mapped at duplicate or triplicate loci among the twelve chromosomes. Extensive conservation in linkage alignment of 13 loci was observed between the lower distal regions of chromosomes 11 and 12 (Nagamura et al., 1995). These conserved regions span 10 cM and 11.8 cM from the distal ends of the linkage maps of chromosomes 11 and 12, respectively. These results suggest that these conserved regions were generated by a duplication of chromosome segments.

In addition, 58 wheat genomic DNA fragments have been mapped on the high-density linkage map in collaboration with Cambridge Laboratory in John Innes Centre, UK. As a result, we have established

Table 1. Number of markers and their categories in the high-density linkage map of rice.

Probe source	Chromosome												Total
	1	2	3	4	5	6	7	8	9	10	11	12	
cDNA (callus)	67	51	59	38	45	41	41	25	25	28	28	19	467
cDNA (root)	63	48	61	32	34	45	38	24	18	15	17	21	416
Genomic	15	19	12	12	9	13	6	8	11	9	12	12	138
Not I linking	18	7	7	8	5	11	6	6	4	5	8	5	90
YAC end	4	3	3	2	4	6	3	0	1	2	3	2	33
Telomere	0	0	0	0	1	0	0	0	0	0	2	1	4
RAPD	10	5	3	8	11	25	7	10	8	5	13	5	110
STS	4	1	2	3	2	3	4	2	4	3	8	1	37
Wheat	9	4	7	4	7	7	4	4	5	0	3	4	58
Others	3	3	4	7	0	5	0	0	0	1	3	4	30
Total	193	141	158	114	118	156	109	79	76	68	97	74	1383
Total length (cM)	192	156	169	129	124	130	129	125	100	86	123	112	1575

that wheat chromosome groups 1, 2, 3, 4, 6 and 7 clearly correspond to rice chromosomes 5, 4 and 7, 1, 3, 2 and 6, respectively. Markers on wheat chromosome group 5 were mapped on rice chromosomes 1, 3, 9, 11 and 12. Surprisingly, in all chromosomes, most of the markers analyzed showed the same linkage ordering between rice and wheat (Kurata et al., 1994a).   This preliminary synteny analysis of rice and wheat revealed high conservation for linkage ordering of DNA markers. Thus, the synteny map and the mapped rice probes will be very useful for the molecular analysis of the corresponding chromosomal regions in wheat. Rice will be a very useful model plant for genome analysis for several cereals, such as wheat, barley, maize and rye.

Tagging Genes for the Traits with Economic and Biological Values Using DNA Markers

One of the application of DNA markers is precise linkage mapping of genes with biological and agronomic value. We have already precisely mapped several genes for morphological and physiological traits, such as *Xa* -1 (bacterial leaf blight resistance), *Se* -1 (photoperiod sensitivity), *Gm* -2 (gall midge resistance; collaboration with Dr. M. Mohan, ICGEB, India), *eg* (extra-glume), *Ph* (phenol staining), *Rc* (brown pericarp and seed coat), *lax* (lax panicle), *dl* (drooping leaf), *lg* (legule less), *la* (lazy growth habit) and *alk* (alkali degeneration) using RFLP markers located on RGP linkage map. We have also embarked on a high resolution mapping of rice blast resistance genes, *Pi-b*, *Pi-ta*, *Pi-z* and *Pi-k*, for a marker-assisted selection and for a map-based cloning.

Many rice geneticists and breeders have also carried out the linkage analyses of genes for the traits with economic importance DNA markers. As a result, many DNA markers have been identified to be linked to genes for resistance to diseases, such as rice blast (Yu et al., 1991; Wang et al., 1994) and bacterial leaf blight (Ronald et al., 1992; Yoshimura et al., 1992), genes for resistance to insects, such as brown plant hopper (Ishii et al., 1994) and gall midge (Mohan et al., 1994), photoperiod sensitivity gene (Mackill et al., 1993), photoperiod sensitive genic male sterility genes (Zhang et al., 1994), semidwarf genes (Ogi et al., 1993; Cho et al., 1994; Liang et al., 1994), scented kernel gene (Ahn et al., 1992), gene for accumulation of glucomannan in endosperm cell wall (Yano et al., 1993), etc.

Mapping Quantitative Traits Loci

In contrast to disease and insect resistance, many important traits in breeding, such as yield, culm length, heading date and eating quality, show continuous variation in progenies. Inheritances of those traits are controlled by several genes. It is difficult to identify those genes, known as quantitative trait loci (QTL), because individual effects of these genes on phenotypes are relatively small. Recent progress in DNA markers and their linkage maps enable us to analyze these individual QTLs (Tanksley, 1993). The concept for detecting QTLs using linked major genes was developed many years ago (Sax, 1923). However, it was difficult to put this concept into practice by using conventional genetic markers. So far, many QTLs have been clarified by using DNA markers in various kinds of crop plants, such as tomato (Paterson et al., 1988; deVicente et al., 1993) and maize (Edwards et al., 1992; Stuber et al., 1992). In rice, QTL analysis with DNA markers has been employed to detect genomic regions conferring cooked-kernel elongation (Ahn et al., 1993) and partial resistance to blast disease (Wang et al., 1994). In these studies, putative genomic regions for such a complicated traits could be identified.

QTL analyses of rice also have been progressing from the beginning of the RGP to clarify the feasibility of isolation of genes at QTL with a map-based cloning system. The F_2 population used in the QTL analysis is same population used for the construction of the high-density linkage map (Figure 1). All data for genotype of markers were available for this QTL analysis. A large number of markers, 857 loci, were employed to identify the QTLs affecting the morphological and physiological traits, such as heading date, culm length, panicle length, *etc*. Traits analyzed and summary of the results of QTL mapping are shown in Table 2. Many QTLs were detected, with a wide range of gene effects on phenotypes, using the computer software, MAPMAKER/QTL (Lander et al., 1987; Lander and Botstein, 1989). High resolution QTL mapping also revealed evidence on the existence of multiple QTLs for the same trait in one chromosomal region and on specific gene interactions between identified individual QTLs, such as epistasis and suppression.

In general, it was difficult to determine the precise location and gene action of individual QTLs in the analysis. In tomato, overlapping substitution lines have been used to precisely map QTLs (Paterson et al., 1991). To overcome these problems, we are constructing well characterized genetic stocks, such as near isogenic lines, carrying one or multiple chromosomal segments of the parental line, Kasalath, in the genetic background of the other parental line, Nipponbare (Yano et al., 1994). Procedure of the construction of these genetic stocks is shown in Figure 2. Whole genome surveys have been carried out to identify the introgressed chromosomal segments using RFLP markers. In this way, we will be able to combine/separate the desired/undesired chromosomal regions containing QTLs in selected individual plants. By using these substitution lines or near-isogenic lines for a given chromosomal segment, it will be possible to handle a given QTL as a single Mendelian factor. Thus we would be able to determine the accurate location of a given QTL on a linkage map, to

Table 2. Summary of putative QTLs detected for eleven traits in F2 population derived from a cross between Nipponbare and Kasalath.

Traits	No. of QTL[1]	% of variance explained[2]		
		Max	Min	Total
Culm length	10	22.9	6.1	90.1
Panicle length	6	13.0	6.6	54.1
Panicle number	5	14.3	7.5	47.6
Number of seeds per panicle	5	13.3	8.3	51.6
Awn length	4	25.1	11.1	85.2
Seed shattering habit	5	68.6	6.9	100
Heading date	5	66.7	1.7	88.4
Seed width[3]	4	41.9	8.1	72.5
Seed length[3]	8	14.6	7.1	72.4
Germinablity at low temperature[4]	1	22.6	—	22.6
Chilling injury of seedling[4]	2	16.0	13.4	29.4

[1] A LOD score of 2.0 was used as the threshold for detecting QTL.

[2] Percent of variance which individual QTL explained in total F2 phenotypic variation were calculated using MAPMAKER.

[3] Traits were scored using seeds derived from F2 plants

[4] Traits were scored using F3 seedlings

clarify the precise gene effects and to evaluate the genotype-by-environment interaction of individual QTLs. Once we succeed to map the genes of interest with high resolution, those genes will be isolated by map-based cloning. Putative locations and linked DNA markers of

QTLs will also make marker-assisted selection feasible in rice breeding.

CONSTRUCTION OF PHYSICAL MAP

The genome size of rice is the smallest, about 430 million base pairs per haploid, among cereals. The relatively small size of the rice genome facilitates constructing a physical map. In RGP, we have also

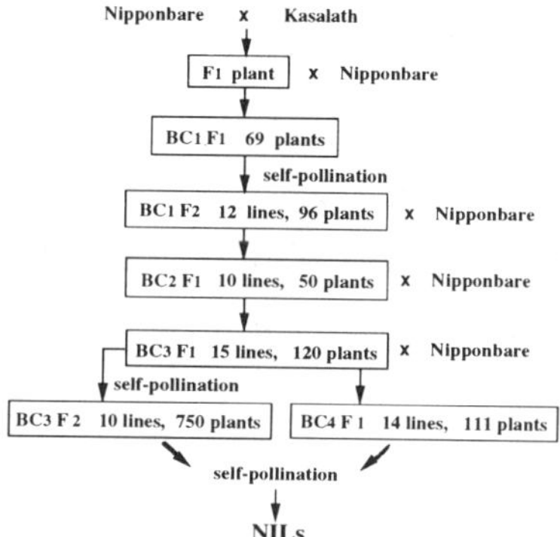

Figure 2. Flowchart of the construction of near isogenic lines (NILs) for the fine mapping of QTLs. In each generations, whole genome surveys using RFLP markers were carried out to select the candidate progenies which were the source of NILs for the putative QTL regions.

embarked on the construction of DNA contigs comprising long and medium size DNA fragments, *i.e.* YAC and cosmid ordered libraries overlapping each other. Main reasons for the construction of a physical map are to make up the fundamental basis of researches for the rice genome structure and to make feasible the isolation of genes for agronomically or scientifically interesting traits by map-based cloning.

Construction of Genomic Library Using YAC and Cosmid

Preparation of genomic libraries carrying large DNA fragments is the first step to construct a physical map. There are several kinds of genomic libraries that can be used for physical mapping, each of them with their own advantages and disadvantages in constructing the libraries and in using them for physical mapping. The three types most commonly applied for physical mapping in other organisms are the libraries constructed with YAC (yeast artificial chromosome), BAC (bacterial artificial chromosome) and cosmid phage vectors. YACs can clone very long genomic DNA fragments, from several hundred kb to over one Mb. The construction of a high quality YAC library and its evaluation is difficult and time consuming work, because of the frequently occurring chimeric clones. After preliminary trials, we were able to construct satisfactory YAC libraries using protoplast cells of *japonica* variety, Nipponbare (Umehara et al., 1995a). The currently used two YAC libraries comprise 7000 clones with an average of 350 kb genomic DNA inserts. These libraries cover *ca.* 5.5 times the rice genome and should cover over 80% of total chromosome length, when all clones are aligned along the chromosomes. About 40% of the clones in the library were chimeric. However, most chimeric clones were observed in the clones with over 400 kb inserts, but not in the clones with smaller inserts. In most cases, it is quite difficult to cover the whole genome with only one kind of genomic clone library. Therefore, RGP has also decided to make cosmid libraries. Cosmid libraries will be very useful not only to fill the gap regions of YAC contigs, but also to divide the long-size YAC clones into several cosmid clones for further analysis, both to isolate target genes and to do detailed physical mapping. A cosmid libraries of genomic DNA of "Nipponbare" has also been constructed for detailed physical mapping (Katayose et al., 1993).

Making Contigs of YAC Clones along with the High-Density Linkage Map

There are several strategies applicable for physical map construction. A complete physical map should be a map which includes all DNA sequences of all 12 chromosomes from one end to the other. For this purpose, it is best to obtain as large genomic DNA fragments as possible and order them along the 12 chromosomes. Collection of the 12 chromosomes independently would be the best way for making chromosome-specific genomic DNA libraries as the source of starting materials for physical mapping. Rice chromosomes, however, cannot be sorted by laser beam chromosome sorting, because of the very small size and continuous variation in length (maybe also in DNA content) among the 12 chromosomes. However, due to the small genome size of rice, it is possible to construct a genomic library of large DNA fragments and to order them directly on the 12 chromosomes. In addition, it was reported that only about 50% of the rice genome is repetitive DNA sequences (Deshpande et al., 1980). This would make it easier in rice to do cloning, selection and ordering of large DNA fragments to build up a physical map using a whole genomic library.

As for rice, we have not yet found any special genome features that could be utilized for efficient physical mapping, except for the small genome size and the low proportion of repetitive sequences. Therefore it seemed to us best to use high-density and high resolution rice genetic map to order YAC clones corresponding to the mapped DNA markers on it. Because we have already constructed a high-density 300 kb interval rice genetic map (Kurata et al., 1994b), it would be possible to cover the whole genome length by selecting and ordering YAC clones with an average length of 350 kb inserts. The practical methods for YAC contig formation used in RGP are shown in

Figure 3. For the screening of YAC clones using RFLP markers, high density colony filters carrying 1500 YAC clones have been made. 7000 clones of the YAC library could be spotted on 5 filters. DNA clones mapped on the high density linkage map were used for the colony hybridization to select YAC clones carrying the sequences of used DNA clone. Once the candidate YAC clones are selected, positive candidate YAC clones are further re-examined by Southern hybridization analysis for detecting marker DNA sequences in them (Figure 3A). In case of STS markers (derived mainly from RAPD

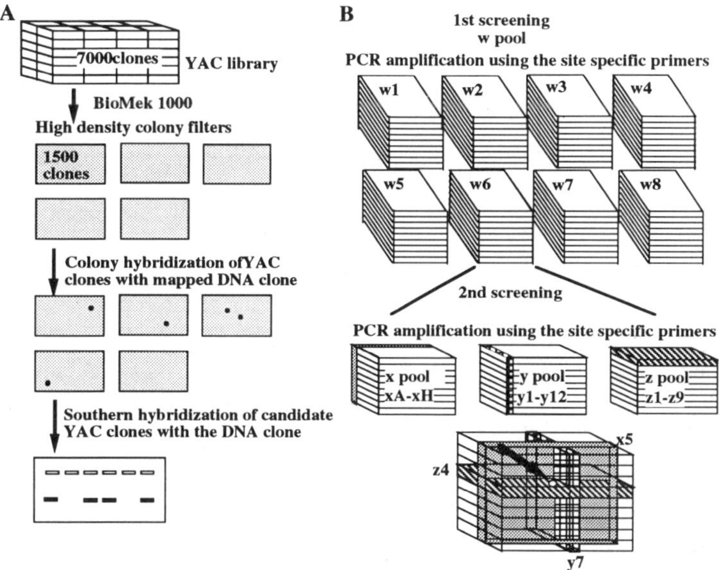

Figure 3. Schematic representation of the methods for a screening of YAC clones to construct ordered contig. A: Screening of YAC clones using the short DNA fragment mapped on the high-density linkage map. B: Screening of YAC clones using the site-specific primer sequences which derived mainly from RAPD markers (STS markers).

markers), all 7000 cloned DNAs are divided into several pool combinations in 96 well microtiter plates for PCR screening with site-specific primer sequences. After PCR screening for the first W pool, 2nd screening for X, Y and Z combination pools are performed to identify the position of the positive YAC clones. When PCR products are amplified in given X, Y and Z pools, respectively, a candidate YAC clone carrying STS marker sequence is identified systematically (Figure 3B).

By using these methods, we have already embarked on the physical map construction for chromosome 1, 3, 4, 6, 7 and 11 (Umehara et al., 1995b). Current status of the screening of YAC clones using DNA clones mapped on the high-density linkage map are shown in Table 3. As of March 1995, 2671 YAC clones have been successfully selected by colony hybridization using 556 DNA clones mapped on the high-density linkage map. Of YAC clones selected, about 80 percent of selected clones have been clarified as to their map locations, while the map location for about 20 percent remained to be clarified. Thus, we

Table 3. Results of the screening of YAC clones using the DNA clones mapped on the high-density linkage map.

Chromosome	Total markers[1]	Used markers[2]	Total YACs[3]	Distinct YACs[4]
Chr. 1	223	74	263	103
Chr. 2	159	—	—	32
Chr. 3	191	122	585	245
Chr. 4	134	45	288	147
Chr. 5	138	76	459	293
Chr. 6	185	132	597	240
Chr. 7	126	93	433	145
Chr. 8	90	—	—	6
Chr. 9	92	—	—	9
Chr. 10	81	—	—	4
Chr. 11	116	20	58	40
Chr. 12	83	—	—	13
Total	1618	556	2671	1277

[1]Number of markers mapped on the high-density linkage map for each chromosome (May 1994).
[2]Number of DNA clones used for the colony hybridization in YAC screening.
[3]Total number of identified YAC clone carrying mapped DNA sequences.
[4]YAC clones which are carrying distinct marker sequences and have been identified their location on the high-density linkage map.

assume that the current physical map covers about one fourth of the genome size. Aiming to introduce more effective methods, trials to use fingerprinting to order YAC clones are also in progress. Isolation of high copy DNA sequences in the rice genome has been carried out in minisatellite or microsatellite assays (Winberg et al., 1993; Zhao et al., 1993), genomic fingerprinting (Ramakrishnan et al., 1994) and analysis for repetitiveness of short nucleotides. Preliminary approaches for this strategy showed a promising results. One of the microsatellites and one short repetitive nucleotide sequence have been found to produce specific multiple banding in distinctive YACs. This should enable great progress in making YAC contigs.

APPLICATION OF RICE PHYSICAL MAP

Completion of the physical map will not only need several kinds of libraries, such as YAC, cosmid and plasmid, but also multiple methods to fill the gaps and to find out rearrangements and other complicated situations in the rice genome. Such combined strategies themselves will reveal the degree and sites of structural complexity in the genome. The resulting physical map will be the main source for multiple analyses of the rice genome structure. Genome organization should be clarified from multiple aspects, for example, by dissecting the whole genome into functional and non-functional segments, surveying for common functional domains and their functions both at the level of gene expression and chromatin/chromosome structure, and also by whole genome sequencing. The rice physical map should be the most valuable source in monocot plants, especially in grass species, for further comparative analyses among them.

Moreover, ordered YAC contigs will provide us an advanced method for gene mapping. In case of genetic mapping of cDNA clones, it may be difficult to locate all of them on the linkage map due to the multiple copy sequences and to the lack of polymorphism. The fine mapping of almost all genes, however, would be possible by using ordered YAC libraries. Actually, we can locate almost all of multicopy and family gene sequences on multiple YAC clones in our system for YAC contig formation. In the future, it would be possible to map most of cDNA clone using ordered YAC clones without the segregation analysis. Such a full expression map will contribute greatly to unraveling genome organization for gene expression and gene evolution in rice and other plants.

PUTTING ALL THE MAPS TOGETHER

During the last two decades the molecular biology of rice has advanced greatly. At present, four kinds of linkage map, which were constructed by using independent probe sources, have already been constructed (McCouch et al., 1988; Saito et al., 1991; Kurata et al., 1994b; Causse et al., 1994). Rice has also recently emerged as a model plant for mapping all cereal genomes and isolating agronomically or scientifically important genes for plant breeding. This is due to its small genome size (430 Mb, the smallest in cereals), the detailed linkage maps becoming available (Kurata et al., 1994b; Causse et al., 1994), as well as the large amount of synteny found between rice and other cereals (Kurata et al., 1994a; Ahn and Tanksley, 1993). On the other hand, for over 70 years, rice geneticists and breeders have clarified the genetic basis of economically and biologically important traits. Due to these efforts, a rice conventional genetic linkage map with more than 150 loci for the morphological and physiological traits has been compiled (Kinoshita, 1994). It is composed of many morphological and physiological genetic marker genes and agronomically important genes, such as disease and insect resistance genes. Integration of this information for the genetic linkage maps will be necessary to facilitate future analysis, such as positional cloning of genes with economic values and the studies on rice genome structure.

For the integration of linkage maps with DNA markers, efforts are in progress in collaboration with Cornell University. We are also developing a consensus framework linkage map using RILs (recombinant inbred lines) which have been constructed at Kyushu University (Tsunematsu et al., 1993). At the present, the framework linkage map of RILs includes about 280 markers of RGP map (Kurata et al., 1994b) and about 100 markers of the linkage map previously constructed in Japan (Saito et al., 1991). Complete integration of several linkage maps will be achieved soon by these efforts. As many research groups can share those RILs, the consensus framework map would facilitate an effective integration of all the information about linkage maps.

For the integration of the conventional linkage maps and those of DNA markers, the linkage map and mapped DNA markers have already been used for tagging genes with agronomic/biological interest, as mentioned before. Comprehensive integration of the linkage map with conventional markers and with DNA markers is in progress and, at present, about 40 conventional genetic markers have been mapped on the RFLP linkage maps (Ideta et al., 1992; Ideta et al., 1993; Ideta et al., 1994). Moreover, DNA markers also enable us to identify the individual QTL for the several morphological and physiological traits as described before. This type of analysis will provide a large amount of information about the chromosomal location and the precise gene effect of many genes involved traits with complicated inheritance.

All the map data, such as physical map, RFLP linkage map and genetic map of phenotypic trait genes, including QTLs, should be put together to generate an informative chromosome map as a comprehensive genome map. In addition, most of the mapped DNA markers have been sequenced partially in RGP (Sasaki et al., 1994; Havukkala et al., 1995). These DNA markers will be very useful to construct a fully integrated genome map. Once a fully integrated genome map is established, all the DNA clones, such as plasmid, cosmid and YAC clones and all their data, such as linkage relationship and sequence data, will be very powerful tools, not only for the analysis of rice genome structure and function, but also for practical rice breeding.

ACKNOWLEDGMENTS

We thank M. Nakagahra, K. Kobayashi for advise and encouragement and Ilkka Havukkala for critical reading of the manuscript. This work was supported by the funds from the Ministry of Agriculture, Forestry and Fisheries of Japan (MAFF) and from the Japan Racing Association (JRA).

REFERENCES

Ahn, S.N., Bollich, C.N., and Tanksley, S.D., 1992, RFLP tagging of a gene for aroma in rice, *Theor. Appl. Genet.* 84:825.

Ahn, S.N., and Tanksley, S.D., 1993, Comparative linkage maps of the rice and maize genomes, *Proc. Natl. Acad. Sci. U.S.A.* 90:7980.

Ahn, S.N., Bollich, C.N., McClung, A.M., and Tanksley, S.D., 1993, RFLP analysis of genomic regions associated with cooked-kernel elongation in rice, *Theor. Appl. Genet.* 87:27.

Causse, M.A., Fulton, T.M., Cho, Y.G., Ahn, N., Chunwongse, J., Wu, K., Xiao, J., Yu, Z., Ronald, P. C., Harrington, S.E., Second, G., McCounch, S.R., and Tanksley, S.D., 1994, Saturated molecular map of the rice genome based on an interspecific backcross population, *Genetics* 138:1251.

Cho, Y.G., Eun, M.Y., McCouch, S.R., and Chae, Y.A., 1994, The semidwarf gene, *sd-1*, of rice (*Oryza sativa* L.). II. Molecular mapping and markers-assisted selection, *Theor. Appl. Genet.* 89:54.

Deshpande V. G., and Ranjekar P. K., 1980, Repetitive DNA in three Graminae species with low DNA content, *Hoppe-Seyler's Z. Physiol. Chem.* 361:1223.

deVicente, M.C., and Tanksley, S.D., 1993, QTL analysis of transgressive segregation in an interspecific tomato cross, *Genetics* 134:585.

Edwards, M.D., Helentjaris, T., Wright, S., and Stuber, C.W., 1992, Molecular-marker-facilitated investigations of quantitative trait loci in maize, *Theor. Appl. Genet.* 83:765.

Havukkala, I., Ichimura, H., Nagamura, Y., and Sasaki, T., 1995, Rice Genome Analysis by ntegration of sequencing and mapping data, *Journal of Biotechnology* (in press).

Ideta, O., Yoshimura, A., Matsumoto, T., Tsunematsu, H., and Iwata, N., 1992, Integration of conventional and RFLP linkage maps in rice. I. Chromosomes 1, 2, 3 and 4, *Rice Genet. Newsl.* 9:128.

Ideta, O., Yoshimura, A., Matsumoto, T., Tsunematsu, H., Satoh, H., and Iwata, N., 1993, Integration of conventional and RFLP linkage maps in rice. II. Chromosomes 6, 9, 10 and 11, *Rice Genet. Newsl.* 10:87.

Ideta, O., Yoshimura, A., Ashikari, M., and Iwata, N., 1994, Integration of conventional and RFLP linkage maps in rice. III. Chromosomes 5, 7, 8 and 12, *Rice Genet. Newsl.* 11:116.

Ishii, T., Brar, D.S., Multani, D.S., and Khush, G.S., 1994, Molecular tagging of genes for brown planthopper resistance and earliness introgressed from *Oryza australiensis* into cultivated rice, *O. sativa, Genome* 37:217.

Katayose, Y., Yoshino, K., and Kurata, N., 1993, Use of cosmid libraries for physical mapping, *Rice Genome* 2(1):6.

Kinoshita,T., 1994, Report of the committee on gene symbolization, nomenclature and linkage map, *Rice Genet. Newsl.* 10:7.

Kishimoto, N., Higo, H., Abe, K., Arai, S., Saito, A., and Higo, K., 1994, Identification of the duplicated segments in rice chromosome 1 and 5 by linkage analysis of cDNA markers of known functions, *Theor. Appl. Genet.* 88:722.

Kurata, N., Moore, G., Nagamura, Y., Foote, T., Yano, M., Minobe, Y., and Gale, M., 1994a, Conservation of genome structure between rice and wheat, *BIO/TECHNOLOGY* 12:276.

Kurata, N., Nagamura, Y., Yamamoto, K., Harushima, Y., Sue, N., Wu, J., Antonio, B.A., Shomura, A., Shimizu, T., Lin, S.Y., Inoue, T., Fukuda, A., Shimano,T.,

Kuboki, Y., Toyama, T., Miyamoto, Y., Kirihara, T., Hayasaka, K., Miyao, A., Monna, L., Zhong, H.S., Tamura, Y., Wang, Z.X., Momma, T., Umehara, Y., Yano, M., Sasaki, T., and Minobe, Y., 1994b, A 300 kilobase interval genetic map of rice including 883 expressed sequences, *Nature Genetics* 8:365.

Lander, E.S., Green, P., Abrahamson, J., Barlow , A., Daly, M.J., Lincoln, S.E., and Newburg, L., 1987, MAPMAKER: An interactive computer package for constructing primary genetic linkage maps of experimental and natural populations, *Genomics* 1:174.

Lander, E.S., and Botstein, D., 1989, Mapping mendelian factors underlying quantitative traits using RFLP linkage maps, *Genetics* 121:185.

Liang, C.Z., Gu. M.H., Pan, X.B., Liang, G.H., and Zhu, L.H., 1994, RFLP tagging of a new semidwarfing gene in rice, *Theor. Appl. Genet.* 88:898.

Mackill, D.J., Salam, M.A., Wang, Z.Y., and Tanksley, S.D., 1993, A major photoperiod-sensitivity gene tagged with RFLP and isozyme markers in rice, *Theor. Appl. Genet.* 85:536.

McCouch, S.R., Kochert, G., Yu, Z.H., Wang, Z.Y., Khush, G.S., Coffman, W.R., and Tanksley S.D., 1988, Molecular mapping of rice chromosome, *Theor. Appl. Genet.* 76:815.

Michelmore,R.W., Paran, I., and Kesseli, R.V., 1991, Identification of markers linked to disease-resistance genes by bulked segregant analysis: A rapid method to detect markers in specific genomic regions by using segregating populations, *Proc. Natl. Acad. Sci. USA* 88:9828.

Mohan, M., Nair, S., Bentur, J.S., Rao, U.P., and Bennet, J., 1994, RFLP and RAPD mapping of the rice *Gm2* gene that confers resistance to biotype 1 of gall midge (*Orseolia oryzae*), *Theor. Appl. Genet.* 87:782.

Murray, M.G., and Thompson, W.F., 1980, Rapid isolation of high molecular weight plant DNA., *Nucleic Acids Res.* 8:4321.

Nagamura, Y., Inoue, T., Antonio, B.A., Shimano,T., Kajiya, H., Shomura, A., Lin, S.Y., Kuboki, Y., Kurata, N., Minobe, Y., Yano, M., and Sasaki, T., 1995, Extensive conservation of duplicate segments between rice chromosome 11 and 12, *Jpn. J. Breed.* (in press)

Ogi, Y., Kato, H., Maruyama, K., Saito, A., and Kikuchi, F., 1993, Identification of RFLP markers closely linked to the semidwarfing gene at the *sd-1* locus in rice., *Japan. J. Breed.* 43:141.

Paterson, A.H., Lander, E.S., Hewitt, J.D., Peterson, S., Lincoln, S.E., and Tanksley, S.D., 1988, Resolution of quanyitative traits into Mendelian factors using a complete linkage map of restriction fragment length polymorphisms, *Nature* 335:721.

Paterson A.H., DeVerna, J.W., Lanini, B., and Tanksley, S.D., 1991, Fine mapping of quantitative trait loci using selected overlapping recombinant chromosomes, in an interspecies cross of tomato, *Genetics* 124:735.

Ramakrishnan, W., Gupta, V. S., and Ranjekar, P. K., 1994, DNA fingerprinting in rice microsatellite and minisatellites, *The proceedings of the 17th meeting of the International Program on Rice Biotechnology. (Rockefeller meeting)* 17:18.

Ronald, P.C., Albano, B., Tabien, R., Abenes, L., Wu, K., McCouch S.R., and Tanksley, S.D., 1992, Genetic and physical analysis of the rice bacterial blight disease resistance locus, *Xa-21, Mol. Gen. Genet.* 236:113.

Saito, A., Yano, M., Kishimoto, N., Nakagahra, M., Yoshimura, A., Saito, K., Kuhara, S., Ukai, Y., Kawase, M., Nagamine, T., Yoshimura, S., Ideta, O., Osawa, R., Hayano, Y., Iwata N., and Sugiura, M., 1991, Linkage map of restriction fragment length polymorphism loci in rice, *Japan. J. Breed.* 41:665.

Sasaki, T., Song, J., Koga-Ban, Y., Matsui, E., Fang, F., Higo, H., Nagasaki, H., Hori, M., Miya, M., Maruyama-Kayano, E., Takiguchi, T., Takasuga, A., Niki,T., Ishimaru, K., Ikeda, H., Yamamoto, Y., Mukai, Y., Ohta, I., Miyadera, N., Havukkala, I., and Minobe, Y., 1994, Toward cataloguing all rice genes : Large-scale sequencing of randomly chosen rice cDNAs from a callus cDNA library, *Plant J.* 6:615.

Sax, K., 1923, The association of size differences with seed-coat pattern and pigmentation in *Phaseolus vulgaris, Genetics* 8:552.

Stuber, C.W., Lincoln, S.E., Wolff, D.W., Helentjaris, T., and Lander, E.S., 1992, Identification of genetic factors contributing to heterosis in a hybrid from two elite maize inbred lines using molecular markers, *Genetics* 823.

Tanksley, S.D., 1993, Mapping polygenes, *Annu. Rev. Genet.* 27:205.

Tsunematsu, H., Hasegawa, H., Yoshimura, S., Yoshimura A., and Iwata, N., 1993, Construction of an RFLP/RAPD linkage map by using recombinant inbred lines, *Rice Genet. Newsl.* 10:89-.

Umehara, Y., Inagaki, A., Tanoue, H., Yasukouchi, Y., Nagamura, Y., Saji, S., Otsuki, Y., Fujimura, T., Kurata, N., and Minobe, Y., 1995a, Construction and characterization of rice YAC library for physical mapping, *Molecular Breeding* 1:79.

Umehara, Y., Tanoue, H., Wang, Z-X., Saji, S., Sue, N., vanHouten, W., Shimokawa, T., Yoshino, K., Ashikawa, I., Kurata, N., and Sasaki, T., 1995b, Construction of a YAC ordered library for 12 rice chromosomes, *Abstract of Plant Genome III* 3:82.

Wang, G., Mackill, D.J., Bonman, J.M., McCouch, S.R., Champoux, M.C., and Nelson R.J., 1994, RFLP mapping of genes conferring complete and partial resistance to blast in a durably resistant rice cultivar, *Genetics* 136:1421.

Winberg, B.C., Zhou, Z. Dallas, J.F., McIntyre, C.L., and Gustafson, J.P., 1993, Characterization of minisatellite sequences from *Oryza sativa*, Genome 36:978.

Yano, M., Zamorski, R., Saito, A., and Shibuya, N., 1993, A dominant gene controlling accumulation of glucomannan in the cell wall of rice endosoerm, *Rice Genet. Newsl.* 10:107.

Yano,M., Harushima, Y., Lin, S. Y., Kuboki, Y., Shomura, A., Shimano, T., Antonio, B.A., Inoue, T., Kajiya, H., Kawamura, Y., Kishida, T., and Nagamura, Y., 1994, Strategy for genetic dissection of quantitative traits into single mendelian factors using DNA makers in rice, *Rice Genome* 3(2):5.

Yoshimura, S., Yoshimura, A., Saito, A., Kishimoto, N., Kawase, M., Yano, M., Nakagahra, M., Ogawa, T., and Iwata, N., 1992, RFLP analysis of introgressed segments in three near-isogenic lines of rice for bacterial blight resistance genes, *Xa-1*, *Xa-3* and *Xa-4*, *Jpn. J. Genet.* 67:29.

Yu, Z.H., Mackill, D.J., Bonman, J.M., and Tanksley, S.D., 1991, Tagging genes for blast resistance in rice via linkage to RFLP markers, *Theor. Appl. Genet.* 81:471.

Zhang, Q., Shen, B.Z., Dai, X.K., Mei, M.H., Saghai Maroof, M.A., and Li, Z.B., 1994, Using bulked extremes and recessive class to map genes for photoperiod-sensitive genic male sterility in rice, *Proc. Natl. Acad. Sci. USA* 91:8675.

Zhao, X., and Kochert, G., 1993, Phylogenetic distribution and genetic mapping of a microsatellite from rice (*Oryza sativa* L.), *Plant. Mol. Biol.* 21:607.

PHYSICAL MAPPING OF THE *Arabidopsis thaliana* GENOME

Melanie Stammers, Renate Schmidt, and Caroline Dean

Department of Molecular Genetics, BBSRC,
John Innes Centre,
Colney Lane,
Norwich NR4 7UH
U.K.

INTRODUCTION

Arabidopsis thaliana has been adopted by the plant community as a model system for studying plants using a molecular approach. *Arabidopsis* is well suited to conventional genetics because it has a short life cycle, is small and has a prolific seed set (Meyerowitz, 1987). It has one of the smallest known genomes with a remarkably low content of interspersed repetitive DNA (Meyerowitz, 1987; Pruitt and Meyerowitz, 1986) making it eminently suitable for molecular studies.

The size of the *Arabidopsis* genome has been the subject of much debate. Methods used to estimate the size of the genome include microspectrophotometry of nuclei (Bennett and Smith, 1976), DNA reassociation analysis (Leutwiler et al., 1984), quantitative gel blot hybridisation (Francis et al., 1990), electron microscopic measurement of chromosome volume (Heslop-Harrison and Schwarzacher, 1990), and flow cytometry (Galbraith et al., 1991; Arumuganathan and Earle, 1991). From these different methods the genome size has been estimated to range from 50 to 150 Mbp. The current concensus of opinion is that the size of the haploid genome is 100Mbp; this small size has obvious advantages for physical mapping and genetic analysis in general.

The complexity of the *Arabidopsis* genome is also under much scrutiny. By studying *Arabidopsis* we hope to learn about the organisation of genomes without the complication of large amounts of repetitive DNA. Several studies conducted to characterise the

Arabidopsis genome have concentrated on quantifying the proportions of repetitive and unique sequences, and the arrangement of these relative to each other. Studies of the reassociation kinetics of the *Arabidopsis* genome (Leutwiler et al., 1984) revealed that rapidly annealing sequences, that is, highly repetitive DNA, makes up 10%, and another 10% is moderately repetitive. Four components of the highly repetitive fraction have been identified. The most abundant sequence is a 180bp repeat which is present in large tandem arrays, this is repeated 4000-6000 times across the genome (Martinez-Zapater et al., 1986). Related to this is a 500bp repeat also found in tandem arrays which is repeated 500 times per genome (Simoens et al., 1988). Together the 180bp and 500bp repeat have been shown to colocalise with the centromeric heterochromatin of all 5 chromosomes (Maluszynska and Heslop-Harrison, 1991). Simoens et al. (1988) also identified the third component a 160bp tandem repeat which is unrelated to the two previous repeats, and of which there are approximately 1500 copies per genome. Lastly, a small 7bp tandem repeat sequence 5'-CCCTAAA-3', associated with the telomeres of all 5 chromosomes which is repeated approximately 350 times at each site was reported by Richards and Ausubel (1988).

Pruitt and Meyerowitz (1986) similarly conducted an experiment to characterise the *Arabidopsis* genome. They analysed the nature of the DNA content of 50 random lambda clones. The majority of these clones (89%), with an average size of 12.8kb, were unique over their entire length. A large proportion of the clones which were found to be repetitive contained rDNA, which was shown to exist mostly in large arrays of tandem repeats. The basic repetitive unit of the rDNA was shown to be 10kb of which there are approximately 570 copies, making up 7 - 8% of the genome and the majority of the moderately repetitive fraction. Little is known about other moderately repetitive sequences; Meyerowitz (1992) suggested they may be transposable elements as is the case in *Drosophila*. The information gained about the frequency of the different classes of DNA sequence was also used to calculate the interspersion pattern of unique and repetitive sequences. The mean unique sequence length was estimated to be 120 - 125kb.

Other aspects of the *Arabidopsis* genome which have been characterised are the G+C content and the methylation status. G+C content was calculated to be 41.4% (Leutwiler et al., 1984), which is a typical value for a flowering plant. The methylation status is however very low, 4.6% (Leutwiler et al., 1984); the general range being from 9.7% for *Brassica oleracea* to 33% for *Secale cereale* (reviewed by Meyerowitz (1992).

The small size of the *Arabidopsis* genome has its disadvantages too; especially concerning cytogenetic studies. Despite this the 5 pairs of *Arabidopsis* chromosomes were first described in 1907 by Laibach. The development of new technology in recent years has improved the situation greatly, and the application of the fluorescent *in situ* hybridisation (FISH) technique is becoming particularly productive.

Figure 1 (adapted from (Schweizer et al., 1987)) shows the current cytogenetic picture of the 5 *Arabidopsis* chromosomes and incorporates information gained from fluorescent *in situ* hybridisation of the cloned repeats and rDNA (Maluszynska and Heslop-Harrison, 1991).

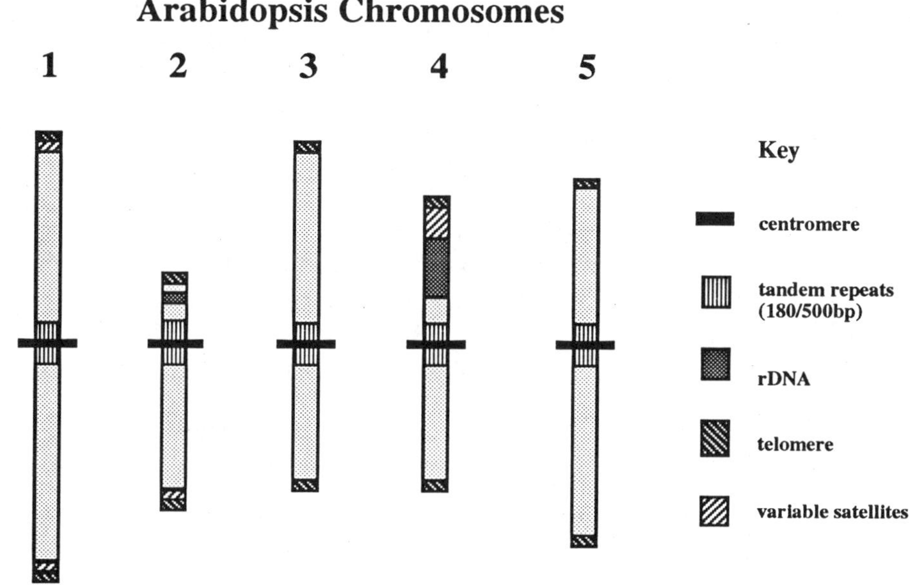

Figure 1. The current cytogenetic view of the 5 Arabidopsis chromosomes

PHYSICAL MAPPING *Arabidopsis*

The cloning of genes by a map-based approach is being used extensively, consequently the completion of the physical map of the *Arabidopsis* genome is a high priority for the *Arabidopsis* research community. The physical mapping of *Arabidopsis* began with a project to cover the whole of the genome in overlapping cosmids (Hauge et al., 1991). This project was modeled on the *C. elegans* genome project which employed a fingerprinting technique to identify overlapping cosmids. In this manner Hauge et al. (1991) screened 17,000 cosmids and from these assembled 750 contigs, which were estimated to cover 90 - 95% of the genome. Efforts were made to close the gaps between the contigs, but it was considered that after fingerprinting 8 - 10 genome equivalents of cosmids the procedure had reached its practical limits.

The larger yeast artificial chromosomes (YACs) have on the whole superceded cosmids as the clones of choice for creating an overlapping library. Several laboratories are now involved in an international effort to construct an overlapping YAC library for *Arabidopsis*. The effort to cover the genome has been shared between these laboratories and several chromosomes are near to completion.

There are four YAC libraries available for this purpose (Creusot et al., 1994; Ecker, 1990; Grill and Somerville, 1991; Ward and Jen, 1990) and together they represent more than 10 genome equivalents.

PHYSICAL MAPPING OF CHROMOSOME 4

Our efforts have concentrated on the construction of an overlapping YAC library for chromosomes 4 and 5. Until recently only 3 of the YAC libraries were available, the EG library (Grill and Somerville, 1991), the EW library (Ward and Jen, 1990) and the yUP library (Ecker, 1990). Cloned inserts in the EG and EW libraries have an average size of 150kb, and inserts in the yUP library average 250kb. A framework of RFLP markers were used to identify YAC clones from which walks were initiated in an attempt to join adjacent RFLP anchored YAC clones. Experiments aimed at chromosome walking were hampered by the high proportion of chimaeric clones in the libraries. The YAC libraries were screened to assess the scale of the problem (Schmidt et al., 1994). Repeated DNA sequences were found to be differentially represented in the EG and EW libraries and YACs carrying the tandemly repeated DNA sequences are often unstable. The libraries were also screened with chloroplast and rDNA sequences and many YACs carrying these sequences were shown to be chimaeric.

Two resources have contributed considerably to the completion of the physical map. These are a new YAC library (Creusot et al., 1994) and a recombinant inbred mapping population (Lister and Dean, 1993). The YAC library which became available in 1994 contains large inserts averaging 450kb and less then 10% chimaeric clones. The recombinant inbred lines are essentially an immortal population for genetic mapping which have been distributed throughout the research community. Markers from the two previously published genetic maps (Chang et al., 1988; Nam et al., 1989) have been mapped onto the recombinant inbred lines and with the publication of the map and distribution of seed, much additional information has been collected. The RIL map has helped the physical mapping by facilitating the ordering of contigs relative to each other and the tight integration of the physical and genetic map makes it an extremely useful tool for map-based cloning.

Using markers alone chromosome 4 fell into 8 contigs with an average size of 2.25Mbp. The remaining gaps were crossed by chromosome walking which showed that several of the contigs were already overlapping, but a marker in the overlap had not been available. Currently the chromosome is covered by just three contigs totalling 16.5Mbp. One of the contigs has been shown to define the short arm of chromosome 4. It hybridises at one end to the rDNA sequeces from the nucleolus organiser region (NOR) situated at the end of chromosome 4 (Copenhaver and Pikaard, 1995), and at the other end to repetitive sequences associated with the centromeric heterochromatin. Thus enabling us to integrate the genetic and physical maps with the cytogenetic map (Figure 2). The long arm is covered by the remaining two contigs which correspond to the majority of the genetic map.

The information we have gained about chromosome 4 enables us to contribute to the debate about the total size of the genome. Measurement of the synaptonemal complex of chromosome 4 has shown that it represents 16.8% of the genome (Albini, 1994). This value excludes the rDNA which does not appear to form a synapyonemal complex. The YAC contigs on chromosome 4 total 16.5Mbp and we estimate the centromeric region to be no less than 1Mbp. Thus, 17.5Mbp represents 16.8% of the genome. Using these values we can estimate that the total genome is approximately 105Mbp.

ANALYSIS OF CHROMOSOME 4

Using the physical map as a framework more can be learned about the organisation of the genome. We have three main areas of interest; recombination, the interspersion pattern of repeated and low copy sequences and sequence analysis.

Recombination

As physical maps of many different organisms ranging from lower to higher eukaryotes are emerging, physical and genetic distances can be compared. The distribution of recombination events is not random across the genome, with recombination lowest at the centromeres and telomeres and greatest midway down the chromosome arm (yeast (Oliver et al., 1992), and in tomato and potato (Tanksley et al., 1992)). In *Arabidopsis* we also have evidence of centromeric repression of recombination and higher rates of recombination in the arms from the physical mapping of chromosome 4. In addition to this global control at the whole chromosome level, there are sequences which appear to stimulate and repress recombination specifically (Simchen and Stamberg, 1969). For example, tomato has a genomic average recombination frequency of 550kb/cM, a 7-fold suppression of this is seen at the *Tm* 2a locus (Ganal et al., 1989) and a 10 fold stimulation at the I_2 locus (Segal et al., 1992). Similarly a 7-fold elevation in the level of recombination has been demonstrated at the *A1* locus in maize (Civardi et al., 1994),

Low definition analysis of recombination on chromosome 4.
We are interested in the sequences associated with large differences in such that a 1kb interval has a recombination frequency of 217kb/cM compared with the genomic average of 1460kb/cM. There are many similar examples in many different organisms including yeast (Symington and Petes, 1988), *C. elegans* (Greenwald et al, 1987), humans (Little et al., 1992), mice (Steinmetz et al., 1986) and rice

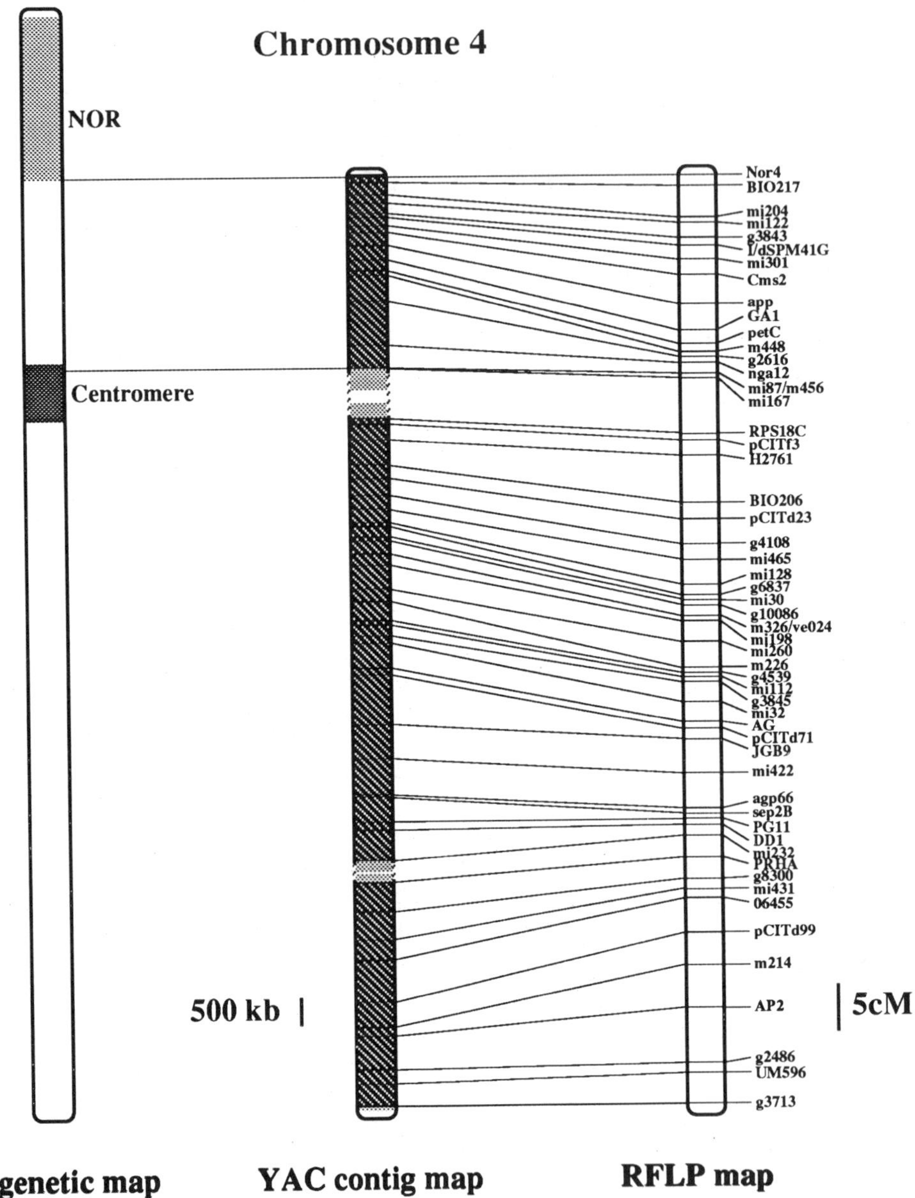

Figure 2. Integration of the physical and genetic maps with the cytogenetic map.

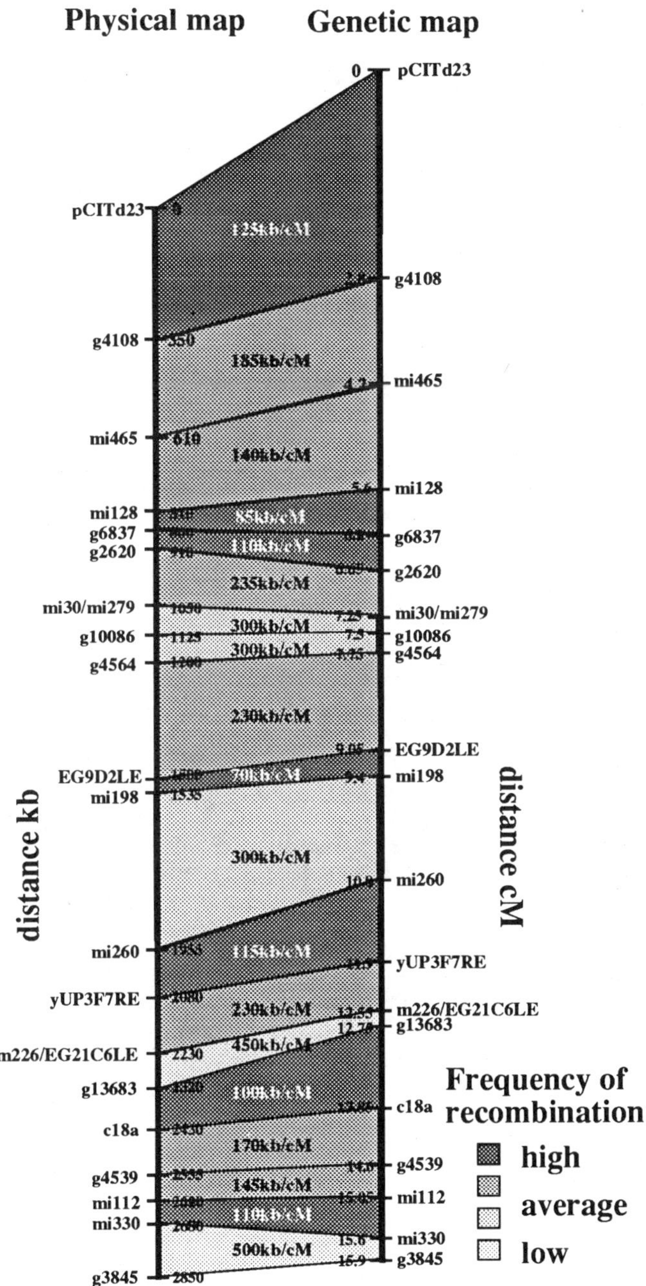

Figure 3. A comparison of genetic and physical distance over a 2.6Mbp region of chromosome 4. Recombination frequency is expressed in kilobases per centiMorgan. The frequency of recombination varies 7-fold between intevals, from 70kb/cM to 500 kb/cM.

79

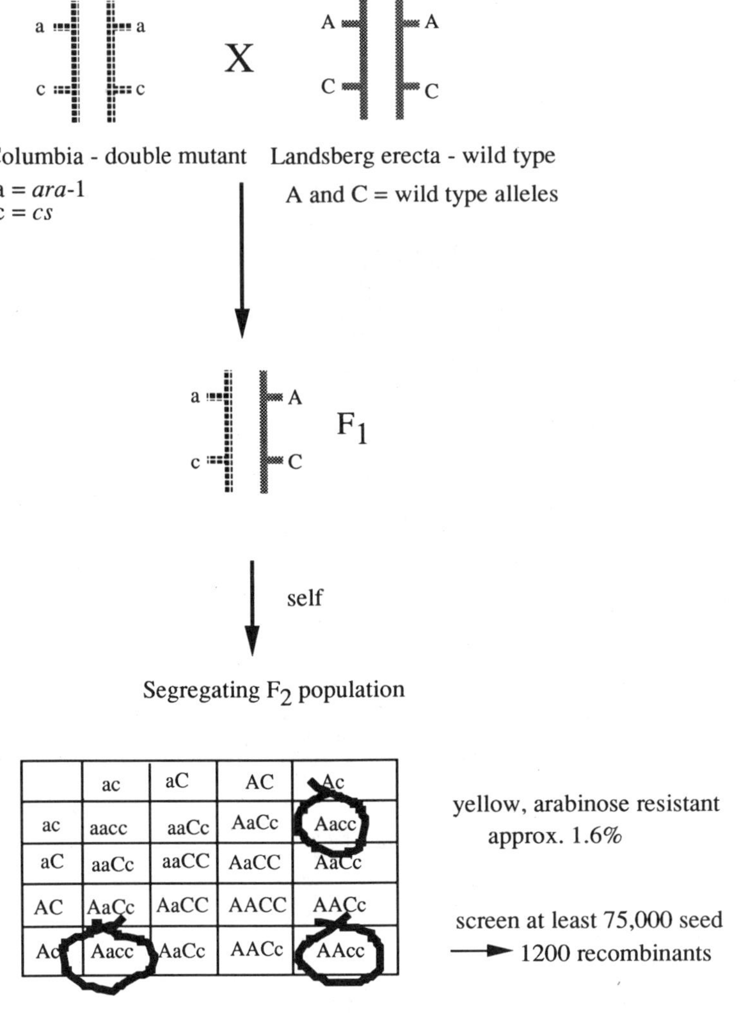

	ac	aC	AC	Ac
ac	aacc	aaCc	AaCc	Aacc
aC	aaCc	aaCC	AaCC	AaCc
AC	AaCc	AaCC	AACC	AACc
Ac	Aacc	AaCc	AACc	AAcc

Figure 4. Scheme for the generation of recombinants between the two phenotypic mutants *ara*-1 and *cs*. A columbia double mutant of *ara*-1 and *cs* is crossed to Landsberg *erecta*. The heterozygous F1 is selfed to produce an F2 segregating population. Yellow, arabinose resistant individuals are selected (Aacc and aacc) from the F2 population, these have a recombination event between *ara*-1 and *cs*.

(Gustafson and Dillé, 1992). It seems there are two levels of control, a global control at the whole chromosome level and on top of this localised control by specific sequences (Simchen and Stamberg, 1969). recombination frequency. In the early stages of the physical mapping project a 2.85Mbp YAC contig was assembled on chromosome 4. The availability of this largephysically mapped region enabled us to compare the physical and genetic maps. Physical distance across the contig could be calculated accurately because of the high redundancy of small YACs of known size and molecular markers including YAC end probes, *Pst*1 genomic clones (Lui et al., in preparation) and cosmids (Nam et al., 1989) were used as RFLPs to construct a genetic map on

300 recombinant inbredlines (RIL) (Lister and Dean, 1993). The large mapping population of 300 RILs, based on a cross between Landsberg *erecta* and Columbia ecotypes, was used to increase the resolution of the map because some of the markers are physically very close. The comparison spanned a total physical distance of 2.85Mbp and a genetic distance of 15.9cM (Figure 3). The average frequency of recombination for the region 180kb/cM which is similar to the genomic average, and recombination was found to vary between intervals 7 fold, from 70kb/cM to 500kb/cM.

High resolution analysis of recombination. A smaller interval of the region on chromosome 4 has been chosen for a more detailed analysis of recombination. The requirements for this type of analysis are a large number of recombinants and an abundancy of markers. In *Arabidopsis* we are fortunate that there are a large number of genetically mapped phenotypic loci. Two phenotypic mutants have been selected, *ara-1* (Dolezal and Cobbett, 1991) and *cs* (Koncz et al., 1990). *ara-1* is a recessive mutation conferring sensitivity to growth on arabinose and *cs* is a recessive mutation affecting chlorophyll levels, this making the plant appear pale green/yellow. *ara-1* maps to a position close to yUP3f7RE (see Figure 4) and *cs* maps south of g3845. A double mutant was constructed in the Columbia ecotype background and this was crossed to Landsberg *erecta* to produce an F_1. The F_1 was allowed to self-fertilize to produce a segregating F_2 population (Figure 4). Recombinants between *ara-1* and *cs* were selected from the F_2 population. Recombinant types which could be selected were pale green/yellow (homozygous at *cs*) and capable of growth on arabinose (either heterozygous for *ara*-1 or homozygous wildtype). At least 75,000 F_2 seed were screened giving 1,200 recombinants. From this we calculated that the genetic distance between *ara-1* and *cs* in this cross is 3cM.

PCR-based markers have been the method of choice to detect the position of recombination events in the interval between *ara* -1 and *cs*. The use of PCR-based markers means no large scale DNA preparations need to be done on large numbers of recombinants. The markers being developed are Cleaved Amplified Polymorhic Sequences (CAPS) (Konieczny and Ausubel, 1993), heteroduplex tests (HDTs) (Chowdhury et al., 1993), and Single Strand Conformational Polymorphism tests (SSCPs) (Orita et al., 1989). A region of the genome to be assayed is amplified by PCR from the recombinant individuals. The nucleotide sequence of this region will be different depending on the parent this portion of the genome was inherited from. These slight differences can be assayed by CAPs by differential cutting of the DNA fragment by a restriction enzyme. Heteroduplex tests and SSCPs rely on conformational changes of the duplex or single-stranded molecules caused by the nucleotide sequence differences. These conformational changes translate into differences in the rate of migration through modified acrylamide gels. The heteroduplex and SSCP tests can potentially detect single nucleotide differences between DNA molecules. A directed approach to developing CAPs markers is being taken by screening cosmids from the region for polymorphism between the parental ecotypes Landsberg *erecta* and Columbia. The cosmid is hybridised to Southern blots with DNA from Landsberg *erecta* and Columbia arranged in pairs and digested with many different restriction enzymes. Polymorphisms which are clearly due to loss or gain of a restriction site can be easily utilised. The appropriate fragment is isolated from the cosmid, cloned and sequenced. A more random approach has been taken to produce heteroduplex and SSCP markers. A cosmid is subcloned and the end 250-300 base pairs of the fragment sequenced. Only small fragments of 100-300bp are required for these tests. The sequence is used to design primers to amplify fragments of the appropriate size which are tested for the heteroduplex and SSCP polymorphisms. The analysis of recombinants is currently in progress.

The interspersion pattern of low copy and repeated sequences

Pruitt and Meyerowitz (1986) analysed the sequence content of 50 lambda clones and from this derived a value of 125kb for the interspersion pattern of low copy and highly repeated DNA sequences if the repeated sequences were randomly distributed. We are conducting an investigation into the nature and location of moderately repetitive elements in the *Arabidopsis* genome. This has involved the examination of a 1Mbp region of chromosome 4 and the analysis of 73 cosmids selected from a library as containing repetitive sequences by virtue of their strong hybridisation to total *Arabidopsis* DNA.

Analysis of the 1Mbp region involved the analysis of YAC end-clones which had been identified as repetitive in a chromosome walking experiment in this region. It transpired that the majority of these were from chimaeric YACs and the repeats identified from a different region of the genome. This whole 1Mbp region appears to be made up exclusively of single or low copy sequences (Helen Thompson, pers. comm.). The 73 cosmids identified as containing repetitive sequences were screened with all other known repeats to eliminate these from the investigation, after which 11 clones remained. A subset of these are moderately repetitive, 50 -100 copies, and are associated with the centromeric heterochromatin. Another small subset are multigene families and a long GA microsatellite has also been found. Another source of information about repeats has come from the building of cosmid contigs for the sequencing program. To date a section of approximately 700kb of chromosome 4 is covered in overlapping cosmids which are to be used as a substrate for sequencing. In building this contig part of which overlaps the 1Mbp region above, 3 regions of dispersed type repeats have been encountered (I. Bancroft, pers. comm.). These data together suggest that most of the highly repetitive sequences reside in centromeric areas and that the mean interspersion pattern of repetitive and unique sequence may be considerably larger than 125kb estimated previously (Pruitt and Meyerowitz, 1986).

Sequence analysis

The most prolific area of sequence analysis of *Arabidopsis* has been the sequencing of directionally cloned anonymous cDNAs, more familiarly known as expressed sequence tags (ESTs). There are projects to sequence ESTs in France (Hofte et al., 1993) and in the U.S. (1033}. More than 14000 *Arabidopsis* ESTs have been sequenced and submitted to the EST database (dbEST), of which approximately 30% show similarity to existing sequences from *Arabidopsis* (Hofte et al., 1993). Homology searches have also shown that approximately 17 % of the ESTs can be assigned a probable function (Newman et al., 1994).

A program to initiate genomic sequencing of *Arabidopsis* has been funded by the European Community. The program is known as ESSA, European Scientists Sequencing *Arabidopsis*. A network of laboratories are contributing to the sequencing effort which is coordinated by Mike Bevan (Norwich, U.K.). The program is in its pilot phase until the end of 1996 when it is hoped to have achieved 1.5Mb of contiguous sequence on chromosome 4, 400kb of sequence from other genomic regions, and the sequencing of cognate cDNAs and ESTs.

The substrate for the genomic sequencing in its initial phase has been cosmid clones (carrying DNA from the Columbia ecotype). Cosmids from the contigs generated by Hauge *et al.* (1991) along with a new cosmid library generated by C. Lister and C. Dean (unpublished results) have been selected by hybridization with whole YACs and then assembled into overlapping arrays by restriction mapping. This resulted in 88% cover (I. Bancroft, pers. comm.). Gaps were filled by generating cosmid libraries from purified YAC clones. This resulted in high redundancy (30 - 40%) libraries facilitating the cloning of 100% of the region. Cosmids covering a 700kb contiguous region have so far been assembled for sequencing.

To analyse the sequence data programs have been used to predict open reading frames, many coincide with homology detected by other searches. At the nucleotide and amino acid level there is homology to known genes and ESTs, with and without gene homology. Analysis of two overlapping cosmids (70kb) from chromosome 4 suggests a gene density of roughly one gene every 5kb (G. Murphy, pers. comm.).

ACKNOWLEDGEMENTS

The authors wish to thank Ian Bancroft, Mike Bevan, George Murphy and Helen Thompson for access to unpublished data. We wish to thank Keith Edwards for helpful discussion and coordinating the Zeneca SRF grant. The integration of the cosmid contigs with the YAC contigs was done in collaboration with Dr. H. Goodman. MS is supported by a SRF grant from Zeneca and RS is supported by a BBSRC grant no PG 208/0608 to CD.

REFERENCES

Albini, S.M., 1994, A karyotype of the *Arabidopsis thaliana* genome derived from synaptonemal complex analysis at prophase I of meiosis, *Plant J.* 5:665.

Arumuganathan, K., and Earle, E.D., 1991, Nuclear DNA content of some important plant species, *Plant. Mol. Biol. Rep.* 9:208.

Bennett, M.D., and Smith, J.B., 1976, Nuclear DNA amounts in angiosperms, *Phil. Trans. R. Soc. Lond.* 274:227.

Chang, C., Bowman, J.L., DeJohn, A.W., Lander, E.S., and Meyerowitz, E.M., 1988, Restriction fragment length polymorphism linkage map for *Arabidopsis thaliana*, *Proc. Natl. Acad. Sci. USA* 85:6856.

Chowdhury, V., Olds, R.J., Lane, D.A., Conard, J., Pabinger, I., Ryan, K., Bauer, K.A., Bhavnani, M., Abildgaard, U., Finazzi, G., Castaman, G., Mannucci, P.M., and Thein, S.L., 1993, Identification of nine novel mutations in type I antithrombin deficiency by heteroduplex screening, *British J. Haem.* 84:656.

Civardi, L., Xia, Y., Edwards, K.J., Schnable, P.S., and Nikolau, B.J., 1994, The relationship between the genetic and physical distances in the cloned *a1-sh2* interval of the *Zea mays* L. genome, *Proc. Natl. Acad. Sci. USA* 91:8268.

Copenhaver, G.P., and Pikaard, C.S., 1995, Genetic and physical mapping of the nucleolus variant rRNA genes at NOR2 and NOR4 using 1- and 2-dimensional pulsed-field gel analysis, *Mol. Cell. Biol.* (In Press)

Creusot, F., Billault, A., Bouchez, D., Caboche, M., Camilleri, C., Chaboute, M.E., Cohen, D., Dron, M., Durr, A., Feltrin, P., Fleck, J., Fovilloux, E., Gigot, C., Lafleuriel, J., Ougen, P., Perrot, V., Picard, G., and Saumier, M., 1994, Construction of a large insert YAC library of *Arabidopsis thaliana, Fourth International Congress of Plant Molecular Biology* no.88.(Abstract)

Dolezal, O., and Cobbett, C.S., 1991, Arabinose kinase-deficient mutant of *Arabidopsis thaliana, Plant Physiol.* 96:1255.

Ecker, J.R., 1990, PFGE and YAC analysis of the *Arabidopsis* genome, *Methods* 1:186.

Francis, D.M., Hulbert, S.H., and Michelmore, R.W., 1990, Genome size and complexity of the obligate fungal pathogen *Bremia lactucae, Exp. Mycol.* 14:299.

Galbraith, D.W., Harkins, K.R., and Knapp, S., 1991, Systemic Endopolyploidy in *Arabidopsis thaliana, Plant Physiol.* 96:985.

Ganal, M.W., Young, N.D., and Tanksley, S.D., 1989, Pulsed field electrophoresis andphysical mapping of large DNA fragments in the *Tm-2a* region of chromosome 9 in tomato, *Mol. Gen. Genet.* 215:395.

Greenwald, I., Coulson, A., Sulston, J., and Priess, J., 1987, Correlation of the physical and genetic maps in the *lin-12* region of *Caenorhabditis elegans, Nucl. Acids Res.* 15:2295.

Grill, E., and Somerville, C., 1991, Construction and characterization of a yeast artificial chromosome library of *Arabidopsis* which is suitable for chromosome walking, *Mol. Gen. Genet.* 226:484.

Gustafson, J.P., and Dille, J.E., 1992, Chromosome location of *Oryza sativa* recombination linkage groups, *Proc. Natl. Acad. Sci. USA* 89:8646.

Hauge, B.M., Giraudat, J., Hanley, S., Hwang, I., Kochi, T., and Goodman, H.M., 1991, Physical Mapping of the *Arabidopsis* genome and its applications, in: *Plant Molecular Biology*, edited by Herman, R.G., ed., Plenum, New York.

Heslop-Harrison, J.S., and Schwarzacher, T., 1990, The ultrastructure of *Arabidopsis thaliana* chromosomes, *Fourth International Conference on Arabidopsis Research* (Abstract)

Hofte, H., Desprez, T., Amselem, J., Chiapello, H., and Caboche, M., 1993, An inventory of 1152 expressed sequence tags obtained by partial sequencing of cDNAs from *Arabidopsis thaliana, Plant J.* 4:1051.

Koncz, C., Mayerhofer, R., KonczKalman, Z., Nawrath, C., Reiss, B., Redei, G.P., and Schell,J., 1990, Isolation of a gene encoding a novel chloroplast protein by T-DNA tagging in *Arabidopsis thaliana, EMBO J.* 9:1337

Konieczny, A., and Ausubel, F.M., 1993, A procedure for mapping *Arabidopsis thaliana* mutations using co-dominant ecotype-specific PCR-based markers, *Plant J.* 4(2):403.

Laibach, F., 1907, Zur fragr nach der individualitat der chromosomen im pflanzenreich, *Beih. Bot. Cbl.* (1 abt.) 22:19.

Leutwiler, L.S., Hough-Evans, B.R., and Meyerowitz, E.M., 1984, The DNA of *Arabidopsis thaliana, Mol. Gen. Genet.* 194:15.

Lister, C., and Dean, C., 1993, Recombinant inbred lines for mapping RFLP and phenotypic markers in *Arabidopsis thaliana, Plant J.* 4:745.

Little, R.D., Pilia, G., Johnson, S., D'Urso, M., and Schlessinger, D., 1992, Yeast artificial artificial chromosomes spanning 8 megabases and 10 -15 centimorgans of human cytogenetic band Xq26, *Proc. Natl. Acad. Sci. USA* 89:177.

Maluszynska, J., and Heslop-Harrison, J.S., 1991, Localization of tandemly repeated DNA sequences in *Arabidopsis thaliana, Plant J.* 1:159.

Martinez-Zapater, J.M., Estelle, M.A., and Somerville, C.R., 1986, A highly repeated DNA sequence un *Arabidopsis thaliana, Mol. Gen. Genet.* 204:417.

Meyerowitz, E.M., 1987, *Arabidopsis thaliana, Ann.Rev.Gen.* 21:93.

Meyerowitz, E.M., 1992, Introduction to the *Arabidopsis* genome, in: *Methods in Arabidopsis Research*, C. Koncz, N-H. Chua, and J. Schell (ed), World Scientific, Singapore.p. 100.

Nam, H.-G., Giraudat, J., den Boer, B., Moonan, F., Loos, W.D.B., Hauge, B.M., and Goodman, H.M., 1989, Restriction fragment length polmorphism linkage map of *Arabidopsis thaliana, Plant Cell* 1:699.

Newman, T., de Bruijn, F., Green, P., Keegstra, K., Kende, H., McIntosh, L., and Ohlrogge, J., 1994, Genes Galore: a summary of methods for accessing results from large-scale partial sequencing of anonymous cDNA clones, *Plant Physiol.* 106:1241.

Oliver, S.G., van der Aart, Q.J.M., Agostoni-Carbone, M.L., Aigle, M., Alberghina, L., et al., 1992, The complete DNA sequence of yeast chromosome 111, *Nature* 357:38.

Orita, M., Iwahana, H., Kanazawa, H., Hayashi, K., and Sekiya, T., 1989, Detection of polymorphisms of human DNA by gel electrophoresis as single-strand conformation polymorphisms, *Proc. Natl. Acad. Sci. USA* 86:2766.

Pruitt, R.E., and Meyerowitz, E.M., 1986, Characterisation of the genome of *Arabidopsis thaliana, J.Mol.Biol* 187:169.

Richards, E.J., and Ausubel, F.M., 1988, Isolation of a higher eukaryotic telomere from *Arabidopsis thaliana, Cell* 53:127.

Schmidt, R., Putterill, J., West, J., Cnops, G., Robson, F., Coupland, G., and Dean, C., 1994, Analysis of clones carrying repeated DNA sequences in two YAC libraries of *Arabidopsis thaliana, Plant J.* 5:735.

Schweizer, D., Ambros, P., Gründler, P., and Varga, F., 1987, Attempts to relate cytological and molecular chromosome data of *Arabidopsis thaliana* to its genetic linkage map, *Arabidopsis Information Service* 25:27.

Segal, G., Sarfatti, M., Schaffer, M.A., Ori, N., Zamir, D., and Fluhr, R., 1992, Correlation of genetic and physical structure in the region surrounding the *I 2 Fusarium oxyspoprum* resistance locus in tomato, *Mol. Gen. Genet.* 231:179.

Simchen, G., and Stamberg, J., 1969, Fine and course controls of genetic recombination, *Nature* 222:329.

Simoens, C.R., Gielen, J., Van Montagu, M., and Inze, D., 1988, Characterization of highly repetitive sequences of *Arabidopsis thaliana, Nucl.Acids Res.* 16:6753.

Steinmetz, M., Stephan, D., and Fischer Lindahl, K., 1986, Gene organisation and recombination hotspots in the murine major histocompatibility complex,

Cell 44:895. Symington, L.S., and Petes, T.D., 1988, Expansions and contractions of the genetic map relative to the physical map of yeast chromosome III, *Mol.Cell Biol.* 8:595.

Tanksley, S.D., Ganal, M.W., Prince, J.P., de Vincente, M.C., Bonierbale, M.W., Broun, P., Fulton, T.M., Giovannoni, J.J., Grandillo, S., Martin, G.B., Messeguer, R., Miller, J.C., Miller, L., Paterson, A.H., Pineda, O., Roder, M.S., Wing, R.A., Wu, W., and Young, N.D., 1992, High density molecular linkage maps of the tomato and potato genomes, *Genetics* 132:1141.

Ward, E.R., and Jen, G.C., 1990, Isolation of single-copy-sequence clones from a yeast artificial chromosome library of randomly-sheared *Arabidopsis thaliana* DNA, *Plant Mol. Biol.* 14:561.

MOLECULAR MARKERS, FOREST GENETICS, AND TREE BREEDING

D.M. O'Malley[1,2], D. Grattapaglia, J.X. Chaparro[1,4], P.L. Wilcox.[1,2], H.V. Amerson[1,2], B.-H. Liu[1,5], R. Whetten[1,2], S. McKeand[2], E.G. Kuhlman[6], S. McCord[1,2], B. Crane[1,2], and R. Sederoff[1,2,3].

Forest Biotechnology Group[1], Department of Forestry[2], Department of Genetics[3], Department of Horticulture[4], Department of Statistics[5], North Carolina State University, Raleigh, NC, 27696-8008, USA, U.S.D.A. Forest Service Laboratories, Athens GA, 30602.[6]

INTRODUCTION

Several years ago, Strauss et al. (1992) thoughtfully evaluated the application of molecular markers in forest tree breeding for marker aided selection. The purpose of their paper was to emphasize the limitations and shortcomings of marker-aided selection particularly in conifers. They argued that studies of quantitative trait loci identified in agronomic crops, which have significant utility (e.g. Stuber, 1992; Stuber et al., 1992), are of little relevance to assessing the potential for marker aided selection in populations of forest trees, and that the near term usefulness of molecular markers for forest tree breeding will be limited. The major barriers to application included cost, the lack of association of markers with traits across breeding populations due to linkage equilibrium, variation in expression of loci affecting quantitative traits due to differences in genetic background, genotype environment interactions, and stability of marker-trait associations over multiple generations. In addition, Strauss et al. (1992) noted that marker-aided selection would be most useful for within family selection, where the economic values of the traits are high, the trait heritabilities are low, and where markers are able to explain much of the genetic variance. However, they argued that important traits in forest trees such as wood volume, are likely to be controlled by large numbers of genes with small effects, and therefore, are unlikely to have useful marker trait associations.

Since that time, considerable advances have been made in technology and theory in the application of molecular markers to the genetics of forest trees. Although the reservations expressed by Strauss et al. (1992) were generally sound, new results lead to a substantially different view of the potential for marker aided selection. In this article, we review recent results that circumvent the limitations of linkage equilibrium, and challenge earlier assumptions about the genetic architecture of valuable traits in forest trees. The major factors that have occurred are: 1) development of PCR based marker systems, 2) the ability to map individual trees, 3) new results that indicate major gene effects are not rare in the genetic architecture of quantitative traits in forest trees, 4) development of theory and computer software to incorporate molecular markers directly into existing breeding programs, and 5) a shift from population breeding in early generations of tree improvement to an emphasis on pedigrees in advanced generations. In contrast with early predictions, we conclude that the application of marker-aided selection to tree breeding programs may be one of the earliest applications of biotechnology to forestry.

SOLVING THE PROBLEM OF LINKAGE EQUILIBRIUM

Linkage equilibrium is the consequence of genetic recombination and chromosome segregation in populations of plants and animals. In such populations, the association of a specific allele at any locus with an allele at another locus, will be lost over generations. Therefore, in unrelated individuals there will be no predictable association of alleles. Any pair of linked heterozygous loci are likely to be found in all possible combinations of linkage phase (coupling or repulsion) and therefore, natural populations of forest trees are not useful for creating genomic maps. Conventional genetic maps are constructed from marker segregation data obtained from families generated by controlled crossing in three generation pedigrees. Some pedigrees of this kind are available in long term breeding programs and have been used successfully for genomic maps, in combination with RFLP techniques. These approaches were the first to study linkage in forest trees (Neale and Williams, 1991; Williams and Neale, 1992; Bradshaw and Stettler, 1992). Genomic maps were laborious and difficult to construct, and marker-trait associations obtained from a single pedigree in a tree species would not be useful generally for the breeding population because different trees could have different linkage relationships for markers and quantitative trait loci. Thus, linkage equilibrium was a conceptual barrier to the application of genomic mapping in forest trees. This conceptual barrier has fallen with the development of methods that could generate genomic maps for many individual trees. These advances were made possible by the development of a new marker system and its application to genetic analysis of conifers using haploid megagametophytes.

APPLICATION OF PCR BASED MARKERS TO FOREST GENETICS

A major technical advance that has facilitated the application of genetic mapping in forest trees is that of polymerase chain reaction (PCR) based markers. In 1990, several researchers developed PCR based systems for the amplification of anonymous fragments from eukaryotic DNA (Williams et al., 1990; Welsh and McClelland, 1990; Caetano-Anolles et al., 1991). The most widely used system, RAPD (Random Amplified Polymorphic DNA) markers, uses single 10 base oligonucleotides to amplify short inverted repeats distributed throughout the genome being assayed. Such inverted repeats are highly polymorphic, and provide large numbers of readily accessible dominant markers. The major advantages of these marker systems is that they can be applied to organisms with compex eukaryotic genomes without any prior information. A typical survey of pine DNA with 100 different arbitrary primers will provide 100 to 200 polymorphic markers. Carefully chosen markers are reliable and repeatable. However, control of reaction parameters and quality of reagents is essential, because amplification is extremely sensitive to initial conditions. Furthermore, the small amounts of DNA required (nanograms per reaction) makes possible the application of RAPD methods to the conifer haploid megagametophyte.

In conifers, the meiotic division of the diploid megaspore mother

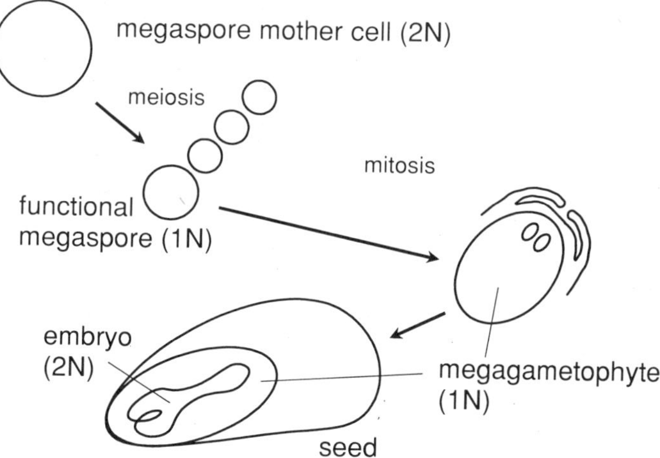

Figure 1. The genetic biology of the haploid megagametophyte of pine. Development of the pine megagametophyte proceeds from a single haploid product of meiosis. The functional megaspore divides by mitosis to produce the megagametophyte, which will develop the archegonia that contain the female gametes. Several gametes could be fertilized by pollen but only one gives rise to the embryo in the mature seed.

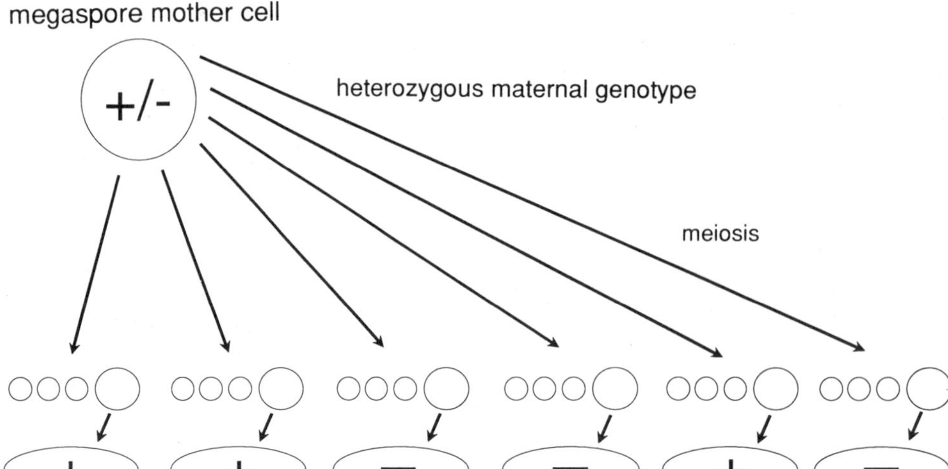

Figure 2. The megagametophyte in each mature seed is genetically equivalent to a haploid progeny plant, thus any heterozygous locus in the seed parent will segregate 1:1 in the megagametophytes, regardless of the pollen contribution.

cell yields four haploid meiotic products. The megagametophyte, a haploid nutritive tissue surrounding the embryo in the mature seed, develops from one of these haploid megaspores (Figure 1). The allelic contribution of the seed parent to the embryo segregates in megagametophytes from that tree (Figure 2). The genotype of each megagametophyte is identical to the maternal gamete that forms the embryo. Genetic analysis of the conifer megagametophyte, using isozymes, has been carried out for many years for population genetics and for linkage analysis (Conkle, 1981). The conifer megagametophyte has been extremely useful for estimation of genetic diversity, heterozygosity, genetic relatedness and for studies of gene flow in natural populations (Wheeler et al., 1982; Millar et al., 1983; Hamrick et al., 1992). The major limitation of a small number of markers were available. Nevertheless, the theory and practice of genetic analysis using the haploid megagametophyte was well established through work on isozymes.

An important feature of megagametophyte analysis is that it is half-sib analysis and independent of the pollen involved in fertilization. Megagametophytes are strictly derived from the products of maternal meiosis, and represent the maternal gametic genotypes exclusively.

The pollen contribution can be ignored and there is no difference in the analysis of open pollinated or control pollinated seeds. If seeds are germinated, the megagametophytes can be collected just before germinants are ready to cast their seed coats. The seed coats still contain the residual megagametophytes inside. It is therefore possible to use the megagametophyte for genetic analysis of the maternal genotype and to grow and test the seedling for a specific phenotype and for its paternal genotype.

For our genetic analysis, we have used RAPD markers almost exclusively. Initially, we used RAPD markers because they could be assayed in megagametophytes, required no prior information, and allowed construction of genomic maps in a relatively short period of time. We also found that they were highly informative in spite of the effect of dominance, because the utility of a marker depends on its position relative to another marker or gene of interest. A close marker linked in coupling to a gene of interest can be extremely useful even though the marker may be dominant. Furthermore, we were able to use automation for the distribution of reagents for RAPD reactions and increase the number of reactions that a single person could carry out per day by a factor of 10. Alternative marker systems, such as microsatellites, and markers based on known sequence will be increasingly important, as sequence information accumulates. New methods such as Amplified Fragment Length Polymorphism (AFLP) (Zabeau, 1993) work well with large complex genomes and have considerable potential for forest trees. Improved systems of marker analysis and automation should further accelerate the construction of genomic maps.

CONSTRUCTION OF A SINGLE TREE MAP FOR LOBLOLLY PINE

In conifers, the tissue from the megagametophyte is the genetic equivalent of a haploid progeny plant from a specific mother tree. A sample of 100 seeds from a single tree can be the equivalent of a mapping population. Each seed is dissected, usually after germination, to remove the megagametophyte, which can be used to prepare several micrograms of genomic DNA. Each seed provides DNA from a different haploid genotype which is analyzed with prescreened primers to obtain genotype information for a sufficient number of RAPD markers to construct a map (Grattapaglia et al., 1991). Markers are prescreened for being polymorphic in the genotype to be mapped, and for repeatability. A similar approach to mapping was taken by Carlson and coworkers (Carlson, 1991; Tuliseram et al., 1992). To begin a genomic map, prescreened primers are used for reactions with a set of seeds that constitute the mapping population of progeny genotypes from a specific tree. RAPD polymorphisms are scored for each primer, based on the presence or absence of specific bands. There is an expectation of 1:1 for a segregating locus that is heterozygous in the maternal parent. We find the frequency of departures from expected segregation ratios to be low. From analysis of each pair of markers, it is possible to determine if there is significant linkage, if so, to assign linkage phase, and to calculate a recombination fraction. A number of mapping programs are available for this analysis including MAPMAKER (Lander et al., 1987),

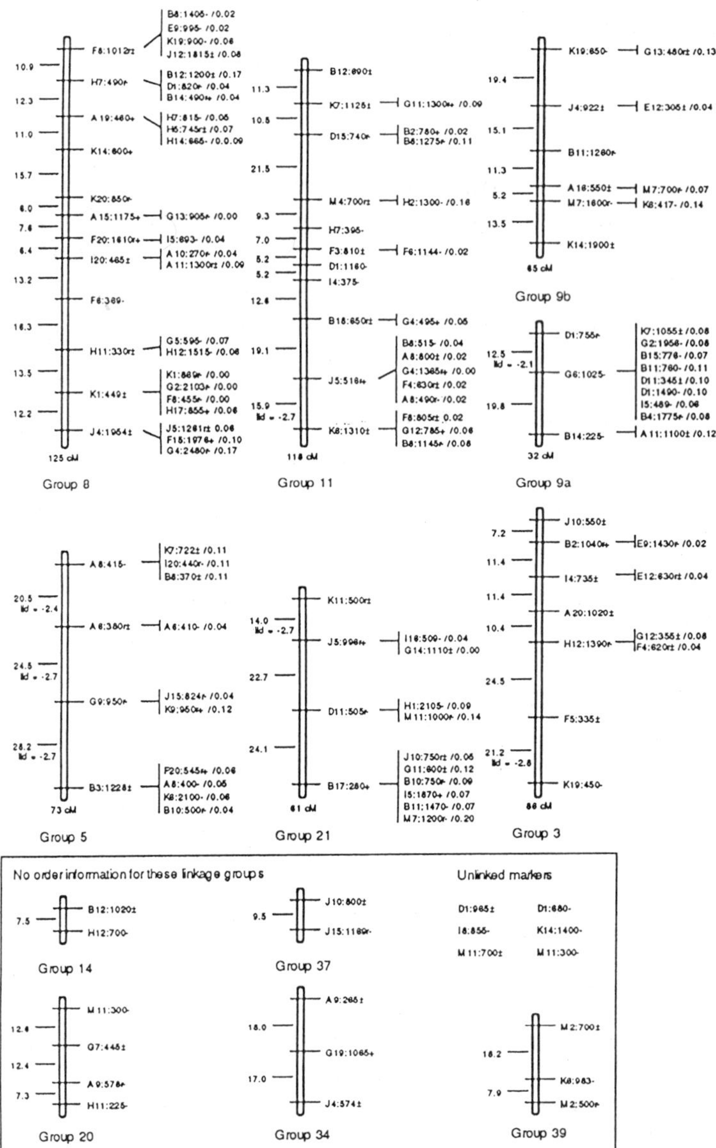

Figure 3. A genomic map of *Pinus taeda* L. (loblolly pine genotype 7-56). The map was constructed from 232 segregating RAPD markers in a mapping sample of 58 megagametophytes from open pollinated seeds of genotype 7-56. The 91 framework markers comprise those that could be ordered with interval support greater than 3 for most linkage groups. Grouping is carried out using a criterion LOD >4, recombination fraction 0.25. The 121 accessory markers are located next to the nearest framework marker. 14 markers remain unordered in 5 more linkage groups. Recombination distance in cM is presented immediately to the left of each linkage map, and the designation for each framework marker is shown immediately to the right. Accessory markers are shown further to the right associated with the closest framework markers. The markers are designated by a code that represents the Operon primer code, the fragment length, the linkage phase (r for repulsion), and a code plus, plus/minus, or minus for decreasing band intensity. Accessory markers also have an additional designation which is the recombination fraction to the closest framework marker. The primary data for the construction of the map has been submitted to the Dendrome project database, (http://s27w007.pswfs.gov).

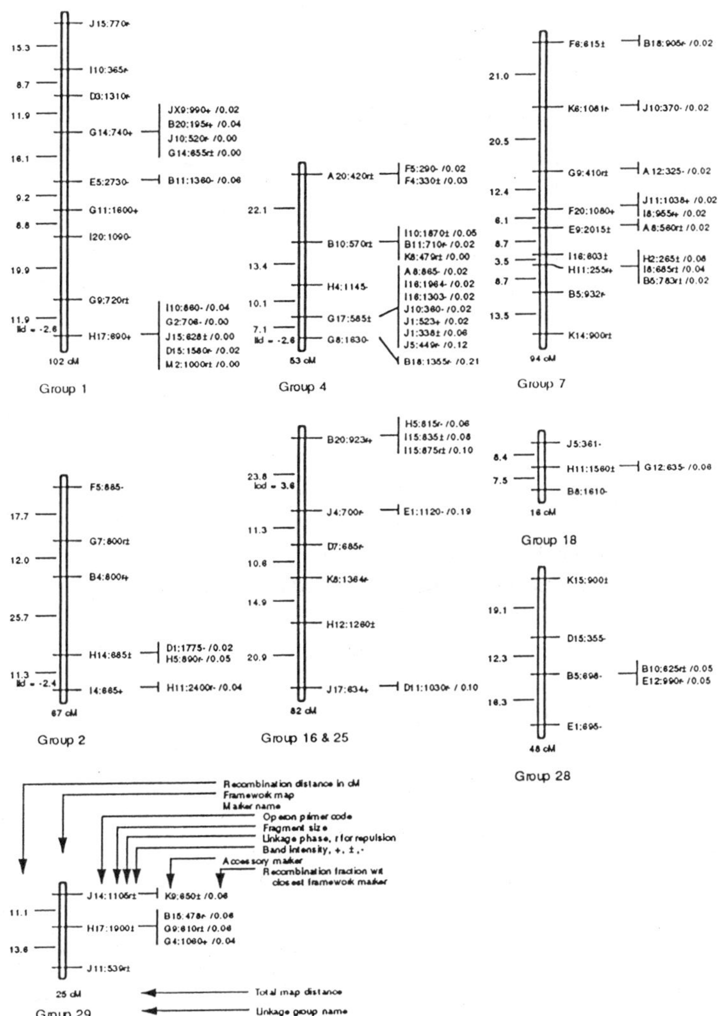

GMENDEL (Liu and Knapp, 1990). In this way we have constructed a linkage map for loblolly pine clone 7-56 from 232 RAPD markers (Figure 3). The map has a total genetic length of 1067 centiMorgans (cM) in ordered linkage groups with an estimated coverage of 85%.

The value of genomic maps depends greatly on the reliability of marker order. We have chosen a "framework map" approach for evaluation of map order (e.g., Plomion et al., 1995). For each subset of three markers within a linkage group, we use a criterion for local locus order that is approximately 1000 times more likely than the next most probable order (interval support > 3). Markers that satisfy this criterion are "framework" markers. Other markers that do not meet the criterion of framework markers, but show significant linkage to framework markers and map to these linkage groups are called accessory markers. The accessory markers are shown adjacent to the major linkage groups along with the recombination fraction with the closest framework marker (Figure 3).

SINGLE TREE MAPS REDUCE THE NEED FOR EXTENDED PEDIGREES

The earliest studies of genetic mapping in forest trees using isozymes, were carried out using megagametophytes, but application of RFLP technology required large amounts of DNA, precluding single megagametophyte analysis. To obtain mapping populations and to ascertain linkage phase (coupling or repulsion) of markers, the first genomic mapping studies used three generation pedigrees (Neale and Williams, 1992; Bradshaw and Stettler, 1992). Extended pedigrees of forest trees are rare. In cultivated crops, varieties are highly inbred and generally homozygous, therefore, heterozygosity is created by crossing inbred lines. Instead of creating heterozygosity through crossing, most forest trees are already highly heterozygous and can be analyzed directly. Where markers can be analyzed in megagametophytes, it is therefore possible to carry out linkage analysis in individual trees. Linkage phase is unknown for any individual tree but is determined readily from the segregation of markers in the megagametophytes. In this way, mapping individual trees solves the problem of linkage equilibrium and greatly reduces the need for extended pedigrees. Extended pedigrees are extremely valuable because it is possible to map the individual trees in the pedigree and to define the effects of specific genes or chromosome regions over generations.

The ability to map individual trees makes accessible genetic analysis of a great many trees that are part of current breeding programs. Mapping of trees using open pollinated seed through half sib analysis may be applied to any specific conifer provided that a sufficient number of seed can be obtained, that adequate amounts of DNA can be obtained from the megagametophytes, and that the tree possesses a reasonably high degree of heterozygosity. These criteria global control at the whole chromosome level and on top of this localised control by specific sequences (Simchen and Stamberg, 1969). recombination frequency. In the early stages of the physical mapping project a 2.85Mbp YAC contig was assembled on chromosome 4. The availability of this largedetailed genetic analysis through half sib mapping. Other gymnosperms such as ginkgo should be amenable to this method also.

A PSEUDOTESTCROSS MAPPING STRATEGY FOR DIPLOIDS

Many species of commercially important forest trees are not conifers and can not be analyzed through genotyping of megagametophytes. However, it has been possible to extend the strategy of mapping individual trees to angiosperms through the use of PCR based markers. The means by which this is possible is through the screening and selection of informative markers that provide the same simple genetic results. The strategy is best illustrated using full sib crosses carried out with two species of *Eucalyptus* (*E. grandis* x *E urophylla*) that can be crossed. Individuals of both species are highly polymorphic. Progeny from such crosses are widely planted in tropical areas of the world. Any particular RAPD marker that is polymorphic in a particular cross could be heterozygous in both parents, or heterozygous in one parent but homozygous null in the other parent. We chose those markers heterozygous in one of the parents and null in the other. With such markers, all progeny show 1:1 segregation from the maternal parent just as in the conifer megagametophyte. DNA can be prepared from seedlings or young progeny plants, and the amounts of DNA obtained are not as limited as they are from megagametophytes of conifers. One set of markers can be used to generate a map from one parent and a second complementary set of markers can be used to create a second map. Using this strategy, we have generated tree-specific maps for *E grandis* and *E. urophylla* (Grattapaglia and Sederoff, 1994). Because this marker strategy uses the same concept as a testcross, but does so through the choice of informative markers in one generation, it is called a pseudotestcross. When applied to both parents simultaneously, it can be called a double pseudotestcross.

The pseudotestcross strategy can be extended further to mapping of individual trees that are open pollinated. This approach is possible when the pollen parents are sufficiently different from the seed parent so that many alleles heterozygous in the female parent are rare in the pollen pool. Under these conditions, the pseudotestcross mating structure also mimics that of the conifer megagametophyte and a map can be created for the maternal parent from a set of progeny plants. Grattapaglia has repeated a map of a previously mapped individual of *E. grandis* using open pollinated progeny crossed to a pollen polymix from *E. urophylla* (Grattapaglia, 1994). In theory, recombination frequencies can be estimated when linked dominant maternal alleles are present in low frequency in the pollen pool. Allele frequencies can be as high as 0.3 in the pollen pool and still be useful for mapping if sufficient progeny can be obtained. This approach further extends the use of open pollinated individuals for mapping. A likelihood function for the such a situation can be formally generalized for any mapping population where the outcrossing rate and the allele frequencies are specified (Liu, 1993). Full sib crosses and half sib mapping are special cases of the general model.

DISSECTION OF COMPLEX TRAITS WITH MOLECULAR MARKERS

Most experiments directed to examine the genetic architecture of quantitative traits in domestic crops do not find that they are determined by a large number of genes with small effects. Typically, a significant component of the genetic variance is associated with specific regions of the genome that can be "tagged" with molecular markers (e.g. Beavis et al., 1991; Goldman et al., 1994; Hayes et al.,

1994; Patterson et al., 1991; Stuber, 1992.). Tanksley and coworkers (see review by Tanksley, 1992) showed that in hybrid breeding, much of the genetic variation in an F$_2$ cross could be attributed to a small number of major genes or chromosomal regions, and that genetic markers could explain much of the phenotypic variation within such an F2 family. In other plants and in animals, crosses between divergently selected lines yielded evidence of QTLs, but the proportion of genetic and phenotypic variation explained by markers appears to be smaller. Groover et al. (1995) found QTLs for wood specific gravity in loblolly pine using divergently selected parents in a three generation outbred pedigree. The QTLs they detected explained 23% of the phenotypic variation. Forest tree populations contain large amounts of genetic variation, but how much variation within families of typical intraspecific crosses can be explained by segregating genetic markers remains to be determined.

Since 1992, considerable progress has been made in forest trees, including conifers regarding the dissection of complex traits and the detection of QTLs. In *Populus*, Bradshaw and Stettler (1995) have analyzed several traits in an inbred F$_2$ from an interspecific cross of *P. tricocarpa* x *P. deltoides*. They mapped QTLs for stem growth and form, and spring leaf flush, evaluated over a two year period. They found one to five QTLs of large effect to be responsible for a large part of the genetic variance for each trait measured. For stem volume, for example, 45% of the genetic variance was controlled by two QTLs. QTLs for syleptic branch leaf area and stem basal area shared similar chromosomal positions suggesting a pleiotropic effect of QTLs affecting stem diameter growth. For spring leaf flush 5 QTLs were found. Strauss et al., (1992) had suggested that interspecific hybrids would be suitable material for such studies.

Similar results have been obtained for *Eucalyptus* in an analysis of a cross between *E. grandis* and *E. urophylla*. The F$_1$ progeny of a cross were analyzed for QTLs following a pseudotestcross mapping experiment (Grattapaglia, 1994; Grattapaglia et al., 1995). QTLs were detected for growth traits in trees at rotation age and for traits involved in vegetative propagation. Ten QTLs were identified for micropropagation response, measured as fresh weight of shoots, six QTLs were identified for stump sprouting ability, and four for rooting ability. These QTLs explained up to 52% of the phenotypic variance for specific traits in this cross. In mature trees, QTLs were detected for height, wood volume, and bark thickness. Three QTLs were detected for wood volume explaining at least 11% of the phenotypic variance, three QTLs were detected for % bark explaining 12 % of the variance (Grattapaglia, 1994). These results establish that QTLs can be detected for traits of commercial value at rotation age and provide strong support for the value of marker-aided selection in forest trees.

In conifers, QTL analysis has been carried out for wood-specific gravity in a three generation pedigree of loblolly pine (Groover et al., 1994). One highly significant region was detected and several additional significant associations were reported as well. Together, the QTLs explained 23% of the phenotypic variance. In addition QTLs have been detected in different species of pines related to early height growth under a variety of conditions (O'Malley, Crane and Plomion, unpublished results).

These experiments allow us to define and distinguish two types of QTLs, those that appear in progeny from a full sib cross and those that appear in a half sib family. Those from a full sib cross are detected in a specific genetic background and may not be important for the population. In contrast, QTLs detected in a half sib family are detected in a genetic background that is more representative of the population. This distinction is analogous to the concept of general combining ability and specific combining ability in plant breeding. We may consider QTLs detected in half sib families as "general combining ability QTLs" and those detected in full sib families as "specific combining ability QTLs".

FUSIFORM RUST; THE GENETIC ARCHITECTURE OF A COMPLEX TRAIT IS ACTUALLY SIMPLE:

Resistance of loblolly pine to fusiform rust disease has long been considered to be a quantitative trait explained by many genes each with small additive effect (Robinson, 1987; Carson and Carson, 1989). Polygenic resistance has been the expected mode of inheritance for long lived forest trees because of the intensity of selection over long periods of time for individual trees. Host resistance to white pine blister rust was found to be due to discrete qualitative inheritance (Kinloch et al., 1970) raising the possibility that other qualitative factors could be important in forest trees (Kinloch and Walkinshaw, 1991). Nance et al. (1992) suggested that molecular marker analysis would be a good method to identify and analyze qualitative factors for resistance in forest trees.

In a study using genomic mapping of host resistance in loblolly pine, Wilcox (1995) and Kuhlman et al. (1995) showed that the resistance of loblolly pine to fusiform rust could be accounted for by a single dominant gene. This result was unexpected because of the extensive work that had been carried out that could not resolve the genetic basis of host resistance. Large variation in phenotypic evaluation of resistance, combined with genetic variation in the pathogen confounded earlier results. When defined fungal inocula were combined with genomic mapping methods, the genetic basis of resistance was resolved. Additional studies showed that the conclusions applied to field level resistance as well as greenhouse tests, and that molecular markers could be found that could predict resistance to rust, in specific families. Such markers may be particularly useful for marker aided selection. These results suggest that many quantitative traits, not yet examined by genomic mapping methods might be easier to analyze that previously expected.

INTEGRATION OF MOLECULAR MARKERS INTO TREE BREEDING

Molecular markers provide a breeding tool for the manipulation of major genes and for variety conversion. It is difficult to precisely analyse the efficiency of breeding methods for traits such as disease resistance, but the use of markers is clearly effective and advantageous given the difficulty of carrying out challenges with the disease organism to assess resistance. The most general application of molecular markers for breeding is in the improvement of efficiency of selection in breeding programs. Marker-aided selection is most efficient for low heritability traits (Lande and Thompson, 1990; Strauss et al., 1992), however, it is not limited to such traits. The heritability of markers is essentially 1. Even for traits with high heritability, markers can increase the efficiency of selection. The efficiency of marker aided selection depends not only on heritability, but also the proportion of additive genetic variance explained by markers, the cost of analysis and the value of the increased gain.

In traditional tree breeding, the additive genetic variance of the population is important for any trait of interest. Half sib families sample more of the genetic variation of the population than a full sib family. QTL effects segregating in a half sib family represent average effects with respect to the breeding population and can be considered as single locus components of breeding value. Breeding value is the average effect of an individual's genes determined from its progeny compared to the entire population (Falconer 1989). Selection on markers tightly associated with loci of high breeding value should therefore increase the efficiency of within family selection that usually is based on phenotypic selection. The efficiency of marker-aided selection increases when multiple traits are considered on the same set of progeny. One set of marker genotypes can then be applied to many traits. The resulting information allows the selection of rare genotypes with the maximum number of desirable QTLs from the progeny set that could not have been identified by phenotype alone. Large numbers of progeny are required to insure that the exceptional genotypes occur in the population, due to the low probability of the segregation of many specific alleles into one individual.

In many breeding programs, a complementary mating design of breeding populations can be combined with QTL analysis. Advanced generation breeding populations in small groups or sublines (McKeand and Bridgwater, 1992), might be particularly suitable. A set of full sib families, in a partial diallele can be used with a set of half sib families generated with a polymix of pollen parents, to look for QTL effects. QTL effects can therefore be detected based on the general combining ability of the parents or from their specific combining ability. The QTLs likely to be the most valuable, that is have the largest single locus breeding value, are likely to be dominant, occur at low frequency in the breeding population, and have large effects. The major factors limiting the application of marker aided selection in forest trees are the costs of genotyping, the large family sizes needed for efficient QTL detection and the lack of knowledge of the number and magnitude of the QTL effects. A better understanding of the genetic architecture of traits that affect height and volume is very important.

INTEGRATION OF QTL ANALYSIS WITH MAPPING GENES OF KNOWN FUNCTIONS:

The success of QTL mapping has implications for molecular biology in addition to the application to quantitative genetics. The identification of a QTL represents the definition of a locus and the association of that locus with a functional role. The QTL therefore defines a functional locus, much as a qualitative mutation does. It will be of considerable interest to identify the functions for QTLs. Several different views on the nature of QTLs have been suggested. Doebley (1993) argues that QTLs are likely to be transcription factors or other genes with pleiotropic effects on development. QTLs could also be associated with complex changes in gene organization, or they could be changes in expressed genes of known function. Large scale cDNA sequencing and mapping efforts will eventually provide information on the function and location of a large fraction of the genes in higher plants essential for normal development, metabolism, and response to environmental stresses. Fine structure mapping of QTLs should lead to the identification of candidate genes associated with QTLs for a wide variety of quantitative functions.

CONCLUSIONS:

1. Experience with agronomic crops is of considerable value for understanding the genetic architecture of quantitative traits in forest trees and for considering the potential value of MAS. QTLs have been detected in the small number of forest trees that have been examined for many traits, and are of interest both for basic science and for applied breeding. There is a great deal of similarity between undomesticated forest trees and agronomic crops in spite of the long history of crop domestication.

2. For breeding programs in forest trees, linkage equilibrium is not a significant barrier to application of genetic mapping when maps are made for individual trees. The high level of genetic diversity and the results of genetic recombination in populations provide the tree geneticist with a rich source of material for genetic analysis or for tree breeding. In contrast with early expectations where marker trait associations will be limited to specific genetic backgrounds, it will be possible to analyze marker trait associations in many different genetic backgrounds and environments. Interactions of QTLs in different genetic backgrounds have now become reasonable objects of genetic study, as well as studies of genotype by environment (GxE) interactions for QTLs, and stability of QTLs over time.

3. Contrary to expectations that markers may find restricted use, it is more likely that a wide variety of traits in forest trees that are of basic or applied interest may be amenable to MAS. Many traits of apparent low heritability are important and the prospects of MAS combined with large families, high selection intensities and within family selection, may prove to be efficient approaches to obtain and deploy rare genotypes in forest tree breeding.

4. Prospects for gene discovery and gene isolation based on quantitative trait analysis continue to improve. QTL analysis can provide a means to identify new functions, difficult to identify by traditional mutation analysis in model plant systems. QTL analysis may be a valuable approach for identifying regulatory interactions, pleiotropy and to identify many genes involved in the development of complex traits. These approaches may be particularly important in forest trees for traits not expressed under controlled conditions or for processes like wood formation, where herbaceous annuals are not good models for study. As large scale gene mapping studies proceed, and as chromosome syntenic relationships are discovered for conifers and other higher plants, quantitative trait analysis will become a more powerful tool for forest tree genetics.

5. The application of MAS to tree breeding is no longer limited by existing technology and therefore may be the earliest use of biotechnology in forest trees. In contrast with genetic engineering where application is limited by the lack of adequate transformation technology, difficulty of vegetative propagation, and government regulatory barriers, MAS may be applied to practical problems in tree breeding in the very near future.

ACKNOWLEDGMENTS

This work was supported by grants from the USDA Plant Genome Program (91-37300-6341 and 92-37300-7549, 92-37300-7548, 93-37300-8839) and through the NCSU Program in Statistical Genetics (N.I.H. GM32518). We are grateful for support from the NCSU Forest Biotechnology Industrial Associates.

REFERENCES

Beavis, W.D., Grant, D., Albertsen, M., and Fincher, F. 1991, Quantitative trait loci for plant height in four maize populations and their associations with quantitative genetic loci, *Theor. Appl. Genet.* 83: 141-145.

Bradshaw, H.D. Jr., and Foster, G.S., 1992, Marker aided selection and propagation systems in trees: advantages of cloning for studying quantitative inheritance, *Can. J. For. Res.* 22: 1044-1049.

Bradshaw, H.D. Jr., and Stettler, R.F., 1995, Molecular genetics of growth and development in Populus: IV. Mapping QTLs with large effects on growth, form and phenology traits in a forest tree, *Genetics* 139: 963-973.

Caetano-Anolles, G., Bassam, B.J., and Gresshoff, P.M., 1991, High resolution DNA amplification fingerprinting using very short arbitrary oligonucleotide primers, *Bio/Technology* 9: 553-557.

Carlson, J.E., Tuliseriam, L.K., Glaubitz, J.C., Luk, V.W.K., and Kauffeldt, C. 1991, Segregation of random amplified DNA markers in F_1 progeny of conifers, *Theor. Appl. Genet.* 87: 805-815.

Carson, S.D., and Carson, M.J., 1989, Breeding for resistance in forest trees: a quantitative approach, *Annu. Rev. Phytopathology* 27: 375-395.

Conkle, M.T., 1981, Isozyme variation and linkage is six conifer species, *in*: "Proceedings of the symposium on isozymes in North American forest trees and forest insects." *Pacific Southwest Expt. Stn.. USDA Genet. Tech. Rep. PSW-48.* pp. 11-17.

Doebley, J., 1993, Genetics, development, and plant evolution. *Current Opin. Genet. Develop.* 6: 865-872.

Falconer, D.S., 1989, "Introduction to Quantitative Genetics", 3rd. edition. Longman Scientific and Technical, Essex, U.K. 438pp.

Goldman, I.L., Rocheford, T.R., and Dudley, J.W., 1994, Molecular markers associated with maize kernel oil concentration in an Illinois high protein x Illinois low protein cross, *Crop Science* 34: 908-915.

Grattapaglia, D., 1994, Genetic mapping of quantitatively inherited economically important traits in Eucalyptus. PhD Dissertation, North Carolina State University, Raleigh, NC, USA.

Grattapaglia, D., Wilcox, P., Chaparro, J.X., O'Malley, D.M., McCord, S., Whetten, R., McIntyre, L., and Sederoff, R., 1991, A RAPD map of loblolly pine in 60 days. *in* "Proceedings of the 3rd International Congress of Plant Molecular Biology", Tucson, AZ, Abstract #2224.

Grattapaglia, D., and Sederoff, R., 1994, Genetic linkage maps of *Eucalyptus grandis* and *Eucalyptus urophylla* using a pseudo-testcross: Mapping strategy and RAPD markers, *Genetics* 137: 1121-1137.

Grattapaglia, D., Bertolucci, F.L., and Sederoff, R., 1995, Genetic mapping of QTLs controlling vegetative propagation in Eucalyptus grandis and E. urophylla using a pseudo-testcross strategy and RAPD markers, *Theor. Appl. Genet.* 90: 933-947.

Groover, A., Devey, M., Fiddler, T., Lee, J., Megraw, T., Mitchell-Olds, T., Sherman, B., Vujcic, C., Williams, C., Neale, D., 1994, Identification of quantitative trait loci influencing wood specific gravity in an outbred pedigree of loblolly pine, *Genetics* 138: 1293-1300.

Hamrick, J.L., Godt, M.J.W., Sherman-Broyles, S.L., 1992, Factors influencing levels of genetic diversity in woody plant species, *New Forests* 6: 95-124.

Hayes, P.M., Liu, B.-H., Knapp, S.J., Chen, F., Jones, B., Blake, T., Frankowiak, J., Rasmusson, D., Sorrells, M., Ullrich, S.E., Wesenberg, D., and Kleinhofs, A., 1994, Quantitative trait locus effects and environmental interactions in a sample of North American barley germplasm, *Theor. Appl. Genet.* 87: 392-401.

Kinloch, B.B., and Walkinshaw, C.H., 1991, Resistance to fusiform rust in Southern pines: how is it inherited? *in* "Proc. 3rd IUFRO Rusts of Pine Working Party Conference, Northern Forestry Center, Forestry Canada. Inf. Rep. NOR-X-317": pp219-228.

Kinloch, B.B., Parks, G.K., and Fowler, C.W., 1970, White pine blister rust: simply inherited resistance in sugar pine, *Science* 167:193-195.

Kuhlman, E.G., Amerson, H.V., and Wilcox, P.L., 1995, Recent research on fusiform rust disease. *in* "Proc. 4th IUFRO Rusts of Pines Working Party Conf.", Tsukuba: 17-21.

Lande, R., and Thompson, R., 1990, Efficiency of marker aided selection in the improvement of quantitative traits, *Genetics* 124: 743-756.

Lander, E.S., Green, P., Abrahamson, J., Baarlow, A., and Daly, M.J. 1987, MAPMAKER,: an interactive computer package for constructing primary genetic linkage maps of experimental and natural populations, *Genomics* 1: 174-181.

Liu, B.-H., and Knapp, S.J., 1990, GMENDEL: a program for Mendelian segregation and linkage analysis of individual or multiple progeny using log-likelihood ratios, *J. Hered.* 81: 407.

Liu, B.-H., Sederoff, R., O'Malley, D., 1993, Linkage mapping using open pollinated populations. *in* "Proc. 22nd Southern Forest Tree Improvement Conf." June 14, 1993, Atlanta GA. p489.

McKeand, S.E., and Bridgwater, F.E., 1992, Third generation breeding strategy for the NCSU-Industry Tree Improvement Program. *in* "Proc. IUFRO Conference S2.02-08, Breeding Tropical Trees", Cartagena, Colombia, pp 234-240.

Moesseler, A., Egger, K.N., and Hughes, G.A., 1992, Low levels of genetic diversity in red pine confirmed by random amplified DNA markers, *Can. J. For. Res.* 22: 1332-1337.

Millar, C.I., 1983, A steep cline in *Pinus muricata.*, *Evolution* 37: 311-319.

Nance, W.L., Tuskan, G.A., Nelson, C.D., and Doudrick, R., 1992, Potential applications of molecular markers for genetic analysis of host pathogen systems in forest trees, *Can. J. For. Res.* 22: 1036-1043.

Neale, D.B., and Williams, C.G., 1991, Restriction fragment length polymorphism mapping in conifers and application to forest genetics and tree improvement, *Can. J. For. Res.* 21: 545-264.

Patterson, A.H., Damon, S., Hewitt, J.D., Zamir, D., Rabinowitch, H.D., Lincoln, S.E., Lander, E.S., and Tanksley, S.D., 1991, Mendelian factors underlying quantitative traits in tomato: Comparison across species and environments, *Genetics* 127: 181-197.

Plomion, C., O'Malley, D.M., Durel, C.E., 1995. Genomic analysis in maritime pine (*Pinus pinaster*): Comparison of two RAPD maps using selfed seeds and open-pollinated seeds of the same individual, Theor. Appl Genet., (in press).

Robinson, R.A., 1987, *in* "Host management in crop pathosystems", Macmillan Publishing Co. N.Y. pp263.

Strauss, S.H., Lande, R., and Namkoong, G., 1992, Limitations of molecular marker aided selection in forest tree breeding, *Can. J. For. Res.* 22: 1050-1061.

Stuber, C.W., 1992, Biochemical and molecular markers in plant breeding. *in* "Plant Breeding Reviews", Vol 9. Dudley, J.W., Hallauer, A.R., and Ryder, M. eds. J. Wiley and Sons, Inc. pp 37-61.

Stuber, C.W., Lincoln, S.E., Wolff, D.W., Helentjaris, T., and Lander, E.S., 1992, Identification of genetic factors contributing to heterosis in a hybrid from two elite maize inbred lines using molecular markers, *Genetics* 132: 823-839.

Tanksley, S.D., 1993, Mapping polygenes, *Ann.Rev. Genet.* 27: 205-233.

Tulseriam, L.K., Glaubitz, J.C., Kiss, G., and Carlson, J.E., 1992. Single tree genetic linkage mapping in conifers using haploid DNA from megagametophytes, *Bio/Technology* 10: 686-690.

Welsh, J., and McClelland, M., 1990, Fingerprinting genomes using PCR with arbitrary primers, *Nucleic Acids Res.* 19: 303-306.

Wheeler, N.C., and Guries, R.P., 1982, Population structure, genic diversity, and morphological variation in *Pinus contorta* Dougl., *Can. J. For. Res.* 12: 595-606.

Wilcox, P., 1995, Genetic dissection of fusiform rust resistance in loblolly pine. PhD Dissertation. North Carolina State University, Raleigh, NC, USA.

Williams, C.G., and Neale, D.B., 1992, Conifer wood quality and marker aided selection: a case study, *Can. J. For. Res.* 22: 1009-1017.

Williams, J.G.K., Kubelik, A.R., Livak, K.J., Rafalski, J.A., and Tingey, S.V., 1990, DNA polymorphisms amplified as arbitrary primers are useful genetic markers, *Nucleic Acids Res.* 18: 6531-6535.

Zabeau, M., 1993, Selective restriction fragment amplification: a general method for DNA fingerprinting. *European Patent Application. Publication No.* 0 534 858 A1.

COMMONALITIES AND CONTRASTS IN THE ORGANIZATION OF THE MAIZE AND SORGHUM NUCLEAR GENOMES

Jeffrey L. Bennetzen, Chang-Nong Liu, Phillip SanMiguel,
Patricia S. Springer, Young-Kwan Jin, Carolyn A. Zanta,
and Zoya Avramova

Department of Biological Sciences
Purdue University
West Lafayette, Indiana 47907

INTRODUCTION

Analysis of plant genome organization has long been the realm of plant geneticists and cytogeneticists. The multipartite (several chromosome) nature of the nuclear genome, heritable and line-specific variations in the cytology or number of chromosomes (Blakeslee, 1922; Randolph and McClintock, 1926; Stadler, 1928; Kostoff, 1929; Philip and Huskins, 1931; McClintock, 1932; Creighton, 1934; Sears, 1939; Swanson, 1940), the linear order of genes whose linkage could be determined by analysis of crossover exchanges in meiosis, the physical exchange of chromosomal segments associated with recombination (Creighton and McClintock, 1931), the properties of telomeres (McClintock, 1941), the behavior of primary and secondary constrictions as centromeres in mitosis and meiosis (Prakken and Muntzing, 1942; Rhoades and Vilkomerson, 1942), the contribution of a particular chromosomal segment (the nucleolar organizer, NOR) to formation of the nucleolus (McClintock, 1934), the existence and preferential transmission of supernumerary (B) chromosomes (Longley, 1927; Darlington and Thomas, 1941; Roman, 1947), the biology of one class of highly repetitive DNA (the knob satellite) (Rhoades and Dempsey, 1966; Peacock et al., 1981), and the properties of a key class of middle repetitive DNAs (transposable elements) (McClintock, 1950) were all identified in plants concurrent with, or prior to, their discovery in other species. Much of this initial work was

Genomes of Plants and Animals: 21st Stadler Genetics Symposium
Edited by J. Perry Gustafson and R. B. Flavell, Plenum Press, New York, 1996

performed with maize, partly due to the early and excellent characterization of its karyotypic properties (Longley, 1924).

Molecular characterization of plant genomes lagged behind that of more intensely studied bacterial, fungal and animal systems. However, renaturation analyses of plant DNA indicated that plants, like animals, have nuclear genomes that are composed of somewhat interspersed single copy, middle repetitive and highly repetitive sequences (Flavell et al., 1974). Two classes of highly repetitive DNA, the ribosomal DNA of the NOR and the knob satellite repeat, were characterized in maize as simply repeated tandem arrays (Phillips et al., 1971; Peacock et al., 1981). Transposable elements of various types were cloned and their contributions to genome organization and function were studied. Many plant genes were cloned, and found to have many of the features also seen in animal genes (introns, promoters with some conserved sequence features necessary for regulation of the genes, etc.). However, beyond the sequences within a few hundred base pairs of genes, little is known regarding plant genome organization. Moreover, other than centromeres and telomeres, little is known of the contribution of higher order genome structure to the expression, transmission, recombination or evolution of plant genomes.

In the last few years, the development of techniques for the mapping and cloning of large (20-2000 kbp) fragments of eukaryotic DNA have allowed investigation of the organization of DNA within multi-gene segments of plant genomes. In addition, DNA marker technology has greatly increased the detail and power of recombinational mapping in plants. Finally, the discovery of common gene content and large regions of recombinational map colinearity in different plant species have provided new tools for gene identification, cloning and analysis, including an evolutionary perspective (Bennetzen and Freeling, 1993).

For the last several years, our laboratory has pursued characterizations of genome composition and organization in plants. We have concentrated on the study of maize (*Zea mays*) and sorghum (*Sorghum bicolor*), two grass species of the tribe *Andropogonae*. Maize and sorghum have undergone about twenty million years of independent descent from a common ancestor (Doebley et al., 1990). A number of genomic features are held in common between these species, although some significant differences are observed. More recent studies indicate that the similarities and differences in the genomes of these species fall into distinctive patterns that may be common to most higher plants. We will summarize some of these new, and previous, findings and provide a first-generation general model for the organization of higher plant genomes.

Commonalities in Maize and Sorghum Nuclear Genomes

Like all higher plants, maize and sorghum have organized their nuclear genomes into multiple chromosomes. Both maize and sorghum have ten pairs of chromosomes in the diploid nucleus, although the maize chromosomes are significantly larger (Laurie and Bennett, 1985). Hybridization of low copy number and gene-containing DNA probes from maize to sorghum has demonstrated that the two species contain essentially the same set of genes and that these genes map into similar linear orders (Hulbert et al., 1990; Binelli et al., 1992; Whitkus et al., 1992; Melake-Berhan et al., 1993; Pereira et al., 1994; Ragab et al., 1994) (Figure 1). Exceptions to the colinearity of these two genomes are observed, but some of these may be due to

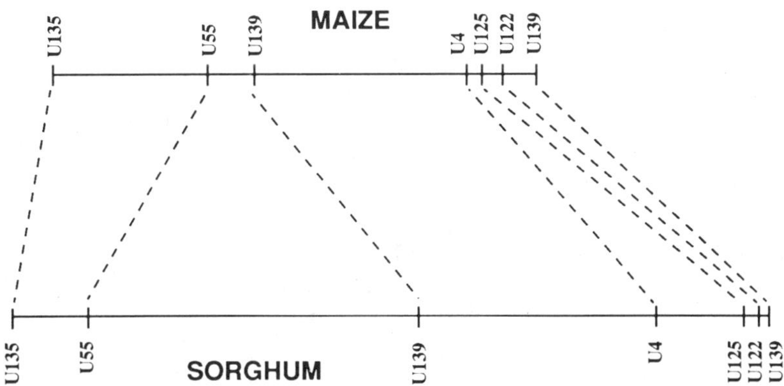

Figure 1. Colinearity of maize and sorghum genomes as determined by comparison of DNA marker linkages. The same DNA markers (all maize DNAs) were mapped in maize and sorghum, and the maps compared (as described in Bennetzen, 1995a). Linkages are drawn to approximate scale, comprising a total of about 120 cM in sorghum and about 75 cM in maize.

inaccuracies/incompleteness in the mapping data, clerical errors, or inappropriate comparisons of different members of the same gene family (Bennetzen, 1995a). Subsequent characterizations have indicated that regional colinearity is observed for very distantly related grasses, like wheat, rice and maize (Ahn et al., 1993).

Genetic maps in both maize and sorghum indicate duplicated blocks of related genes on different chromosomes (Helentjaris et al., 1988; Hulbert et al., 1990; Chittenden et al., 1994), suggesting that these two species have undergone tetraploidization (or numerous segmental duplications) at some time in their distant pasts. The ancient polyploid interpretation is also supported by the existence of n=5 chromosome relatives of sorghum (Doggett, 1988).

Comparisons of linkages across colinear chromosome segments indicates some differences in local recombination frequencies between maize and sorghum. However, the data sets for all of these studies are relatively small, and nearly-equivalent variations in linkage values are observed in within-species comparisons for different mapping populations. Hence, more quantitatively-significant experimentation is needed to investigate this important and interesting point. In general, though, a simplifying extrapolation suggests that both genomes will contain about 1800-2200 cM distributed into their ten linkage groups.

Some repetitive DNAs of maize and sorghum are very similar. The tandemly-repeated ribosomal DNA of sorghum is structurally very similar to that of maize (Springer et al., 1989). Some transposable elements are conserved between the two species; small inverted repeat elements of the *tourist* and *stowaway* families are associated with both maize and sorghum genes (Bureau and Wessler, 1994a; 1994b) and we have cloned and sequenced an *Ac* homologue from sorghum (unpub. obs.). Moreover, like maize, we have found that sorghum also has middle repetitive retrotransposons (Bennetzen, 1995b).

As in maize (Antequera and Bird, 1988; Bennetzen et al., 1994), a very high percentage of the CG and CNG sequences in the sorghum nuclear genome are apparently 5-methylated at cytosines (Figure 2 and unpub. obs.). Our previous analyses have indicated that these methylated blocks in maize are composed primarily (or exclusively) of mixed classes of interspersed repetitive DNA (Bennetzen et al., 1994; Springer et al., 1994). As in maize, the methylated blocks in sorghum mostly range in size from 20 kbp to 200 kbp (Figure 2). The size range of these repetitive DNA blocks correlates with the size range of chromatin loops attached to the nuclear scaffold, and we have some data indicating that scaffold attachment sites are found preferentially at the boundaries between repetitive and single copy blocks (Avramova et al., 1995).

Differences in Maize and Sorghum Nuclear Genomes

The maize and sorghum genomes differ greatly in size; the maize nucleus contains about 3.5 times more DNA than does the sorghum nucleus (Arumuganathan and Earle, 1991; Michaelson et al., 1991), as reflected in the smaller size of sorghum chromosomes (Laurie and Bennett, 1985). There is also some variation in the copy numbers of gene families between the two species, with sorghum usually having either the same or a slightly lower copy number (Hulbert et al., 1990; Whitkus et al., 1992). There are exceptions to this rule, however, as in the case where the sorghum homologues of maize *Adh2* (Whitkus et al., 1992) and *Adh1* (our unpub. obs.) exist in twice as many copies as they do in maize. Since probably over 80% of the maize genome is made up of repetitive DNAs (Hake and Walbot, 1980), this difference in gene family copy number should not contribute significantly to the overall difference in genome size. Hence, most of the difference in genome size should be caused by variation in the amount of repetitive DNA in the two species. The available data suggest that the sorghum

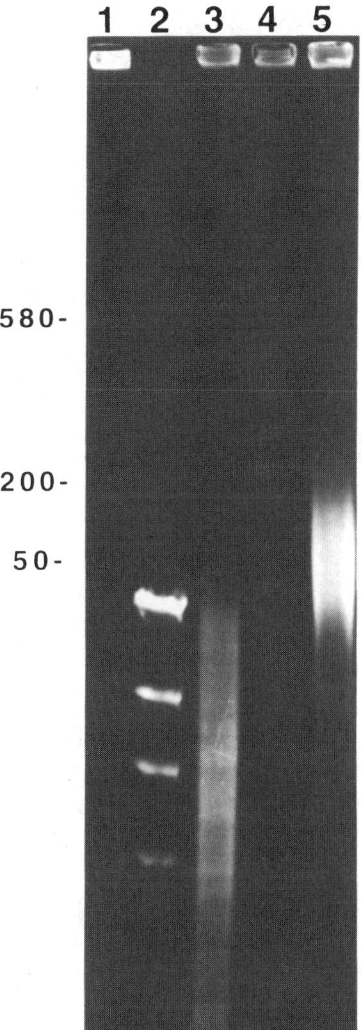

Figure 2. Pulsed field gel analysis of sorghum genomic DNA. Lane 1: lambda ladder. Lane 2: lambda DNA digested with *Hind*III. Lane 3: sorghum DNA digested with *Hind*III. Lane 4: sorghum DNA digested with *Hae*III, a restriction enzyme that cleaves at GGCC. Lane 5: sorghum DNA digested with *Hpa*II, a restriction enzyme that cleaves at CCGG and is inhibited by 5-methylation of cytosines at its recognition sequence. The picture is of an ethidium bromide stained gel, with many of the fragments (including all those in lane 4) having run off of the gel due to their small size. Numbers to the left are in kbp, and are derived from lambda ladder or yeast chromosome size standards. DNA preparation, digestions and gel analysis were all as previously described (Bennetzen et al., 1994).

genome contains about 30% of its DNA in the repetitive category. At a chromosomal level, the karyotype of the small sorghum chromosomes has not indicated large satellite clusters like the knobs of maize (Longley, 1924; Rhoades and Dempsey, 1966) and B chromosomes and some sorghum relatives [Longley, 1927; Darlington and Thomas, 1941]) (K. Schertz, pers. comm.).

If the major factor accounting for the difference in the genome sizes of maize and sorghum is the percentage of repetitive DNA, then it is interesting that the methylated repetitive DNA blocks in both species mostly fall into a similar size range of 20-200 kbp (Figure 2: Bennetzen et al., 1994). This result indicates that, if sorghum does not have smaller repetitive blocks, then it must have fewer such repetitive blocks.

Despite extensive colinearity, specific rearrangements do differentiate the maize and sorghum genomes (Hulbert et al., 1990; Binelli et al., 1992; Whitkus et al., 1992; Melake-Berhan et al., 1993; Pereira et al., 1994; Ragab et al., 1994). The genetic maps of sorghum using maize-comparable clones are not yet sufficiently detailed to guarantee that all of the rearrangements have been detected, but it is clear that the majority of the events are inversions, duplications and/or translocations involving large segments of single arms.

With the notable exception of the highly conserved rDNA tandem repeat (Springer et al., 1989), most of the highly repetitive DNAs present in maize do not detectably hybridize to sorghum (Bennetzen et al., 1994) and the converse is also true (unpub. obs.). For the dispersed repetitive elements that we have analyzed, all appear to be mobile DNAs in both sorghum and maize (unpub. obs.). Hence, although the specific repeats may be different between maize and sorghum, their general type seems to be the same.

It is not clear whether a lack of cross-hybridization between most highly reiterated DNAs of maize and sorghum actually means that these highly repeated elements are not present in both species. It is also possible that these elements have evolved to a degree that limits their detection by cross-hybridization. Evidence for both of these processes exists. A species-specific repetitive element could be derived by horizontal transfer, which is well established in animal systems (Abad et al., 1989; Kidwell, 1992; Kim et al., 1994) and appears likely in plants. The rapid evolution of mobile repetitive DNAs within maize (Bennetzen and Springer, 1994; Jin and Bennetzen, 1994; Purugganan and Wessler, 1994; Bennetzen, 1995b) or between species (MacRae and Clegg, 1992) has been well documented, and often makes the relatedness of these elements difficult to discern/confirm unless one utilizes sequence comparisons.

The similar gene content and colinearity of maize and sorghum genomes, matched with the smaller genome size (and presumed lower repetitive DNA content) of sorghum, suggests that sorghum genes may be closer together than the same genes would be in maize. We have begun a test of this hypothesis in the areas around the *A1* and *Sh2* loci of these species. In maize, these two genes are separated by about 140 kb of mostly repetitive DNA (Civardi et al., 1994). Using a maize *Sh2* probe, we have cloned the homologous gene from sorghum on a bacterial artificial chromosome (BAC). Preliminary analyses indicate that an *A1* homologue is also on this BAC, and that these two genes are within 20 kbp of each other (SanMiguel, Woo, Wing, and Bennetzen, unpub. obs.). These results suggest, of course, that map-based cloning techniques like chromosome walking would be most easily accomplished in a small-genome grass, even if the targeted gene is in a large-genome grass (Bennetzen and Freeling, 1993).

A Model of Grass Genome Organization: Constants and Variables

Detailed molecular analyses of maize genome organization in regions of 20-500 kbp (Bennetzen et al., 1994; Civardi et al., 1994; Springer et al., 1994) indicate that, on average, relatively short blocks of unmethylated single copy DNA (i.e., genes) will be flanked by long (5-200 kbp) blocks of mixed classes of repetitive/methylated DNA. Our preliminary analyses of sorghum genome organization suggest a similar organization, as do initial studies of other grass genomes (Moore et al., 1993). Figure 3 presents a crude, first-generation model for the general organization of a standard region of a grass genome. Within a given chromosomal segment in mature somatic tissues, gene order and unmethylated state would be constants. Other proposed constants would be the presence, repeat-type composition, methylated state and size range of repetitive blocks (shaded). Variables would be the particular repeats, the placement of repeat blocks, and the distance between genes.

Prospects

The ability to analyze comparable regions of related genomes will now permit investigations into the nature and evolution of higher plant genome organization. The biological significance of various

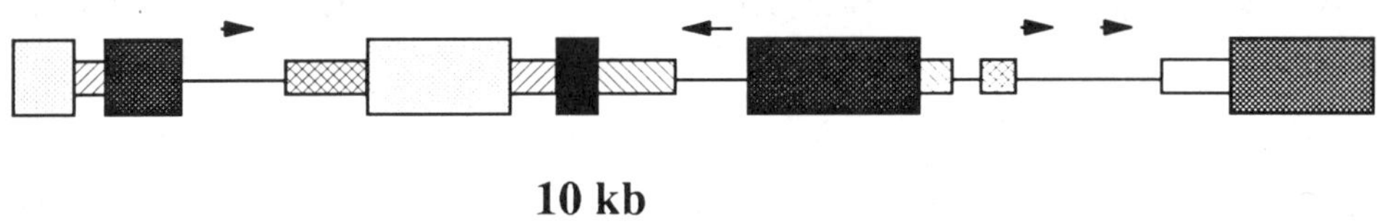

Figure 3. A simple model for genome organization in grasses with large genomes. Narrow lines indicate single copy, unmethylated DNAs. Narrow boxes denote middle repetitive, often methylated DNAs. Wide boxes indicate highly repetitive, methylated DNAs. Each box exhibits a different fill as an indication of the different mixtures of repetitive DNAs that will be found within it. Arrows represent the transcripts specified by genes found in the single copy regions. Grasses with smaller genomes should have more instances where genes are not separated by large blocks of repetitive DNA.

components will be tested by the degree to which they are conserved between species. The colinearity of different grass genomes suggests a selected function for gene order, perhaps related
to the regulation of gene expression. The nature and existence of position-effect (Henikoff, 1990) and sense-suppression (Flavell, 1994; Jorgensen, 1995) phenomena suggest that the higher order organization of genomes can significantly affect gene function.

The comparison of the placement of scaffold attachment regions (Avramova et al., 1995), and other proposed "insulator" functions (Kellum and Shedl, 1991), will be particularly noteworthy between grass species (like maize) where single genes are usually bounded by large blocks of repetitive (presumably heterochromatic) DNA and species (like sorghum) where different genes may often not be separated by repetitive DNA blocks. Local comparisons of recombination rates and locations between species, particularly in cases where species differ in the presence of repetitive blocks between genes, will provide direct evidence regarding gene-specific biases for the initiation and/or resolution of recombinational events.

Detailed comparative mapping of similar genetic functions (either qualitative or quantitative) from maize and sorghum will provide important information on the genetic basis of commonalities and differences in basic grass physiology and development (Bennetzen and Freeling, 1993). Such studies will also identify maize genes of particular interest that may be cloned by map-based techniques employing the smaller sorghum genome. The synergy of such parallel studies, extrapolated across the full range of grasses (and perhaps beyond), will open up a new era in highly simplified and powerful plant genome analysis and manipulation (Bennetzen and Freeling, 1993; Bennetzen, 1995a).

ACKNOWLEDGEMENTS

We thank K. Schertz, R. Wing, and S.-S. Woo for permission to cite unpublished observations. This research and preparation of this manuscript were supported by a grant from the USDA (94-37300-0299). P. SanMiguel is supported by an NSF training grant in Plant Genetics.

REFERENCES

Abad, P.C., Vaury, C., Pelisson, A., Chaboissier, M.-C., Busseau, I., and Bucheton, A., 1989, A LINE element, the I factor of *Drosophila teissieri*, is able to transpose in other *Drosophila* species, *Proc. Natl. Acad. Sci. USA* 86:8887.

Ahn, S., Anderson, J.A., Sorrells, M.E., and Tanksley, S.D., 1993, Homoeologous relationships of rice, wheat and maize chromosomes, *Mol. Gen. Genet.* 241:483.

Antequera, F., and Bird, A., 1988, Unmethylated CpG islands associated with the genes in higher plant DNA, *EMBO J.* 7:2295.

Arumuganathan, K., and Earle, E.D., 1991, Nuclear DNA content of some important plant species, *Plant Mol. Biol. Rep.* 9:208.

Avramova, Z., SanMiguel, P., Georgieva, E., and Bennetzen, J.L., 1995, Matrix attachment regions and transcribed sequences within a long chromosomal continuum containing maize *Adh1*, *Plant Cell*, in press.

Bennetzen, J.L., 1995a, The use of comparative genome mapping in the identification, cloning and manipulation of important plant genes, *in*: "The Impact of Plant Molecular Genetics," B.W.S. Sobral, ed., Birkhauser, Boston, in press.

Bennetzen, J.L., 1995b, The contributions of retroelements to plant genome structure, function, and evolution, *Trends Micro.*, in press.

Bennetzen, J.L., and Freeling, M., 1993, Grasses as a single genetic system: genome composition, collinearity and compatibility, *Trends Genet.* 9:259.

Bennetzen, J.L., Schrick, K., Springer, P.S., Brown, W.E., and SanMiguel, P., 1994, Active maize genes are unmodified and flanked by diverse classes of modified, highly repetitive DNA, *Genome* 37:565.

Bennetzen, J.L., and Springer, P.S., 1994, The generation of Mutator transposable element subfamilies in maize, *Theor. Appl. Genet.* 87:657.

Binelli. G., Gianfranceschi, L, Pe, M.E., Taramino, G., Busso, C., Stenhouse, J., and Ottaviano, E., 1992, Similarity of maize and sorghum genomes as revealed by maize RFLP probes, *Theor. Appl. Genet.* 84:10.

Blakeslee, A.F., 1922, Variation in Datura due to changes in chromosome number, *Amer. Nat.* 66:16.

Bureau, T.E., and Wessler, S.R., 1994a, Mobile inverted-repeat elements of the *Tourist* family are associated with the genes of many cereal grasses, *Proc. Natl. Acad. Sci. USA* 91:1411.

Bureau, T.E., and Wessler, S.R., 1994b, *Stowaway*: A new family of inverted repeat elements associated with the genes of monocotyledenous and dicotyledenous plants, *Plant Cell* 6:907.

Chittenden, L.M., Schertz, K.F., Lin, Y.-R., Wing, R.A., and Paterson, A.H., 1994, A detailed RFLP map of *Sorghum bicolor* X *S. propinquum*, suitable for high-density mapping, suggests ancestral duplication of *Sorghum* chromosomes or chromosomal segments, *Theor. Appl. Genet.*, 87:925.

Civardi, L, Xia, Y., Edwards, K.J., Schnable, P.S., and Nikolau, B.J., 1994, The relationship between genetic and physical distances in the clones *a1-sh2* interval of the *Zea mays* L. genome, *Proc. Natl. Acad. Sci. USA* 91:8268.

Creighton, H.B., 1934, Three cases of deficiency in chromosome 9 in *Zea mays*, *Proc. Natl. Acad. Sci. USA* 20:111.

Creighton, H.B., and McClintock, B., 1931, A correlation of cytological and genetical crossing-over in *Zea mays*, *Proc. Natl. Acad. Sci. USA* 17:492.

Darlington, C.D., and Thomas, P.T., 1941, Morbid mitosis and the activity of inert chromosomes in Sorghum, *Proc. Roy. Soc., London, B*, 130:127.

Doebley, J., Durbin, M., Golenberg, E.M., Clegg, M.T., and Ma, D.P., 1990, Evolutionary analysis of the large subunit of carboxylase (*rbc*L) nucleotide sequence among the grasses (Gramineae), *Evolution* 44:1097.

Doggett, H., 1988, "Sorghum," John Wiley and Sons, New York.

Flavell, R.B., 1994, Inactivation of gene expression in plants as a consequence of specific sequence duplication, *Proc. Natl. Acad. Sci. USA* 91:3490.

Flavell, R.B., Bennett, M.D., Smith, J.B., and Smith, D.B. 1974, Genome size and proportion of repeated nucleotide sequence DNA in plants, *Biochem. Genet.* 12:257.

Hake, S, and Walbot, V., 1980, The genome of Zea mays, its organization and homology to related grasses, *Chromosoma* 79:251.

Helentjaris, T., Weber, D.L., and Wright, S., 1988, Identification of the genomic locations of duplicate nucleotide sequences in maize by analysis of restriction fragment length polymorphisms, *Genetics* 118:353.

Henikoff, S., 1990, Position effect variegation after 60 years, *Trends Genet.* 6:422.

Hulbert, S.H., Richter, T.E., Axtell, J.D., and Bennetzen, J.L., 1990, Genetic mapping and characterization of sorghum and related crops by means of maize DNA probes, *Proc. Natl. Acad. Sci. USA* 87:4251.

Jin, Y.-K., and Bennetzen, J.L., 1994, Integration and nonrandom mutation of a plasma membrane proton ATPase gene fragment within the *Bs1* retroelement of maize, *Plant Cell* 6:1177.

Jorgensen, R.A., 1995, Cosuppression, flower color patterns, and metastable gene expression states, *Science* 268:686.

Kellum, R. and Shedl, P., 1991, A position-effect assay for boundaries of higher order chromosomal domains, *Cell* 64:941.

Kidwell, M.G., 1992, Horizontal transfer of *P* elements and other short inverted repeat transposons, *Genetica* 86:275.

Kim, A., Terzian, C., Santamaria, P., Pelisson, A., Prud'homme, N., and Bucheton, A., 1994, Retroviruses in invertebrates: the *gypsy* retrotransposon is apparently an infectious retrovirus of *Drosophila melanogaster*, *Proc. Natl. Acad. Sci. USA* 91:1285.

Kostoff, D., 1929, An androgenic Nicotiana haploid, *Zeitschr. Zellforsch.* 9:640.

Laurie, D.A., and Bennett, M.D., 1985, Nuclear DNA content in the genera *Zea* and *Sorghum*. Intergeneric, interspecific and intraspecific variation, *Heredity* 55:307.

Longley, A.E., 1924, Chromosomes in maize and maize relatives, *J. Agr. Res.* 28:673.

Longley, A.E., 1927, Supernumerary chromosomes in *Zea mays*, *J. Agr. Res.* 35:769.

MacRae, A.F., and Clegg, M.T., 1992, Evolution of *Ac* and *Ds1* elements in select grasses (Poaceae), Genetica 86:55.

McClintock, B., 1932, A correlation of ring-shaped chromosomes with variegation in *Zea mays*, *Proc. Natl. Acad. Sci. USA* 18:677.

McClintock, B., 1934, The relation of a particular chromosomal element to the development of the nucleoli in *Zea mays*, *Zeitschrift fur Zellforschung und mikroskopische Anatomie* 21:294.

McClintock, B., 1941, The stability of broken ends of chromosomes in maize, *Genetics* 26:234.

McClintock, B., 1950, The origin and behavior of mutable loci in maize. *Proc. Natl. Acad. Sci. USA* 36:344.

Melake-Berhan, A., Hulbert, S.H., Butler, L.G., and Bennetzen, J.L., 1993, Structure and evolution of the genomes of *Sorghum bicolor* and *Zea mays*, *Theor. Appl. Genet.* 86:598.

Michaelson, M.J., Price, H.J., Ellison, J.R., and Johnston, J.S., 1991, Comparison of plant DNA contents determined by feulgen microspectrophotometry and laser flow cytometry, *Am. J. Bot.* 78:183.

Moore, G., Abbo, S., Cheung, W., Foote, T., Gale, M., Koebner, R., Leitch, A., Leitch, I., Money, T., Stanscombe, P., Yano, M., and Flavell, R., 1993, Key features of cereal genome organisation as revealed by the use of cytosine methylation-sensitive restriction endonucleases, *Genomics* 15:472.

Peacock, W.J., Dennis, E.S., Rhoades, M.M., and Pryor, A.J., 1981, Highly repeated DNA sequence limited to knob heterochromatin in maize, *Proc. Natl. Acad. Sci. USA* 78:4490.

Pereira, M.G., Lee, M., Bramel-Cox, P., Woodman, W., Doebley, J., Whitkus, R., 1994, Construction of an RFLP map in sorghum and comparative mapping in maize, *Genome* 37:236.

Philip, J., and Huskins, C.L., 1931, The cytology of *Mathiola incana* R. Br., especially in relation to the inheritance of double flowers, *J. Genet.* 24:359.

Phillips, R.L., Kleese, R.A., and Wang, S.S., 1971, The nucleolus organizer region of maize (*Zea mays* L.): chromosomal site of DNA complementary to ribosomal RNA, *Chromosoma* 36:79.

Prakken, R., and Muntzing, A., 1942, A meiotic peculiarity in rye, simulating a terminal centromere, *Hereditas* 28:441.

Purugganan, M.D., and Wessler, S.R., 1994, Molecular evolution of *magellan*, a maize Ty3/*gypsy*-like retrotransposon, *Proc. Natl. Acad. Sci. USA* 91:11674.

Ragab, R.A., Dronavalli, S., Saghai Maroof, M.A., and Yu, Y.G., 1994, Construction
of a sorghum RFLP map using sorghum and maize DNA probes, *Genome*
37:590.
Randolph, L.F., and McClintock, B., 1926, Polyploidy in *Zea mays* L., *Amer. Nat.*
60:99.
Rhoades, M.M., and Dempsey, E., 1966, The effect of abnormal chromosome 10 on
preferential segregation and crossing over in maize, *Genetics* 53:989.
Rhoades, M.M., and Vilkomerson, H., 1942, On the anaphase movement of
chromosomes, *Proc. Natl. Acad. Sci. USA*, 28:433.
Roman, H., 1947, Mitotic nondisjunction in the case of interchanges involving
the B-type chromosome in maize, *Genetics* 32:391.
Sears, E.R., 1939, Cytogenetic studies with polyploid species of wheat. I.
Chromosomal aberrations in the progeny of a haploid of *Triticum
vulgare*, *Genetics*, 24:509.
Springer, P.S., Edwards, K.J., and Bennetzen, J.L., 1994, DNA class organization on
maize *Adh1* yeast artificial chromosomes, *Proc. Natl. Acad. Sci. USA*
91:863.
Springer, P.S., Zimmer, E.A., and Bennetzen, J.L., 1989, Genomic organization of
the ribosomal DNA of sorghum and its close relatives, *Theor. Appl. Genet.*
77:844.
Stadler, L.J., 1928, Mutations in barley induced by X-rays and radium, *Science*
68:186.
Swanson, C.P., 1940, The distribution of inversions in Tradescantia, *Genetics*
25:438.
Whitkus, R., Doebley, J., and Lee, M., 1992, Comparative genome mapping of
sorghum and maize, *Genetics* 132:1119.

PROSPECTS FOR COMPARATIVE GENOME ANALYSES AMONG MAMMALS

Leslie A. Lyons and Stephen J. O'Brien

Laboratory of Viral Carcinogenesis
Frederick Cancer Research and Development Center
National Cancer Institute
Frederick, MD   21702-1201

INTRODUCTION

The discovery of hypervariable microsatellite markers and the technical advances of marker typing has facilitated efforts towards the construction of genetic linkage maps in several diverse vertebrate species. These mapping projects have advanced the exploration of each organism's genome and its biology. But less effort has been focused on the mapping of coding genes, genetic markers which allow comparisons of genome organization across species. The establishment of universal Type I (coding gene) genetic markers, which can be mapped in all species, will allow true comparisons of genome organization across species and will provide the framework for comparative genetic research (O'Brien, 1991). Comparative genetics allows the gene poor species to consider the information of gene rich species, exponentially increasing the perspective available for disease and genetic trait analyses. In developing the genetic map of the cat (O'Brien and Nash; Nash and O'Brien; Gilbert et al., 1988; O'Brien et al., 1988), we have focused on the mapping of Type I markers in order to capture the information locked in the genome of other species and to assist the use of the feline genome in evolutionary and inherited disease research (O'Brien et al., 1993).

Genomes of Plants and Animals: 21st Stadler Genetics Symposium
Edited by J. Perry Gustafson and R. B. Flavell, Plenum Press, New York, 1996.

THE FELINE GENOME AND COMPARATIVE GENETICS

The past decade has witnessed a gradual development of the genetic map of the cat. The karyotypes of felids are highly conserved, 2n=19 for most felids, and the chromosome pairs are easily distinguishable by standard G-banding*See*See*See* methods (Wurster-Hill and Centerwall, 1982; Modi and O'Brien 1988). The diverse size and morphology of the feline chromosomes facilitated the characterization and usage of a Chinese hamster x cat and a mouse x cat somatic cell hybrid panel. These hybrid panels have permitted the syntenic mapping of nearly 100 isozymes, oncogenes, and other coding genes on the feline chromosomes (Marshall et al., 1995). Comparisons of the syntenic maps of cats and cows have suggested that the cat and cow genomes are 3-4 times less rearranged than mice, when compared to the structure of the human genome (O'Brien et al., 1988). This difference laid the ground work and enthusiasm for comparative genetic studies across several diverse species. These crude syntenic maps and karyotypic investigations quickly led to evolutionary considerations of genome organization and the role of genomic rearrangements in species development, as well as the possible cross-species inferences of biological processes.

The investigation of inherited disease is an important area of comparative genetics that would benefit from a cat gene map. The location for the genes causing disease should be predictable in one species when they are mapped to a chromosomal segment of homology in another. Homologous disease gene locations are not always certain in mouse-human comparisons due rearrangement of the two genomes into over 120 chromosomal segments of conserved homology (Nadeau et al., 1991; Copeland et al., 1993). Homologous disease gene locations are often found to be near rearrangement breakpoints between mouse and human genomes, while the fewer rearrangements in the feline make chromosomal predictions more reliable. For example, Figure 1 depicts the location for polycystic kidney disease type 2 locus on human chromosome 4. This human chromosomal segment is in a region homologous to two mouse chromosomes, 5 and 3. The same region is conserved as a larger block of homology in the cat genome on cat chromosome B1; thus mapping efforts and gene localization techniques could concentrate on that single chromosome in the cat. The cat also provides a comparative model for cardiomyopathies, retinal atrophies, viral immunodeficiencies, and other inherited and acquired diseases (Barnett and Curtis 1985; Hardy et al., 1980; Pedersen and Floyd 1985; Pedersen 1987). The diverse morphology of the domesticated feline breeds is a potential resource for the elucidation of polygenic interactions of developmental and quantitative traits. Therefore, the development of additional mammalian species as disease models will offer valuable and unique insight into gene interaction and organismal biology.

The discovery of feline-human genome conservation from the rudimentary syntenic map of the cat has fueled the investigation of

the cat genome. The feline genome project's major objectives are to
provide a carnivore representative for genetic evolutionary analyses
and to aid in the advancement of human and feline health issues.
These objectives have reached beyond syntenic relationships and
justify the construction of a genetic linkage map for the cat.

THE FELINE GENOME PROJECT

The Laboratory of Viral Carcinogenesis has committed to
an organized effort to develop the genetic map of the cat. The project
is designed to utilize the biology of the cat in several research areas

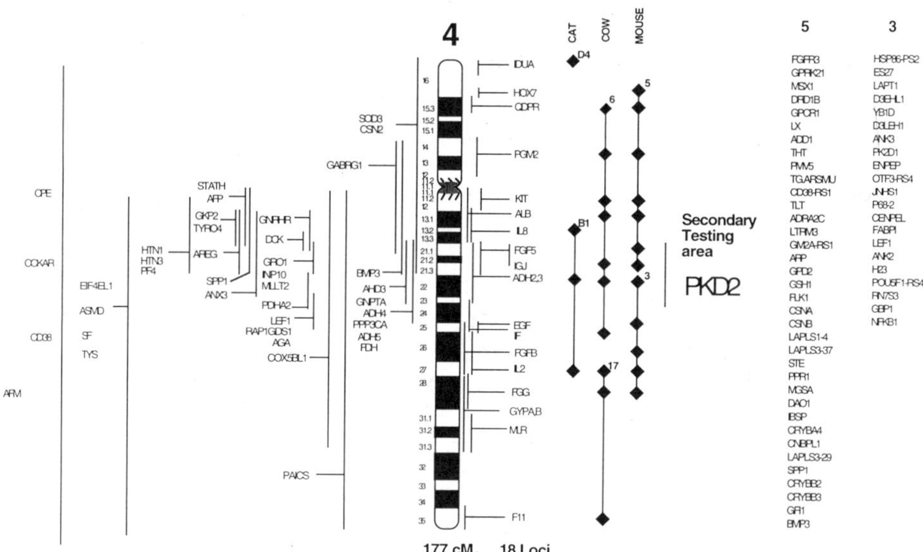

Figure 1. Comparative Mapping: Gene Rich to Gene Poor. (Left) Human genes
mapping to 4q21-q23 from HGM11. (Right) Mouse genes mapping to mouse
chromosomes 5 and 3 in regions homologous to human 4q21-q23 from Mouse
Genome Database. Diamonds denote genes mapped in corresponding species
with species chromosome number indicated. Loci to right of idiogram represent
comparative anchor loci.

that are described, but which so far have little genetic information.
The cat is an excellent model for host-virus interactions. Feline
leukemia virus is an excellent model for viral carcinogenesis (Hardy et
al., 1980). Feline immunodeficiency virus (FIV) causes a disease

progression similar to human immunodeficiency virus (HIV) in a percentage of infected domestic cats, but is apparently non-pathogenic in larger felids (O'Brien 1986). The recent radiation and the popularity of the 37 extant felids makes cats ideal for evolutionary and population studies (O'Brien 1986; Wayne et al., 1989). Domestic cats are common household pets and non-domestic feline species are attractive zoo exhibits. The human fascination with the feline has assisted the investigation of the cat's biology by a growing field of scientists, veterinarians, breeders, and reproductive-physiologists. The charismatic nature of the cat has led to the establishment of over thirty fancy cat breeds throughout the world. Fancy cat breeding has led to the detection of inherited disease models and has facilitated the establishment of cat colonies segregating for a variety of genetic diseases. Over two dozen inherited diseases which have analogous phenotypes in humans have been described in cats (Nicholas 1996) (Table 1). Our laboratory has had an inherent interest in viral and

Table 1. Hereditary Human Diseases with Feline Models.

Disease	Affected Protein	Gene symbol	Human Locus	Feline Locus
Albinism*	Tyrosinase	TYR	11q14-q12	D1
Amyloidosis*	Cystatin	CST†	20p	
Cardiomyopathy				
Chediak-Higashi Syndrome*	(unknown)	CHS	1q43	
Corneal Dystrophy	(unknown)	CDGG1	5q22-q33.3	
Diabetes Mellitus, insulin dependent*	Insulin	IDDM	6p21.3	
Diabetes Mellitus, non-insulin depend*	Islet amyloid polypeptide	NIDDM	1 2	B4
Ehlers-Danlos Syndrome	Collagen			
Ehlers-Danlos Type VII	Procollagen protease			
Endocardial Fibroelastosis	(unknown)	EFE	Xq28	
Fabry Disease	Alpha-galactosidase	GLA	Xq22	X
Factor XII	Factor XII	F12	5q33-qter	
Fucosidosis	Alpha-L-fucosidase	FUCA1	1p34	C1
Globoid cell leukodystrophy*	Galactosylceraminidase	GALC	14q24.3-q32	
Glycogen Storage Disease Type II	Alpha-1,4-glucosidase	GAA	17q23	D1
Glycogen Storage Disease Type IV	Glycogen branching enzyme	GBE1	3p12	
GM1 Gangliosidosis	Beta-galactosidase	GLB1	3p21.33	B3
GM2 Gangliosidosis (Sandhoff's)	Beta-hexosminidase A	HEXB	5q13	(B3)§
Hemophilia A	Factor VIII	F8C	Xq28	
Hemophilia B	Factor IX	F9	Xq27.1-q27.2	
Hyperlipoproteinaemia*	Cholesterol metabolism			
Mannosidosis	Alpha-mannosidase	MANB	19cen-q12	(B3)§
Mucopolysaccharidosis Type I	Alpha-L-iduronidase	IDUA	4p16.3	D4
Mucopolysaccharidosis Type VI	Arylsulfatase B	ARSB	5q11-q13	
Neuronal ceroid lipofuscinosis*	(unknown)	CLN†		
Niemann-Pick Type A	Sphingomyelin phosphodiesterase-1	SMPD1	11p15.4-p15	
Niemann-Pick Type C*	(unknown)	NPC	18q11-q12	
Ornithine aminotransferase deficiency	Ornithine aminotransferase	OAT	10q26	
Polycystic Kidney Disease	(unknown)	PKD1	16p13.31-p13	
Progressive Retinal Atrophy				
Testicular Feminization*	Androgen receptor	AR	Xq11-q12	X
Von Willebrand Disease*	Coagulation factor VIII	F8VWF	12p13.3	

*diseases with murine models   †several
§ HEXA-B3; MANA-B3
Current collaborations in bold.

evolutionary felid studies, and with the further development of the genetic map of the cat, we have initiated collaborations to investigate the genetic basis of many feline diseases.

The reference family for the cat genome linkage project is a feline interspecific backcross pedigree. This pedigree has been constructed by mating female domestic cats to male Asian leopard cats, *Prionailurus bengalenesis*. The leopard cat is within the size range of domestic cats, but has an intractable temperament. Leopard cats will occasionally naturally mate with domestic cats, if reared together from kittenhood, but mature leopard cats require assisted reproduction methods for the production of hybrid cats (Lyons et al., 1996). Our pedigree, which is currently 50 backcross offspring, has been developed by both natural and assisted reproduction. Approximately one-third of the backcross offspring have been produced by artificial insemination techniques, whereas the remaining two-thirds of the pedigree samples have been collected from cat breeders. Hybrids between the leopard cat and the domestic have been common in the United States for over 20 years. A fancy cat breed has been developed from these hybrids and is commonly known as the Bengal. Fourth generation Bengals qualify for championship competition and represent one of the few breeds which are bred for their spotted coats.

The F1 males of the interspecific feline cross follow Haldane's rule and are sterile (Haldane 1922). Female F1's can be mated to either male domestic or leopard cats. Backcross males, from either cross, are also sterile. This sterility of the F1 males prohibits the production of F2 offspring and reciprocal backcrosses, although male fertility has been noted in subsequent generations. We have investigated the physiological basis of sterility in F1 males. Histological and morphological examination of the F1 male testis reveals Sertoli cell degeneration and the complete lack of spermatogenesis. These hybrid cats exhibit an earlier arrest in spermatogenesis than the F1 males of the interspecies mouse cross, *Mus spretus* x *mus domesticus* (Bonhomme 1978). Mouse hybrids, as well as horse x donkey hybrids (Chandley et al., 1974), display arrested spermatogenesis after meiosis I. Subsequent generations of the interspecies cat hybrids are fertile, thus the correlation of fertility/sterility with leopard/domestic cat gene inheritance should reveal the segregation of specific genes involved with reproduction. Reproductive isolation is one of the mechanisms leading to speciation. The identification of the genes leading to reproductive isolation in the cat could help elucidate the process of species development in the Felidae.

It is recognized that the hybridization of less divergent species of cats would likely provide a more prolific cross, but sterility studies, interesting viral studies, and polymorphism information content may be compromised with other such crosses. The availability of pure African and/or European wildcats is also more limited than access to leopard cats. Genetic differences between the leopard cat and the domestic cat may provide valuable insight to viral diseases and other inherited traits. Leopard cats are resistant to FeLV (Hardy et al., 1980) and do not have the endogenous retrovirus, RD-114 (Reeves and

O'Brien 1984). The cross between the domestic and the leopard cat allows the investigation of these unique feline viruses. As non-responders, the leopard cats also facilitate the mapping of the dominant gene controlling the response to catnip.

The 8-10 million years of evolutionary divergence between the domestic and leopard cat (Collier and O'Brien 1985) permit the development of an extremely efficient pedigree for genetic linkage studies. Over 95% of domestic cat specific microsatellites primers amplify in both cat species (Menotii-Raymond and O'Brien 1995). Southern blot analyses of coding genes reveal unique band differences between the two cat species with screening only 3-4 restriction enzymes. Thus, this completely phase-known and completely heterozygous pedigree provides an extremely efficient family for the mapping of both Type I, coding gene markers, and Type II, microsatellite markers. As few as 61 offspring can effectively produce a 5 cM genetic map (Lyons et al., 1994). Our present pedigree of 50 backcross offspring would be sufficient to establish a 10 cM map for the cat. Thus, although not highly prolific, the domestic x leopard cat cross is most robust and versatile for genetic studies of felines.

Marker Development

The advancement of microsatellite typing techniques has greatly improved the feasibility of rapidly and efficiently developing a genetic linkage map of Type II markers for any species. Yet microsatellite markers are not very effective for across species comparisons, even within species of the same order. Type I markers are efficacious for such comparisons, but inefficient and time consuming typing methods have led to the paucity of coding gene markers on all genome maps, with the exception of the mouse (Coopland et al., 1993). Without the advantage of a diverse genetic cross for a reference pedigree, the tedious detection of polymorphism and the low heterozygosity of coding genes has impeded Type I marker mapping for all species.

Our laboratory has focused on the advancement of Type I marker usage for mammalian genome projects. In collaboration with members of the comparative mapping community, 321 genes were selected as anchor loci (O'Brien et al., 1993). The genes selected as anchor loci had been previously mapped in more than one species, were publically available as DNA clones, and were evenly spaced by using human cytogenetic and mouse linkage data. Preference was given to genes shown to be involved in disease and reproduction and genes which flanked known blocks of homology when compared to the human genome. Thus, a set of markers that could be mapped by all species mapping efforts, has been established to provide a framework for genome structure comparisons across mammalian orders.

Even though a common set of comparative anchor markers has been established, time consuming typing methods and low heterozygosity levels remain major drawbacks for mapping efforts. We have explored techniques for establishing polymerase chain reaction (PCR) based typing of universal Type I markers in any mammal species, in a manner which will also increase the likelihood of polymorphism detection for these highly conserved markers. Portions of the following procedure have been automated to enable the rapid construction of a set of primers for a given Type I gene which amplify the corresponding gene in diverse species (Lyons et al., 1996).

The initial step in our procedure is to identify all the published sequences for a particular human gene. We start by searching Genbank for the gene accession numbers, searching with human gene symbols and gene names. After verifing that the accession numbers are the correct gene, we then attempt to build a contiguous single gene sequence from identified sequences, by splicing together exons and 5' or 3' sequences into one larger gene sequence. This "contig" sequence is then used to search the vertebrate DNA sequence database to identify all non-human species sequenced for the same gene. For each species, the sequences which are the longest and have the greatest homology to the human gene are selected for alignment. The gene sequences from the various species are aligned and any known intron/exon boundary information is noted. The consensus sequence in downloaded into a primer designing program for the selection of primers. Primer parameters are held constant for all the genes. The gene primers are selected from the middle of adjacent exons, in the areas with the highest conservation amongst the aligned species. Primers are designed to amplify the intron region between the two exons. This placement ensures primers to be well conserved across species, but amplifies intron regions which have little selection pressures and should accumulate mutations and DNA polymorphism more rapidly than exon regions of the same gene.

Our goal is to develop universal PCR primers for a group of comparative mapping anchor loci which can be assayed by PCR methods in any mammal species and will have 5-10 cM speciation across the murine and human genomes, respectively. Over 400 gene specific primers have been tested for gene recognition in selected mammalian species (dog, seal, horse, deer, sheep, pig, cow, bat, echidna, Tammar wallaby, mouse, hamster, rabbit and humans). Preliminary results reveal that the primer design is robust and that these species can be amplified with over half of the primers using PCR conditions that were optimized empirically for cat genomic DNA. A concerted attempt to place these Type I loci in species from several mammalian orders offers the prospect of deriving comparative inference from across the mammalian radiations.

Comparative Map Usage

Effective comparative genetics depends on availability of both a Type I and a Type II genetic map (O'Brien 1991). Since the microsatellites developed from domestic cats can be mapped in the leopard cat (Menotti-Raymond and O'Brien 1995), the feline interspecific pedigree will produce an integrated Type I and Type II marker map. Both marker categories are required for effective disease gene homology determinations. Once a disease phenotype has been identified in two different species, gene homology must be determined. Verification of the homologous gene as the locus responsible for the disease phenotype in different species is necessary for development of clinical diagnostics and therapies. Effective treatments developed in the cat could be used in humans for the genetically homologous disease. For example, the gene for polycystic kidney disease (PKD1) has been recent identified and mapped to human chromosome 16p (Intern. Polycycstic Kidney Disease Consortium 1995) (Figure 2). An autosomal dominant form of PKD has been described in cats (Biller et al., 1995) but its map location has not been determined. The comparative mapping anchor loci for human 16p are *HMBA1*, *PRM1*, *IL4R*, and *PRKCB1*. The feline interspecific backcross pedigree will have these 4 genes mapped as well as linked microsatellites. Once it is shown that this region is conserved as a large block between humans and cats, the cat microsatellites linked to the human 16p markers can be screened in the feline PKD pedigree for linkage to the PKD disease phenotype. Linkage would strongly suggest disease gene homology (Figure 2). True homology cannot be determined until the exact causative mutations are discovered in each species.

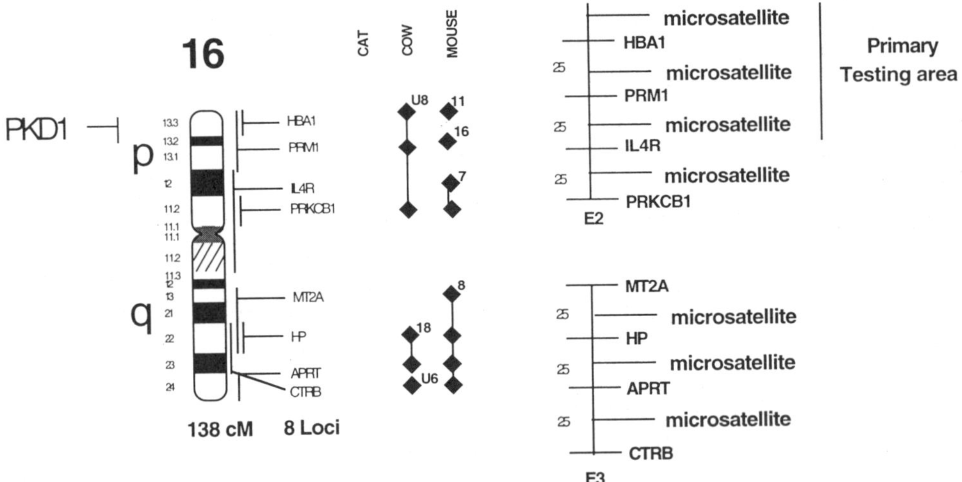

Figure 2. Comparative Mapping of Polycystic Kidney Disease. known felinehuman homology blocks allow rapid disease homology determination. Microsatellites linked to the Type I markers in the region where PKD1 is mapped can be analyzed in feline PKD families to determine homology.

Comparative genetics allows the exploitation of the gene rich species, mice and humans, by the gene poor species. Once comparative mapping anchor loci have been mapped in different species, then all the genetic information identified for those species is interpretable for the other species. As presented in Figure 2, if the feline PKD is not homologous to human PKD1, the region homologous to human chromosome 4 should then be searched since a second locus for polycystic kidney disease has been mapped to this region in humans, although gene has not yet been identified. Human 4q21-q23 is maintained as a single conserved unit in the cat. If PKD maps to the homologous segment on cat chromosome B1, then the genes located in the region 4q21-q23 of human and in homologous segment of cat chromosome B1 become possible candidates.

Certain linages of Abyssinian cats have an inherited form of progressive retinal atrophy (PRA) (Barnett and Curtis 1985). Humans also have progressive retinal atrophies but neither specific genes nor the homologous disease between humans and cats have been identified. The cat disease pedigree can be screened with Type II markers for linkage to the feline PRA. Once linkage has been established, the neighboring Type I markers can be identified. These markers now become candidate genes. These linked Type I markers allow the identification of homologous chromosome segments in the human and mouse. The plethora of genes mapped in these two species within the homologous gene regions provides a wide selection of

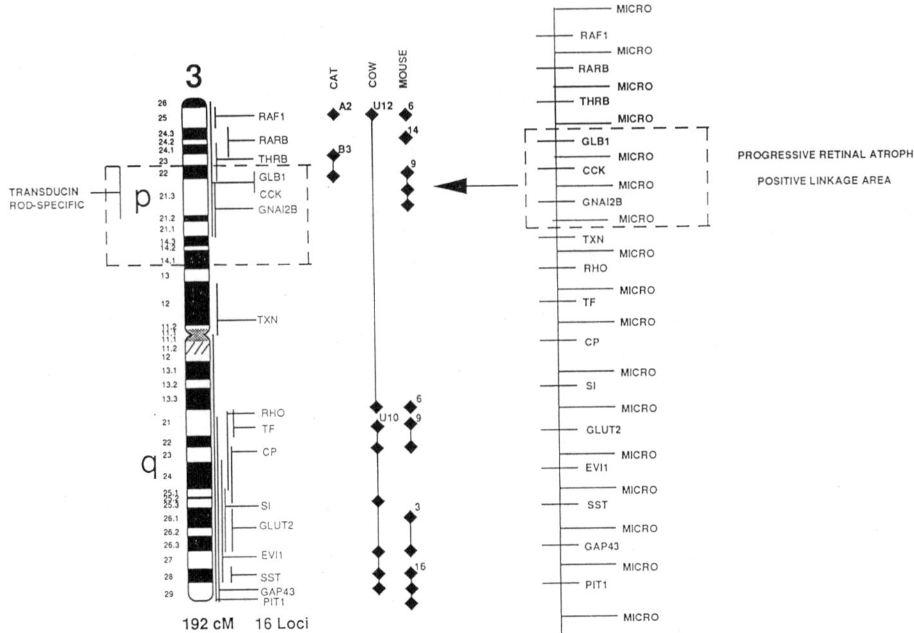

Figure 3. Reverse Comparative Genetics. Hypothetical model for the mapping of a progressive retinal atrophy in the cat. Type II markers showing linkage to the disease phenotype can be used to find the homologous regions in a gene rich species. Type I markers linked to the Type II markers allow human and/or mouse regions to be identified for candidate gene selection.

candidate gene possibilities (Figure 3). Every gene does not have to be mapped in all species, just enough loci to delineate the blocks of conservation. This reverse comparative approach may identify a retinal specific gene, mapped in either mouse or humans, which now would become the prime candidate for the feline and human PRA.

SUMMARY

Comparative genomics has progressed beyond the role of mapping homologous genes. As the genomes of more diverse species become mapped with both Type I and Type II markers, the mechanisms of biological interactions will begin to unravel. Type II markers have helped identify genes involved with economically important traits, such as meat and milk quality and quantity in agriculturally important species. The genetic map of the mouse is rich in both Type I and Type II markers, making the mouse the principle animal model in diverse fields of biological research. As Type I marker maps become more developed in diverse species such as cows, cats, deer, and wallabies, comparative mappers will begin to complement the comparative inference begun by human and murine gene mapping. Second and third models for human inherited disease will be developed in these additional species, as well as models for genes controlling development and complex traits. Eventually, comparative mapping may become a genomic "web" of information, allowing each species to be as rich as the other, exponentially more so than when each stands alone.

REFERENCES

Barnett, K.C., and Curtis, R., 1985, Autosomal dominant progressive retinal atrophy in Abyssinian cats, *J Heredity* 76:168.

Biller, D.S., Chew, D.J., and DiBartola, S.P., 1990, Polycystic kidney disease in a family of Persian cats, *JAVMA* 196:1288.

Bonhomme, F., Martin, S., and Thaler, L, 1978, Hybridation en laboratoire de *Mus musculus L* et *Mus spretus Lastaste, Experientia* 34:1140.

Chandley, A.C., Jones, R.C., Dott, H.M., Allen, W.R., and Short, R.V., 1974, Meiosis in interspecific equine hybrids. I The male mule (*Equus asinus x E. caballus*) and hinny (*E. caballus x E. asinus), Cytogenet. Cell Genet.* 13:330.

Collier, G.E., and O'Brien, S.J., 1985, A molecular phylogeny of Felidae: Immunological distance, *Evolution* 39:473.

Copeland, N.G., Jenkins, N.A., Gilbert, D.J., Eppig, J.T., Maltais, L.J., Miller, J.C., Dietrich, W.F., Weaver, A., Lincoln, S.E., Steen, R.G., Stein, L.D., Nadeau, J.H., and Lander, E.S., 1993, A genetic linkage map of the mouse: Current applications and future prospects, *Science* 262:57.

Gilbert, D.A., O'Brien, J.S., and O'Brien, S.J., 1988, Chromosomal mapping of lysosomal enzyme structural genes in the domestic cat. *Genomics* 2:329.

Haldane, J.B.S., 1922, Sex ratio and nisexual sterility in hybrid animals, *J. Genetics* 12:101.

Hardy, W.D., Essex, M., and McClelland, A.J., eds., 1908, Feline Leukemia Virus, Elsevier/North Holland, New York, pp. 586.

Lyons, L.A., Menotti-Raymond, M., and O'Brien, S.J., 1994, Comparative Genomics: The next generation, *Animal Biotechnology* 5:103.

Lyons, L.A., Menotti-Raymond, M., and O'Brien, S.J., 1994, Comparative Genomics: The next generation, *Animal Biotechnology* 5:103.

Lyons, L.A., Laughlin, T.F., Copeland, N.G., Jenkins, N.A., Womack, J.E., and O'Brien, S.J., 1996, Comparative anchor tagged sites for integrative mapping of mammalian genomes, (In preparation).

Lyons, L.A., Brown, J.L., Munsen, L, Tuthill, D., Howard, J.G., Wildt, D.E., and O'Brien, S.J., 1996, Reproductive assessment of sterile male F1 domestic cat x leopard cat hybrids (*Felis catus x Prionailurus bengalensis,* (In preparation).

Marshall Graves, J.A., Wakefield, M.J., Peters, J., Searle, A.G., Womack, J.E., and O'Brien, S.J., 1995, Report of the committee on comparative gene mapping, *in:*"Human Gene Mapping, 1994: A Compendium", Cuticchia, A.J. ed., Johns Hopkins University Press, Baltimore, pp. 962.

Menotti-Raymond, M.A., and O'Brien, S.J., 1995, Evolutionary conservation of ten microsatellite loci in four species of Felidae, *J. Hered.* 86:319.

Modi, W.S., and O'Brien, S.J., 1988, Quantitative cladistic analyses of chromosomal banding data among species in three orders of mammals: Hominoid primates, felids and arvicolid rodents, *in:*" Chromosome Structure and Function", Gustafson, J. P. and Appels, R. eds., Plenum Press, New York pp.215.

Nadeau, J.H., Davisson, M.T., Doolittle, D.P., Grant, P., Hillyard, A.L., Kosowsky, M., and Roderick, T.H., 1991, Comparative map for mice and humans, *Mamm. Genome* 1:S461.

Nash, W.G., and O'Brien, S.J., 1982, Conserved regions of homologous G-banded chromosomes between orders in mammalian evolution: Carnivores and primates, *Proc. Natl. Acad. Sci.*, USA 79:6631.

Nicholas, F., 1996, Online Mendelian Inheritance in Animals (OMIA), http://morgan.angis.su.oz.au.

O'Brien, S. J., 1986, Molecular genetics in the domestic cat and its relatives, *Trends Genet.* 2:137.

O'Brien, S.J., 1991, Mammalian genome mapping: Lessons and prospects, *Current Opinion in Genetics and Development* 1:105.

O'Brien, S.J., and Nash, W.G., 1982, Genetic mapping in mammals: Chromosome map of the domestic cat, *Science* 216:257.

O'Brien, S.J., Seuanez, H.N., and Womack, J.E., 1988, Mammalian genome organization: An evolutionary view, *Annual Review of Genetics,* 22:323.

O'Brien, S.J., Womack, J.E., Lyons, L.A., Moore, K.J., Jenkins, N.A., and Copeland, N.G., 1993, Anchored reference loci for comparative genome mapping in mammals, Nature Genet. 3:103.

Pedersen, N.C., 1987, Coronavirus diseases (coronavirus enteritis, feline infectious peritonitis), in:" Diseases of the Cat", Holzworth, J. ed., W.B. Saunders, Philadelphia, pp. 193.

Pedersen, N.C., and Floyd, K., 1985, Experimental studies with three new strains of feline infectious peritonitis virus: FIPV-UCD2, FIPV-UCD3 and FIPV-UCD4, *Comp. Cont. Educ. Prac. Vet.* 7:1001.

Reeves, R.H., and O'Brien, S.J., 1984, Molecular genetic characterization of the RD-114 gene family of endogenous feline retroviral sequences, *J. Virol.* 52:164.

The International Polycystic Kidney Disease Consortium, 1995, Polycystic kidney disease: The complete structure of the PKD1 gene and its protein, *Cell* 81:289.

Wayne, R.K., Benveniste, R.E., Janczewski, D.N., and O'Brien, S.J., 1989, Molecular and biochemical evolution of the Carnivora, *in:*"Carnivore Behavior, Ecology and Evolution", Gittleman, J. L ed., New York, Cornell University, pp. 465.

Wurster-Hill, D.H., and Centerwall, W.R., 1982, The interrelationships of chromosome banding patterns in canids, mustelids, hyena, and felids, *Cytogenet. Cell Genet.* 34:178.

GENOME ANALYSIS IN FARM ANIMALS

James E. Womack

Department of Veterinary Pathobiology
Texas A&M University
College Station, TX 77843-4467

INTRODUCTION

Genome research provides one of the exciting new frontiers in the life sciences. It should not be surprising that the agricultural sciences, long concerned with genetic improvement, are beginning to invest in building maps and localizing important traits to chromosomal regions. The following is a brief overview of the progress and problems of genome analysis in farm animals.

WHY MAP FARM ANIMAL GENOMES

To Study Chromosomal Evolution

Biologists are naturally curious about how genomes are organized, how this organization evolved, and how it might influence gene expression. Consequently, a driving force in animal genome mapping is to answer these questions and consequently, advance the fundamental biological knowledge of the species. Comparative gene mapping, mapping homologous genes in multiple species, provides more information about chromosomal evolution between distant species than we can retrieve with today's best cytogenetic technologies. It has therefore become a major tool in the study of chromosomal evolution. The well-funded and well-organized human genome initiative will provide the prototypic mammalian gene map. The map of the laboratory mouse will be the first yard stick for evaluating genomic evolution. Other mammals, including the livestock

Genomes of Plants and Animals: 21st Stadler Genetics Symposium
Edited by J. Perry Gustafson and R. B. Flavell, Plenum Press, New York, 1996

species, have important roles to play, however, in unraveling the pathways of chromosomal rearrangement that have accompanied mammalian evolution. It is already clear that some mammals such as cattle and cats have genomes that are more conserved relative to human than is the mouse genome. Continued comparative mapping in domestic animals will make a valuable contribution to the big picture of mammalian genome evolution.

To Permit Marker-Assisted Selection

The more practical goal of marker-assisted selection (MAS) drives gene mapping in many laboratories. Genetic markers of positive alleles for economic trait loci (ETL), including quantitative trait loci (QTL), have the potential to enhance the rate and efficiency of genetic gain through selective breeding (Soller and Beckmann, 1982; Weller et al., 1990; Smith and Simpson, 1986). The ideal marker for a selective breeding program would be the gene actually responsible for the trait. The candidate gene approach to marker identification requires a basic knowledge of the physiological processes underlying the trait and extensive screening for variation in candidate genes related to these processes. Ideally, sequence variation related to the trait can be incorporated directly into the marker assay. As an excellent example, Shuster et al., (1992) identified the genetic defect responsible for leukocyte adhesion deficiency (LAD) in Holstein cattle as a missense mutation coding amino acid 128 in CD18. The mutant and normal alleles can be distinguished by the polymerase chain reaction (PCR) providing the ultimate genetic marker of this economic trait locus.

The physiological basis and consequently the genes for most economic traits are unknown. ETL, even QTL, can be genetically mapped, however (Lander and Botstein, 1989; Paterson et al., 1989; Georges et al., 1993a; Georges et al., 1993b, Andersson, et al., 1994, Georges et al., 1995). Markers mapped in close proximity to ETL can be used to assist selection if the recombination frequency (map distance) is sufficiently small and the phase of marker and ETL alleles is known. Efficiency of MAS can be dramatically increased by identifying markers flanking the ETL since recombination in the region spanned by two markers can be detected. A major goal of animal gene mapping is therefore a map of highly polymorphic markers spaced at intervals of 20 cM (1cM = 1% recombination) or less across every cattle chromosome. These markers can be used to map QTL and, under appropriate breeding protocols, may also be used for MAS.

Linked markers can be used in selection only with knowledge of phase of the marker allele and the ETL allele. This is not a problem in defined pedigrees where the phase is initially known and map distances are small. Linked markers cannot be used in the absence of pedigree information, however, unless the marker and trait are in linkage disequilibrium.

To Facilitate Positional Cloning

A third goal of gene mapping is to identify and clone genes responsible for ETL. The term "reverse genetics" has been replaced by "positional cloning" or "map based cloning" to denote the application of map information to clone a gene responsible for a trait in the absence of information about the biochemical basis of the trait. Success of positional cloning has been experienced in searches for several human disease genes, exemplified by the cloning of the cystic fibrosis gene (Rommens et al., 1989). The task of cloning genes for ETL in livestock is a formidable one, however, for a number of reasons. Our maps are not as dense as those of the human. Large insert libraries for livestock are only beginning to be developed and used. Chromosomal deletions of economically important genes have not been identified and propagated. Our task is further complicated by the quantitative nature of most of our traits of interest and the paucity of research support world-wide relative to the human genome initiative. Alternative strategies to conventional positional cloning must be developed.

METHODS FOR ANIMAL GENOME MAPPING

Synteny Mapping

Synteny means on the same strand, or in genetic terminology, on the same chromosome. A synteny map is simply a list of genes that are known to reside on the same chromosome in a particular species. An unfortunate misuse of the word synteny has crept into recent genetic literature. Such statements as "...a greater degree of synteny between cattle and human than between cattle and mouse,..." or "...a limited degree of synteny has been demonstrated between human, mouse and swine..." are genetically meaningless. "Synteny conservation" as used by Nadeau (1989) to describe the location of two or more homologous genes on the same chromosome in different species was the likely intent of the word "synteny" in the above statements.

Somatic cell genetics is the common method for building synteny maps. Hybrid somatic cells can be constructed such that the chromosomes of one progenitor species are preferentially lost. Each hybrid clone will retain a partial genome of that species along with the complete genome of the other which is usually a transformed rodent cell line. Since chromosome loss is more or less random, each clone will retain a different subset of chromosomes from the species being mapped. Analysis of pairs of genes in a panel of hybrid cell lines will reveal concordance or discordance of their retention. Concordance of retention is evidence for the location of two genes on the same chromosomes. Conversely, discordance of retention is evidence for asynteny, their location on different chromosomes. Gene products or

DNA sequences may be mapped by synteny analysis so long as the presence or absence of the gene or gene product of the targeted species can be ascertained against the fully retained rodent genomic background. Enzyme electrophoresis, Southern blotting with unique sequence probes, and PCR amplification with species discriminating primers have all been effective analytical tools for synteny mapping.

Somatic cell genetics does not typically assign markers to specific chromosomal sites or even to chromosomal regions. Consequently, genes on a synteny map are usually unordered. Somatic cell methods employing rearranged chromosomes are an exception to this generalization and have been used very effectively to order genes in the human map. Unfortunately, panels of hybrid cell lines with rearranged chromosomes have not yet been employed in mapping genomes of domestic animals.

In Situ Hybridization

Whole genomes, repetitive elements, and unique sequences have all been successfully localized to chromosomal sites by *in situ* hybridization. This technique employs the attachment of a visually detectable marker to a DNA probe followed by hybridization of the probe to denatured DNA of an otherwise intact chromosome. The specificity of hybridization is determined by the uniqueness of the probe. While radioactive probes dominated the early application of this technology, fluorescent probes are now generally used. In her review of fluorescence *in situ* hybridization (FISH) Trask (1991) notes the following advantages of FISH over isotopic labeling. It provides superior spatial resolution and generally requires the visualization of fewer labeled chromosomes. It is faster and the probe is more stable. The sensitivities are similar, each requiring a few kilobase (kb) pairs of uninterrupted sequence as the hybridizing probe. Schemes have been developed which allow multiple probes with different color signals to be used on the same chromosomes.

Cosmids, cloning vectors that accommodate large DNA sequences (up to 45 kb), have been effectively utilized as probes for FISH. Since these large inserts often contain repetitive DNA, the target DNA must first be prehybridized with unlabeled total genomic DNA. This method has been effectively used to anchor linkage maps to chromosomes by hybridizing cosmids that contain highly polymorphic markers used in linkage mapping.

The above technologies result in physical maps which describe the physical relationships of loci to the chromosomes on which they reside. Higher resolution physical maps which define markers in contiguous clones (contig maps) are forthcoming in livestock but will most likely span small genomic regions of special interest rather than whole chromosomes as is targeted in the human genome initiative and is prerequisite to total genome sequencing.

Linkage Mapping

Linkage maps are defined in biological rather than physical terms, a map unit representing one percent recombination in meiosis. Since linkage is measurable only in gametic products, linkage mapping requires detection of maternal and paternal alleles in gametes of heterozygous individuals. A linkage map is then made from the percentage of recombinants in the parental arrangements of alleles of two loci on a chromosome. The segregation of three or more loci permits ordering of genes on the map since double recombinants are rare relative to single recombinants.

Markers for mapping purposes have been catagorized as Type I or II by O'Brien (1982). Type I markers are expressed sequences (genes) and are usually conserved from one species to another. Thus, they are the material for comparative gene mapping. Unfortunately they are usually not highly polymorphic and therefore difficult to incorporate into linkage maps. Type II markers are highly polymorphic and are more widely used for linkage mapping. The value of these markers lies in their high level of polymorphism. They are not necessarily expressed genes but often anonymous stretches of DNA.

The polymorphic DNA markers that dominate the linkage maps of most plant and animal species are one of two general classes. Restriction fragment length polymorphisms (RFLPs) identify alleles on the basis of polymorphism in the presence or absence of specific restriction sites. Genomic DNA is digested with a particular restriction enzyme, separated by electrophoresis and blotted onto a solid membrane (Southern, 1975). Hybridization to a labeled probe often reveals polymorphism in DNA fragment sizes. One major advantage of RFLPs as linkage markers is that they often define polymorphisms in or around Type I markers, depending on the probe used for detection. They are therefore useful for comparative mapping and also provide a potential source of variation in candidate genes for economic traits. Blotting and probe development are laborious and time consuming, however, and blotting exhausts DNA resources much faster than does PCR based methods.

Microsatellites (Weber and May, 1989; Fries et al., 1990) have become the Type II marker of choice in linkage mapping in most animal species. These islands of di-, tri-, or tetranucleotide repeats are proving to be ubiquitous throughout animal genomes. The number of repeats is usually the polymorphic entity, with a half dozen or more alleles not uncommon in a breeding population. Microsatellites are defined by unique flanking sequences which serve as primers for PCR amplification and also as tags of unique sites in the respective genome.

Other types of markers on linkage map include enzyme polymorphisms, blood group antigens, and phenotypic traits like coat color or horn development. These all add interest and biological significance to the map. Moreover, they provide potential for enhanced comparative mapping across species.

Linkage mapping requires observable meiotic products, usually in the form of offspring from individuals segregating the markers discussed above. Large full sib-ships or half sib-ships are ideal for linkage analysis in farm animals. Since a comprehensive map requires large numbers of markers scored on a common set of meiotic products, sharing of family material among laboratories is the most effective arrangement for generating a linkage map. Sets of shared reference families are available for most species including cattle (Barendse et al., 1994). Offspring are not essential for linkage analysis, however, since multiple DNA loci can now be analyzed for allelic variation directly on meiotic products. Van Eijk, et al. (1993) successfully ordered three loci on bovine Chr 23 by PCR after primer extension preamplification (PEP) of individual sperm.

CURRENT STATUS OF ANIMAL GENOME MAPS

Physical Mapping

More than 400 Type I loci have been mapped in cattle (Fries, et al., 1993; O'Brien et al., 1993) primarily through somatic cell genetics. Of the 100 or so chromosomal assignments in the sheep map (Echard et al., 1994), most were made by synteny mapping. More than 80 of these are Type I markers. While these synteny mapped markers indicate the boundaries of chromosomal conservation relative to the map rich genomes of mice and humans and show extensive sheep-cattle conservation, they provide an incomplete comparative map. Conservation of gene order is not addressed by the synteny map.

In situ hybridization, especially FISH has been used effectively to address the order of Type I loci, to assign syntenic groups to specific chromosomes, and to anchor the rapidly growing linkage map to chromosomes (Fries et al., 1993; Solinas-Toldo et al., 1993; Iannuzzi et al., 1993, Gallagher et al., 1993). There are presently more than 50 *in situ* localizations of unique sequences on cattle chromosomes. All bovine syntenic groups are now assigned to specific chromosomes and the bovine linkage map is anchored at 35 sites on 26 chromosomes. Thirty-four *in situ* assignments in sheep were reviewed by Echard et al. (1994). At least 80 chromosomal assignments have been made in pigs, most by *in situ* hybridization (Andersson et al., 1993). Sixty of these are Type I markers.

Linkage Mapping

The published bovine linkage map of 200 markers (Barendse et al., 1993) has grown to almost 500 markers in less than a year (Barendse, personal communication). This was made possible in large measure by international cooperation and the use of a common set of reference families. Combined with the independent development of other maps (Bishop et al., 1994) there are probably more than 800 markers presently assigned to cattle linkage groups. The goal of 20 cM resolution has clearly been achieved over at least 95% of the total genome. The international map includes 65 type I markers. A pig linkage map of 128 markers, 60 of which are informative for comparative mapping was published by Ellegren et al. (1994). Rohrer et al. (1994) published a 383 marker map of the pig, almost all of which were microsatellites. Two reference families of chickens have each produced linkage maps in excess of 200 markers (Burt et al., 1995).

Comparative Mapping

Mammalian genomes are highly conserved. Since domestic animals will likely always be "map poor" relative to the "map rich" genomes of humans and mice, it behooves us to engage in comparative mapping.

Approximately 400 loci have been mapped in both cattle and humans. Most of these have also been mapped in mice. While extensive conservation of synteny has been observed between cattle and humans (Womack and Moll, 1986; Threadgill et al., 1991), conservation of linkage (conservation of gene order) may not be so prevalent. Barendse et al. (1993) incorporated a sufficient number of Type I loci into the linkage map to demonstrate the presence of several rearrangements of gene order within conserved syntenies.

The sheep and cattle maps, as should be expected, are highly conserved with only three apparent differences (Echard et al., 1994). Developing these maps in parallel should enhance and economize the efforts targeted to each map.

To make the process of comparative gene mapping in mammals more efficient, O'Brien et al. (1993) proposed a list of comparative anchor reference loci (CARL). These 321 Type I markers were selected to span the human genome at approximately 10 cM intervals and to include the anchor loci for human and mouse maps. Concerted efforts to map domestic animal homologues of these loci, by synteny, cytogenetic and linkage approaches, will greatly facilitate comparative mapping and the use of the growing human and mouse database for animal improvement.

A recent breakthrough in comparative gene mapping is heterologous chromosomal painting on ZOO-FISH painting. Rettenberger et al. (1995) painted the pig genome with 22 human autosome specific libraries. Solinas-Toldo et al. (in press) have done the same experiments including the X chromosome in cattle. These studies delineate the boundaries of chromosomal conservation at the cytogenetic level and are thus far highly consistent with the results of comparative synteny mapping. Like synteny mapping they do not address conservation of gene order.

Economic Traits

A number traits of economic significance are beginning to appear on animal genome maps. In addition to bovine LAD (Schuster et al., 1992; Threadgill et al., 1991), uridine monophosphate synthase deficiency (UMPS) is also mapped to a specific site on chromosome 1 (Schwenger et al., 1993; Ryan et al., 1994; Barendse et al., 1993). BoLA is associated with susceptibility to leukemia virus infection (Lewin et al., 1988). Georges et al. (1993a) have linked the polled locus to microsatellites on chromosome 1. The weaver disease maps to markers on chromosome 4 (Georges et al., 1993b) and also appears to be associated with a quantitative trait for improved milk production. Variation around the prolatin gene on chromosome 23 (Cowan et al., 1990) is related to milk production in certain Holstein sire families and Georges et al. (1995) have used mapped microsatellites to locate an additional five QTL for milk production. Mapping the Booroola fecundity gene (Montogmery et al., 1993) and a hyper muscling gene called callipyge (Cockett et al., 1994) are recent successes with the genomic approach to ETL mapping in sheep. Growth and fatness QTL were mapped in pigs in a cross involving the European wild boar (Andersson et al., 1994 and previously the candidate gene approach identified the genetic defect for porcine malignant hyperthermia (Otsu et al., 1991). The number of ETL on animal genome maps is rapidly expanding.

FUTURE DIRECTIONS

Sufficient numbers of mapped markers now exist in cattle, sheep, pigs and chickens for extensive genome coverage of families segregating ETL. Unfortunately, different ETL require different segregating families. These resource families are usually expensive to develop and maintain. Nonetheless, the families are an integral and necessary step in the ultimate application of the gene map to economic improvement. The mapping of ETL to 10-20 cM regions of a chromosome may be followed by high resolution mapping to ultimately identify and clone the responsible genes. Chromosome specific libraries will aid this process.

The increased emphasis on identifying and mapping all the expressed genes of mice and humans places even greater need for comparative mapping in domestic animals. The mapping of an ETL to a 10 cM region could eventually result in a list of only 100-200 candidate genes directly from the human and mouse maps. The efficient use of comparative maps will require on increased emphasis on placing Type I markers on animal linkage maps. Such a "comparative positional cloning" approach will undoubtedly benefit from the growing number of cDNAs being mapped in humans and strategies to capture cDNAs from specific chromosomal regions.

REFERENCES

Andersson, L., Haley, C.S., Ellegren, H., Knott, S.A., Johansson, M., Andersson, K., Andersson-Eklund, L., Edfors-Lilja, I., Fredholm, M., Hansson, I., Håkansson, J., and Lundström, K., 1994, Genetic mapping of quantitative trait loci for growth and fatness in pigs, *Science* 236:1771.

Barendse, W., Armitage, S.M., Ryan, A.M., Moore, S.S., Clayton, D., Georges, M., and Womack, J.E., 1993, A genetic map of DNA loci in bovine chromosome 1, *Genomics* 18:602.

Barendse, W., Armitage, S.M., Kossarek, L.M., Shalom, A., Kirkpatrick, B.W., Ryan, A.M., Clayton, D., Li, L., Neibergs, H.L., Zhang, N., Grosse, W.M., Weiss, J., Creighton, P., McCarthy, F., Ron, M., Teale, A.J., Fries, R., McGraw, R.A., Moore, S.S., Georges, M., Soller, M., Womack, J.E., and Hetzel, D.J.S., 1994, A genetic linkage map of the bovine genome, *Nature Genetics* 6:227.

Bishop, M.D., Kappes, S.M., Keele, J.W., Stone, R.T., Sunden, S.L.F., Hawkins, G.A., Solinas Toldo, S., Fries, R., Grosz, M.D., Yoo, J., and Beattie, C.W., 1994, A genetic linkage map of cattle, *Genetics* 136:619.

Burt, D.W., Bumstead, N., Bitgood, J.J., Ponce de Leon, F.A., and Crittenden, L.B., 1995, Chicken genome mapping: a new era in avian genetics, *Trends in Genetics*, 11:161-201.

Cowan, C.M., Dentine, M.R. Ax, R.L., and Schuler, L.A., 1990, Structural variation around prolactin gene linked to quantitative traits in an elite Holstein sire family, *Theor. Appl. Genet.* 79:577.

Cockett, N.E., Jackson, S.P., Shay, T.L., Neilson, D., Moore, S.S. Steele, M.R., Barendse, W., Green, R.D., and Georges, M., 1994, Chromosomal location of the callipyge gene in sheep (*Ovis aries*) using bovine DNA markers, *Proc. Natl. Acad. Sci. USA* 91:3019.

Echard, G., Broad, T.E., Hill, D., and Pearce, P, 1994, Present status of the ovine gene map (*Ovis aries*); comparison with the bovine map (*Bos taurus*), *Mammal. Genome*, 5:324.

Ellegren, H., Chowdhary, B.P., Johansson, M., Marklund, L., Fredholm, M., Gustavsson, I., and Andersson, L., 1994, A primary linkage map of the porcine genome reveals a low rate of genetic recombination, *Genetics* 137:1089.

Fries, R., Eggen, A., and Stranzinger, G., 1990, The bovine genome contains polymorphic microsatellites, *Genomics* 8:403.

Fries, R., Eggen, A., and Womack, J.E., 1993, The bovine genome map, *Mammal. Genome* 4:405.

Gallagher, D.S., Threadgill, D.W., Ryan, A.M., Womack, J.E., and Irwin, D.M., 1993, Physical mapping of the lysozyme gene family in cattle, *Mammal. Genome* 4:368.

Georges, M., Drinkwater, R., King, T., Mishra, A., Moore, S.S., Nielsen, D., Sargeant, L.S., Sorensen, A., Steele, M.R., Zhao, X., Womack, J.E., and Hetzel, J., 1993a, Microsatellite mapping of a gene affecting horn development in *Bos taurus*, *Nature Genetics* 4:206.

Georges, M., Dietz, A.B., Mishra, A., Nielsen, D., Sargeant, L., Sorensen, A., Steele, M.R., Zhao, X., Leipold, H., and Womack, J.E., 1993b, Microsatellite mapping of the gene causing weaver disease in cattle will allow the study of an associated quantitative trait locus, *Proc. Natl. Acad. Sci. USA* 90:1058.

Georges, M., Nielsen, D., Mackinnon, M., Mishra A., Okimoto, R., Pasquino, A.T., Sargeant, L.S., Sorensen, A., Steele, M.R., Zhao, X., Womack, J.E., and Hoeschelel, I., 1995, Mapping quantitative trait loci controlling milk production in dairy cattle by exploiting progeny testing, *Genetics* 139:907.

Iannuzzi, L., Di Meo, G.P., Gallagher, D.S., Ryan, A.M., Ferrara, L., and Womack, J.E., 1993, Chromosomal localization of omega and trophoblast interferon genes in goat and sheep by fluorescent insitu hybridization, *J. Hered.* 84:301.

Lander, E.S., and Botstein, D., 1989, Mapping mendelian factors underlying quantitative traits using RFLP linkage maps, *Genetics* 121:185.

Lewin, H.A., Wu, M.-C., Stewart, J.A., and Nolan, T.J., 1988, Association between Bol*LA* and subclinical bovine leukemia virus infection in a herd of Holstein-Friesian cows, *Immuno.* 27:338.

Montgomery, G.W., Crawford, A.M., Penty, J.M., Dodds, K.G., Ede, A.J., Henry, H.M., Pierson, C.A., Lord, E.A., Galloway, S.M., Schmack, A.E., Sise, J.A., Swarbrick, P.A., Hanrahan, V., Buchanan, F.C., and Hill, D.F., 1993, The ovine Booroola fecundity gene (*FecB*) is linked to markers from a region of human chromosome 4q, *Nature Genetics* 4:410.

Nadeau, J.H., 1989, Maps of linkage and synteny homologies between mouse and man, *Trends in Genet.* 5:82.

O'Brien, S.J., 1992, Mammalian genome mapping; lessons and prospects, *Curr. Opinion Genetics Develop.* 1:105.

O'Brien, S.J., Womack, J.E., Lyons, L.A., Moore, K.J., Jenkins, N.A., and Copeland, N.G., 1993, Anchored reference loci for comparative genome mapping in mammals, *Nature Genetics* 3:103.

Otsu, K., Khanna, V.K., Archibald, A.L., and MacLennan, D.H., 1991, Cosegregation of porcine malignant hyperthermia and a probable causal mutation in the skeletal muscle ryanodine receptor gene in backcross families, *Genomics* 11:744.

Paterson, A.H., Lander, E.S., Hewitt, J.D., Peterson, S., Lincoln, S.E., and Tanksley, S.D., 1989, Resolution of quantitative traits into Mendelian factors by using complete linkage map of restriction fragment length polymorphisms, *Nature* 335:721.

Rettenberger, G., Klett, C., Zechner, U., Kunz, J., Vogel, W., and Hameister, H., 1995, Visualization of the conservation of synteny between humans and pigs by heterologous chromosomal painting, *Genomic* 26:372.

Rohrer, G.A., Alexander, L.J., Keele, J.W., Smith, T.P., and Beattie, C.W., 1994, A microsatellite linkage map of the porcine genome, *Genetics*, 136:231.

Rommens, J.M., Iannuzzi, M.C., Kerem., B.-S., Drumm, M.L., Melmer, G., Dean, M., Rozmahel, R., Cole, J.L., Kennedy, D., Hidaka, N., Zsiga, M., Buchwald, M., Riordan, J.R., Tsui , L-C., and Collins, F.S., 1989, Identification of the cystic fibrosis gene: chromosome walking and jumping, *Science* 245:1059.

Ryan, A.M., Gallagher, D.S., Jr., Schöber, S., Schwenger, B., and Womack, J.E., 1994, Somatic cell mapping and in situ localization of the bovine uridine monophosphate synthase gene (UMPS), *Mammal. Genome* 5:46.

Schwenger, B., Schöber, S. and Detlef, S., 1993, DUMPS cattle carry a point mutation in the uridine monophosphate synthase gene, *Genomics* 16:241.

Shuster, D.E., Kehrli, Jr., M.E., Ackermann, M.R., and Gilbert, R.O., 1992, Identification and prevalence of a genetic defect that causes leukocyte adhesion deficiency in Holstein cattle, *Proc. Natl. Acad. Sci. USA*, 89:9225.

Solinas-Toldo, S., Fries, R., Steffen, P., Neibergs, H.L., Barendse, W., Womack, J.E., Hetzel, D.J.S., and Stranzinger, G., 1993, Physically mapped, cosmid-derived microsatellite markers as anchor loci on bovine chromosomes, *Mammal. Genome* 4:720.

Soller, M., and Beckmann, J.S., 1982, Restriction fragment length polymorphisms and genetic improvement, Proc. 2nd World Congress Genetics Applied to Livestock Production 6:396.

Southern, E., 1975, Detection of specific sequences among DNA fragments separated by gel electrophoresis, *J. Mol. Biol.* 98:503.

Smith, C., and Simpson, S.P., 1986, The use of genetic polymorphism in livestock improvement, *J. Anim. Breed. Genetics* 103:205.

Threadgill, D.S., Krause, J.P., Krawetz, S.A., and Womack, J.E., 1991, Evidence for the evolutionary origin of human chromosome 21 from comparative gene mapping in the cow and mouse, *Proc. Natl. Acad. Sci. USA*, 88:154.

Trask, B.J., 1991, Fluorescence *in situ* hybridization, *Trends in Genetics* 7:150.

Van Eijk, M.J.T., Russ, I., and Lewin, H.A., 1993, Order of bovine *DRB3, DYA,* and *PRL* determined by sperm typing, *Mammal. Genome* 4:113.

Weber, J.L., and May, P.E., 1989, Abundant class of human DNA polymorphisms which can be typed using the polymerase chain reaction, *Amer. J. Human Genetics* 44:388.

Weller, J.L., Kashi, Y., and Soller, M., 1990, Power of daughter and grandaughter designs for determining linkage between marker loci and quantitative trait loci in dairy cattle, *J. Dairy Sci.* 73:2525.

Womack, J.E., and Moll, Y.D., 1986, A gene map of the cow: Conservation of linkage with mouse and man, *J. Hered.* 77:2.

SOYBEAN GENOME ORGANIZATION: EVOLUTION OF A LEGUME GENOME

Randy Shoemaker[1], Terry Olson[2], and Vladimir Kanazin[2]

[1]USDA-ARS-FCR
[2]Department of Agronomy
Iowa State University
Ames, Iowa 50011

INTRODUCTION

The Leguminosae contains approximately 650 genera and over 18,000 species, making it the third largest family of flowering plants (Polhill and Raven, 1981). Of the three subfamilies, the Papilionoideae is the largest and includes the most diverse and economically important legumes. The soybean (*Glycine max* L. Merr.) is the most agronomically important member of this group.

The soybean, in most instances, acts genetically as a diploid and we normally regard 2n=40 as the diploid chromosome number. However, most genera of the *Phaseolae* have a genome complement of 2n=22. It has been suggested that *Glycine* was probably derived from a diploid ancestor (n=11) which underwent aneuploid loss to n=10 and subsequent polyploidization to yield the present 2n=2x=40 genome size (Lackey, 1980).

Evolutionary studies and haploid genome analysis have also suggested that soybean is an ancient tetraploid (Hadley and Hymowitz, 1973). Observations that members of soybean multi-gene families contain pairs of more closely related genes supports this hypothesis (Grandbastien et al., 1986; Hightower and Meagher, 1985; Lee and Verma, 1984; Nielsen et al., 1989), as does the presence of many examples of duplicated qualitative genes (Palmer and Kilen, 1987) and the finding that >90% of random genomic fragments used as probes in the construction of soybean RFLP maps detect two or more fragments (Shoemaker et al., 1995).

There is a strong tendency for polyploid genomes to evolve into a diploid state through sequence diversification (Leipold and Schmidtke, 1982). Major changes in the amount of nuclear DNA and in the structure of chromosomes also occur during this evolutionary process (Stebin, 1966; Ohno, 1970). Major genome restructuring occurs as genetic rearrangements gradually accumulate and restore the genome to the diploid state (Ohno, 1970). This is the evolutionary process of diploidization.

Polyploidization results in a duplication of an entire genome but duplications of only a chromosomal region or segment can occur as well. This type of chromosomal duplication occurs during the diploidization process. Both polyploidization and regional or segmental duplications result in `paralogous' regions of chromosomes (regions of chromosomes that, after having been duplicated, contain homoeologous loci). Discovery and analysis of paralogous genomic regions in a diploidizing tetraploid can provide a wealth of information about the sequence divergence and structural rearrangements associated with the evolutionary dynamics of a tetraploid genome.

COMPARATIVE MAPPING

That intra- rather than inter-chromosomal events may have played a larger part in the evolution of plant genomes has been indicated by the finding of large syntenic regions between the genomes of distantly related species such as rice, maize, wheat (Ahn et al., 1993; Ahn and Tanksley, 1993) and maize and sorghum (Whitkus et al., 1992; Pereira et al., 1993; Hulbert et al., 1990). Such regions of conserved linkage would not be present if inter-chromosomal recombination has been a frequent agent of genome divergence. Similarly, the major differences observed between the genomes of the closely related tomato and potato genomes were the presence of intra-chromosomal rearrangements (inversions) (Bonierbale et al., 1988; Tanksley et al., 1992). The arrangement of the pepper genome relative to the genome of tomato, both of which belong to the Solanacease family indicates that significant rearrangements of gene order have occurred, but that genetic loci have remained on the same linkage groups (Tanksley et al., 1988; Prince et al., 1992). However, that inter-chromosomal events can play a significant role in the evolution of some plant genomes is shown by the frequent chromosomal rearrangements within the genus *Brassica* (Kianian and Quiros, 1992; Slocum et al., 1990).

Map comparisons of two legumes, *V. radiata* (mungbean) and *V. unguiculata* (cowpea), demonstrated some rearrangement in the order of markers on a linkage group, but also showed general linkage conservation (Menancio-Hautea et al., 1993). Comparative mapping between pea (*Pisum*) and lentil (*Lens*) however, showed blocks of linkage groups common to one member, scattered throughout the linkage groups of the other member (Weeden et al., 1993; Simon et al.,

1993). Boutin et al. (1995) conducted a comparative mapping study using RFLP maps of mungbean (2n=22), common bean (*Phaseolus vulgaris*, 2n=22), and soybean (*Glycine max*, 2n=40). Mungbean and common bean genome conservation was shown to be similar to that observed between tomato and potato (Bonierbale et al., 1988) where genomes were composed primarily of conserved linkage blocks. The average size of these conserved linkage blocks was large, averaging 36.6 cM. The longest conserved block between mung bean and common bean was 103.5 cM. When comparing soybean to mungbean and common bean, however, only short and dispersed linkage blocks were conserved, e.g. segments of up to 16, 12, 11, and ten different linkage groups from soybean were found on separate single linkage groups of mungbean and segments of up to nine, eight, and seven different linkage groups from soybean were found on separate single linkage groups of common bean (Boutin et al., 1995). The average size of these conserved linkage blocks was relatively small, averaging 12.2 cM and 13.9 cM, respectively, although conserved blocks as large as 37.8 cM were observed.

Although some of the conserved linkage blocks between soybean and mungbean or common bean appear to be of comparable genetic sizes between the two genomes, many do not. Even when showing evidence of colinearity, many blocks demonstrate large differences in genetic distance between markers. For example, a conserved block on soybean linkage group I spans 65 cM but spans only 14 cM of mungbean linkage group 4 and a 30 cM block on mungbean linkage group 5 spans 126 cM on soybean linkage group H. Similar patterns can be observed between soybean and common bean (Boutin et al., 1995). This suggests that either recombination rates vary greatly between genomic regions of the different genomes or physical distances between markers in soybean have been greatly enlarged in some regions.

CONSERVATION OF DUPLICATED REGIONS OF THE SOYBEAN GENOME

Previous studies have suggested that within soybean multi-gene families (actins, leghemoglobins and seed storage proteins), pairs of more closely related genes can be identified (Grandbastien et al., 1986; Lee and Verma, 1984; Neilsen et al., 1989). Calvo et al. (in review) demonstrated that although sequence identity was very high among all members of clones of the *Em*-like gene family for soybean late embryogenic abundant genes, the clones could be divided into two classes based on more similar sequence identities throughout the gene. These two classes also demonstrated somewhat different patterns of expression during seed development.

It has also been shown that some members of gene pairs may exhibit significant conservation of flanking sequences (Grandbastien et al., 1986; Lee and Verma, 1984; Neilsen et al., 1989). That conservation can extend over a region of more than 10 kb was revealed by heteroduplex analysis of 42 kb regions encompassing the glycinin loci *gy1gy2* and *gy3* (Nielsen et al., 1989). Lambda clones containing corresponding regions of the two loci displayed non-continuous regions of hybridization over 30-52% of their length, indicative of an intersepesion of homologous and non-homologous sequences. Thus, these loci had originated as part of a duplication exceeding 42 kb in size. However, since this region contained at least four other genes in addition to the glycinin genes, it was not possible to determine whether the extensive flanking sequence homology was due to selection pressure on the gene sequences or was indicative of a high degree of sequence conservation between duplicated regions of the soybean genome.

A detailed analysis of flanking sequence homology between regions of the soybean genome containing duplicated copies of coding sequences was conducted by Polzin et al. (in review). Lambda clones were isolated that contained sequences homologous to soybean late embryogenesis (*Sle1-4*) genes as well as homologous to sequences of A071, a soybean genomic clone of a gene of unknown function. Cross-hybridizations among the lambda clones of regions flanking the coding regions indicated that sequence homology between *sle* loci extended from only a few hundred bases to several kilobases. Within the cross-hybridizing regions no rearrangements as evidenced by non-colinearity of probe hybridization were observed. Comparisons of sequence homology between lambda clones of A071 gave similar results to those observed for the late embryogenic gene family; cross-hybridization did not extend much beyond the coding sequence homologous regions. Also, hybridization within the regions of the A071 lambda clones that cross-hybridized, was colinear.

Small low-copy subclones of lambda clones containing the A071 and late embryogenic abundant gene sequences were selected to be used as RFLP probes and were placed on the soybean molecular genetic map (Figures 1 and 2). These subclones were selected from regions that did not cross-hybridize between lambda clones. These results indicated that subclones showing no cross-hybridization between lambda clones, still sometimes mapped together on the RFLP map (Polzin et al., in review). These results indicated that subsequent divergence of genomic regions following duplication of the genome had oftentimes not disrupted the linkages between loci and flanking sequences, but had altered the distance between them. The implications for utilizing these conserved linkages include chromosome landing and regional saturation.

The finding that there have been changes in the physical distance between the loci and their flanking sequences is consistent with either intra-chromosomal rearrangements or differential accumulation of repetitive sequence. Potential examples of both processes can be identified for the *sle* loci. For example, the accumulation of repetitive sequence upstream of *sle1* (data not shown) may have increased the distance between this locus and the cosegregating duplicate copy of the *sle4-5* locus. In contrast, it is unlikely that the difference in distance between *sle4-13* and *sle1* (24 cM) (Figure 2) is due simply to accumulation of repetitive sequence. Instead, it is likely that intra-chromosomal rearrangements have increased the distance between these two loci. Also, no repetitive sequence was present between the *sle4* locus and the region of the *sle4* clone which would contain the *sle1-2* sequence if the *sle4* and the *sle1-2* sequences had remained the same distance apart. Thus, the accumulation of repetitive sequence cannot account for the change in distance observed between loci in all cases. Therefore, intra-chromosomal rearrangements have almost certainly played a significant role in the divergence of duplicated regions of the soybean genome.

Our results suggest that rather than containing long tracts of conserved sequence, the duplicated regions of the soybean genome

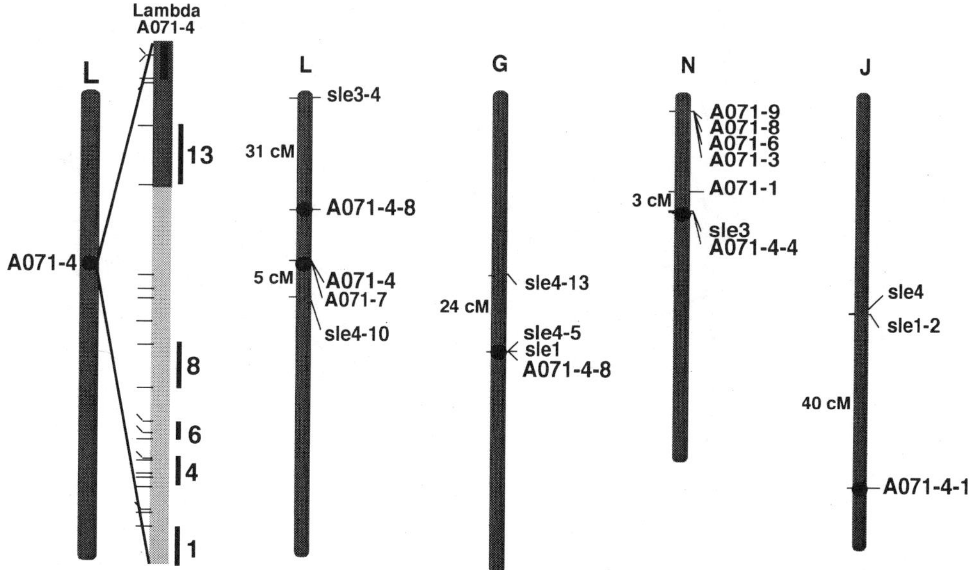

Figure 1. Small low-copy subclones of lambda clones containing the A071-4 locus (from Linkage group L) and their relative map position on the soybean genetic map. Subclones are taken from regions of the lambda clones shown not to cross-hybridize to other lambda clones containing A071-related sequences.

consist of islands of conserved sequence scattered throughout a "sea" of non-homologous DNA. Most, or all, of these "islands" likely correspond to genes and a few kilobases of immediately flanking sequence. Thus, at the molecular level, conservation of the duplicated regions of the soybean genome appears to be limited to the conservation of coding regions   Three mechanisms could explain both the non-homologous flanking regions and the differences in repetitive sequence distribution found between the duplicated loci: 1) differential accumulation of repetitive sequences; and 2) intra-chromosomal rearrangements, deletions or insertions or 3) inter-chromosomal recombinations (translocations), both of which would juxtapose new sequences (including repetitive sequences) near one

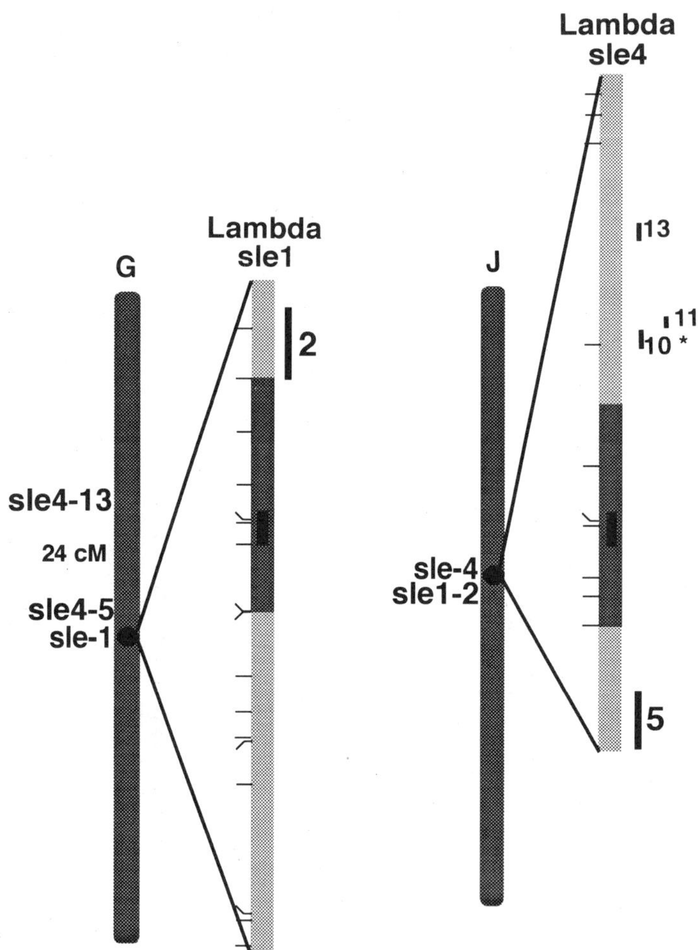

Figure 2. Small low-copy subclones of lambda clones containing the Sle1 and Sle4 coding sequences (from Linkage Groups G and J) and their relative map position on the soybean genetic map. Subclones are taken from regions of the lambda clones shown not to cross-hybridize to other lambda clones containing Sle-related sequences.

member of a duplicated locus. The linkage conservation described above suggests that inter-chromosomal recombinations have played little or no recent role in the divergence of these regions because such events would have destroyed genetic linkage.

PARALOGY IN THE SOYBEAN GENOME

One byproduct of RFLP map construction has been the ability to estimate the degree of duplication within plant genomes. These estimates can vary considerably. Estimates of locus duplication based upon the number of fragments to which a RFLP probe hybridizes, have been estimated at 26% in rice (McCouch et al., 1988) to 80% in Brassica (Song et al., 1991). In our laboratory, similar associations of RFLP fragments to mapped loci, and band counting experiments using multiple enzymes have suggested that more than 90% of non-repetitive soybean sequences may be present in two or more copies. This is a significantly higher level of duplication of sequences than has been reported for other supposed ancient tetraploids such as maize (28.6%) (Helentjaris et al., 1988; Ahn and Tanksley, 1993). Out of 549 loci on the soybean molecular genetic map, 338 match to some type of mapped duplicated loci. The observation that approximately 60% of RFLP probes detects three or more fragments, and probably therefore, three or more loci in soybean, suggests that this higher level of duplication is not simply due to greater conservation of duplicated loci from ancestral genomes, but that much of the genome has undergone genome duplications in addition to the original tetraploidization event. Integration of genetic maps from soybean has allowed the detection of many more duplicate loci than was possible from a single population. Nearly 650 loci and twenty-five linkage groups are resolved after joining several independent maps. These linkage groups contain from 3 to 55 loci and span from 8.7 to 141.7 cM. Each linkage group contains loci (from 1 to 33) that are duplicated on other linkage groups based upon detection of loci using standard hybridization conditions and RFLP mapping. These duplicated loci reside on from 1 to 17 (average 9.2) separate groups (Table 1). The number of putative homoeologous segments detected within the genome, based on the co-occurrence of three or more duplicated loci on separate groups, range from two to ten, with an average of approximately 4.2. These data suggest that either an additional round of genome tetraploidization has occurred with the soybean, or more likely, the genome has undergone many events of segmental or regional duplications previous to or following the original polyploidization.

Because most soybean probes detect multiple fragments it is important to discern the locus identity of each of these in order to develop a clear picture of the arrangement of duplicated segments within the picture of the entire genome. The incidence of polymorphism between any two soybean genotypes is relatively low

Table 1. Examples of soybean linkage groups and coincidence of duplicate loci. Putative homoeologous linkage groups are based on the co-occurrence of three of more loci with the tested linkage groups.

Tested Linkage Group	Total cM	Total Loci	No. Loci Detecting Dup. Loci	No. Groups Detected	Putative Homoeologies To Tested LG
a1	90.0	22	15	8	1
a2	106.9	42	21	13	3
b1	98.6	37	23	10	1
d1	131.7	39	18	16	8
g	141.1	55	33	17	5
n	96.5	32	22	13	9
s	14.5	4	1	1	unknown

so that the picture observed from any single map is usually quite fragmentary. However, different parental combinations can provide mapping information for different combinations of loci using the same set of probes. By using specific probe/enzyme combinations and known mapped polymorphism to detect anchoring loci it is possible to create independent maps and join them using available software (Stam, 1993). This allows us to generate a combined map containing many more duplicate loci. The picture observed from these combined maps is oftentimes quite informative (Figure 3).

Many instances are observed where several markers on a linkage group define putative homoeologous segments on other linkage groups. One example is shown in Figure 3. In some instances these paralogous linkages are collinear (linkage groups `e' and `k'), and in others, rearrangements such as inversions (linkage groups `a2' and `e') and possible translocations are seen.

The incidence and frequency of paralogous regions within the soybean genome suggests an evolutionary history much more complex than would be expected simply from a single tetraploidization event. Many probes used in the construction of the molecular map detect several duplicate loci. In addition to the probes exampled in Figure 3 which detect loci on the soybean genetic map, and which define possible homoeologous regions, many more map as dispersed duplications. An example of these relationships among linkage groups is shown in Figure 4.

Two mechanisms associated with genome duplications can be readily observed through analysis of RFLP maps: total genome duplication (tetraploidization) and regional or segmental duplications. Both processes can be detected by the presence of duplicated loci. Lundin (1993) suggested that regional duplication is the probable mechanism for the evolution of clustered gene families. Following

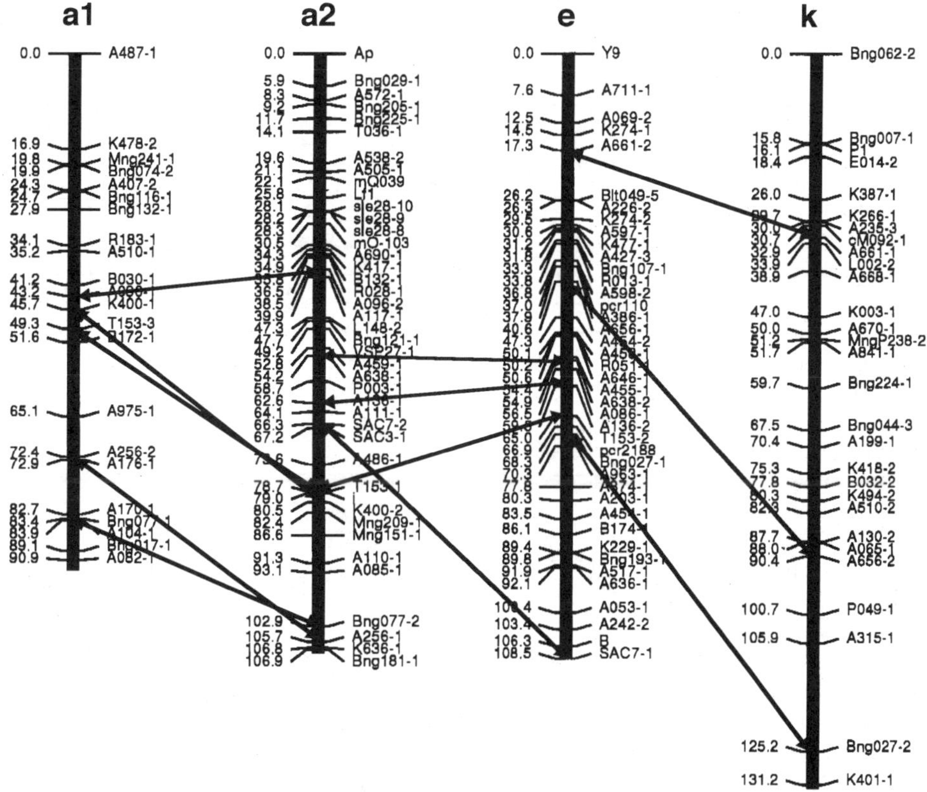

Figure 3. Soybean linkage groups. Lines drawn between linkage groups connect duplicate loci. Numbers to the left of each linkage groups indicate genetic distances in centimorgans and numbers to the right of each linkage group indicate loci designations.

regional duplication the two loci (segments) are on the same chromosome. However, subsequent events can result in the two segments ending-up on different chromosomes. In contrast to segmental duplication, the polyploidization event results in duplication of the entire genome and results in the presence of many duplicate loci whose members are almost always present on different chromosomes (Lundin, 1993). Detailed analysis of RFLP map structures can provide much information concerning these duplication events in a plant's history.

Early RFLP mapping analysis of the soybean genome provided the detection of collinear duplicated loci suggesting that duplicated Q and paralogous regions of the genome had been detected (Keim et al., 1992). However, only one region of extensive paralogoy was detected and was detected by only a few duplicate loci. In contrast, many

Linkage Group I

Marker							
BLT010-2							f
I8-2			g				
A450-2	j						
B046-1	j						
E014-1							k
B046-1	j						
L050-2				b1	c2	l	
B162-2		n					
A071-6		n					
A071-3		n					
A071-5(i)	j	n	g				
A071-4		n					
A071-1		n					
L204-1				b1			
28-10-2							a2
A132-2				b1			h
L050-3				b1	c2	l	
Bng095-1		n					
A533-1		n					
A363-1	j						
A802-2		n					

Linkage Group G

Marker								
A655-2		c2						
K069-1	b1							
Bng225-2								a2
A121-2		c2						
A426-3		c2	n					
MP238-1							k	
A427-2				b2				e
B151-1					g			
I8-3								l
Sle-1			n			d1		
40-5-1								j
B151-2					g			
K011-4		c2				d1		i
T005-2				b2				
A245-2								f
L002-1							k	
A235-1						d1		
Bng069-1								p
A378-1	b1							
A586-2								un

Figure 4. Soybean genetic markers from linkage groups G and L. The markers are listed in the order in which they appear on the linkage group. The letter designations within columns indicate other linkage groups to which the markers also hybridize and map.

linkage groups contained duplicate loci from many other linkage groups. These results suggested that the soybean genome may have undergone intense rearrangement and reorganization following tetraploidization and may not retain significant collinear paralogous regions as seen in other tetraploids. However, as has been observed this appearance of scrambling' of the soybean genome, although true to some extent, likely only reflects our inability to map adequate numbers of duplicate loci to resolve the regions of paralogy.

REFERENCES

Ahn, S., Anderson, J.A., Sorrells M.E., and Traksley, S.D., 1993, Homoeologous relationships of rice, wheat and maize chromosomes, *Mol Gen Genet.* 241:482.

Ahn, S., and Tanksley, S.D., 1993, Comparative linkage maps of the rice and maize genomes, *Proc Natl Acad Sci. USA* 90:7980.

Bonierbale,M.W., Plaisted, R.L., and Tanksley, S.D., 1988, RFLP maps based on a common set of clones reveal modes of chromosomal evolution in potato and tomato, *Genetics* 120:1095.

Boutin, S., Young, N., Olson, T., Yu, Z-H., Shoemaker, R., and Vallejos, E., 1995, Genome conservation among three legume genera detected with DNA markers, *Genome* (in press).

Calvo, E.S., Wurtele, E., and Shoemaker, R., 1995, Cloning, mapping, and analyses of expression of the Em-like gene family in soybean (*Glycine max* (L.) Merr.), *Plant Mol Biol.* (in review).

Grandbastien, M.A., Berry-Lowe, S., Shirley, B.W., and Meagher, R., 1986, Two soybean ribulose-1,5-bisphosphate carboxylase small subunit genes share extensive homology even in distant flanking sequences, *Plant Mol Biol.* 7:451.

Hadley, H.H., and Hymowitz, T., 1973, Speciation and cytogenetics, *in*: "Soybeans: Improvement, Production, and Uses," B.E. Caldwell, ed., American Society of Agronomy, Madison, Wisconsin.

Helentjaris, T., Weber, D., and Wright, S., 1988, Identification of the genomic locations of duplicate nucleotide sequences in maize by analysis of restriction fragment length polymorphisms, *Genetics* 118:353.

Hightower, R.C., and Meagher, R.B., 1985, Divergence and differential expression of soybean actin genes, *EMBO J.* 4:1.

Hulbert, S.H., Richter, T., Axtell, J., and Bennetzen, J., 1990, Genetic mapping and characterization of sorghum and related crops by means of maize probes, *Proc Natl Acad Sci. USA* 87:1990.

Kianian, S.F., and Quiros, C., 1992, Generation of a *Brassica oleracea* composite RFLP map: linkage arrangements among various populations and evolutionary implications, *Theor Appl. Genet.* 84:544.

Lackey, J.A., 1980, Chromosome numbers in the Phaseoleae (Fabaceae:Faboideae) and their relation to taxonomy, *Amer J Bot.* 67:595.

Lee, J.S., and Verma, D.S., 1984, Structure and chromosomal arrangement of leghemoglobin genes in kidney bean suggest divergence in soybean leghemoglobin gene loci following tetraploidization, *EMBO J.* 12:2745.

Leipold, M., and Schmidtke, J., 1982, Gene expression in phylogenetically polyploid organisms, *in*: "Genome Evolution," G. Dover, and R. Flavell, eds., Academic Press, New York

Lundin, L.G., 1993, Evolution of the vertebrate genome as reflected in paralogous chromosomal regions in man and the house mouse, *Genomics* 16:1.

McCouch, S.R., Kochert, G., Yu, Z-H., Wang, Z.Y., Khush, G., Coffman, W.R., and Tanksley, S., 1988, Molecular mapping ofrice chromosomes, *Theor Appl Genet.* 76:815.

Menancio-Hautea, D., Fatokun, C., Kumar, L., Danesh, D., and Young, N., 1993, Comparative genome analysis of mungbean (*Vigna radiata* L. Wilczek) and cowpea (*V. unguiculata* L. Walpers) using RFLP mapping data, *Theor Appl Genet.* 86:797.

Nielsen, N., Dickinson, C., Cho, T.-J., Thanh, V.H., Scallon, B.J., Fischer, R.L., Sims, T.L., Drews, G.N., and Goldberg, R.B., 1989, Characterization of the glycinin gene family in soybean, *Plant Cell* 1:313.

Ohno, S., 1970, "Evolution by Gene Duplication", Springer Verlag, New York.

Palmer, R.G., and Kilen, T.C., 1987, Qualitative genetics and cytogenetics, *in*: "Soybeans: Improvement, Production, and Uses," J.R. Wilcox, ed., American Society of Agronomy, Inc., Madison, Wisconsin.

Pereira, M.G., Lee, M., Bramel-Cox, P., Woodman, W., Doebley, J., and Whitkus, R., 1993, Construction of an RFLP map in sorghum and comparative mapping in maize, *Genome* 37:236.

Polhill, R.M., and Raven, P.H., 1981, "Advances in Legume Systematics", Royal Botanic Gardens, Kew.

Prince, J.P., Pochard, E., and Tanksley, S.D., 1992, Construction of a molecular linkage map of pepper and a comparison of synteny with tomato, *Genome* 36:404.

Shoemaker, R.C., Polzin, K., Lorenzen, L.L., and Specht, J., 1995, Molecular genetic mapping of soybean, *in*: "Soybean: Genetics, Molecular Biology and Biotechnology," D.P.S. Verma and R. Shoemaker, eds., C.A.B. International, Wallingford, UK.

Simon, C.J., Tahir, M., and Muehlbauer, F.J., 1993, Linkage map of lentil (Lens culinaris) 2N=14. *in*: "Genetic Maps of Complex Genomes" S. O'Brien, ed., Cold Spring Harbor, New York.

Slocum, M.K., Figdore, S., Kennard, W., Suzuki, J., and Osborn, T., 1990, Linkage arrangement of restriction fragment length polymorphism loci in *Brassica oleracea, Theor Appl Genet.* 80:57.

Song, K.M., Suzuki, J.Y., Slocum, M., Williams, P., and Osborn, T.C., 1991, A linkage map of *Brassica rapa* (syn. campestries) based on restriction fragment length polymorphism loci, *Theor Appl Genet.* 82:296.

Stam, P., 1993, Construction of integrated genetic linkage maps by means of a new computer package: Join Map, *The Plant J.* 3:739.

Stebbins, G.L., 1966, Chromosomal variation and evolution, *Science* 152:1463.

Tanksley, S.D., Lapitan, N.L., and Prince, J.P., 1988, Conservation of gene repertoire but not gene order in pepper and tomato. *Proc Natl Acad Sci. USA* 85:6419.

Tanksley, S.D., Ganal, M., Prince, J., de Vicente, M., Bonierbale, M., Broun, P., Fulton, T., Giovannoni, J.J., Grandillo, S., Martin, G.B., Messeguer, R., Wu, W., and Young, N., 1992, High density molecular linkage maps of the tomato and potato genomes, *Genetics* 132:1141.

Weeden, N., Muehlbauer, F., and Ladizinsky, G., 1993, Estensive conservation of linkage relationships between pea and lentil genetic maps, *J Hered.* 83:123.

Whitkus, R., Doebley, J., and Lee, M. 1992, Comparative genome mapping of sorghum and maize. *Genetics* 132:1119.

COMPARATIVE CEREAL GENOME ANALYSIS-RECONSTRUCTING THE ANCESTRAL CEREAL GENOME

Graham Moore

Cereals Research Department, John Innes Centre, Colney Lane, Norwich NR4 7UH, United Kingdom

INTRODUCTION

More than 60 million years ago speciation occurred giving rise to the cereals which eventually became the major crops grow today (Wolfe et al., 1989; Martin et al., 1989). In the last 50 years, many of these cereals have been studied largely in isolation, with little exchange of information between the breeders or scientists working on them. With the advent of plant Biotechnology, there is unlikely to be sufficient resources to study individual crops like maize, sorghum, rice, barley, rye, wheat, millet, sugarcane in any great detail, as compared to the resources devoted to the programmes on yeast, human and Arabidopsis. Yet in the next 50 years we will require as much cereal grain as has been consumed since the beginning of agriculture. Clearly it is going to be important to use the information generated from the study of these cereals more efficiently. Comparative genome analysis can provide just such a framework for collation of all this information.

Since their speciation, cereals such as wheat, barley, maize and rice possess different chromosome numbers and genome sizes. The rice genome is 4.3 x108 bp, while the three genomes of hexaploid bread wheat total 1.7 x1010 bp in size. From genome studies, it is now possible to consider several levels of genome organisation. The lowest organisation is the kilobase and below level includes genes and repetitive sequences. The sub-megabase level comprises of repetitive sequences organised into units which have undergone amplification. The multi-megabase level comprises the non-random organisation of repetitive sequences in satellites or compartments along a

Genomes of Plants and Animals: 21st Stadler Genetics Symposium
Edited by J. Perry Gustafson and R. B. Flavell, Plenum Press, New York, 1996

chromosome. The sub-chromosomal level comprises the linkage blocks containing genes whose linkage is conserved throughout many cereal genomes. Then finally the chromosomes whose number, gene and recombination distribution varies between species.

This review is to be structured in a similar fashion. Comparative cereal genome analysis should provide sets of markers for selection of traits in many crops, methods for gene isolation, collation of data, elements effecting gene expression and distribution of recombination.

KILOBASE LEVEL

Genes

Analysis of gene sequences from dicot and monocots indicates that the average G+C content of monocot genes is higher than that of dicot genes with the majority having a G+C content of 60-70% while dicot genes have an average around 46%. The reason for this difference rests with the use of different synonymous codons. In fact cereals and mammals share the most commonly used codon for 13 of the 18 amino acids while dicots share only 7 of 18 amino acids. It is therefore not trival to detect monocot genes by Southern blot analysis using dicot genes as probes. The Japanese Rice Genome Programme has sequenced a large number of rice cDNAs making it possible to compare the protein homologues of these genes with those derived from the *Arabidopsis* cDNA sequencing programme, thus allowing for cross-mapping of cereals and dicots. From cross-mapping studies between the cereals, it has been noted that the majority of rice cDNAs tested revealed homoeologous sequences on Southern blots of wheat, barley, millet, and rye DNA suggesting strong sequence conservation.

In the case of animal genomes, landmarks have been identified which allow a gene to be identified with relative ease within a long stretch of DNA (Bird, 1987). These large genomes have a CpG content which is lower than the 4% expected from its 40% G+C rich genome. However regions (CpG islands) localised at the 5'end (promoter) of genes have been shown to have the expected CpG content. The CpG dinucleotides in these islands are unmethylated and provide recognition sites for restriction enzymes such as NotI. 75% of NotI sites are located within islands and more than half of CpG islands are coincident with genes. In contrast to these genomes, wheat genomes exhibit no CpG shortage (Swartz et al., 1962; Russell et al., 1971). In plants, 5-methylcytosine is not just confined to CpG dinucleotides, but is also present in over 80% of the trinucleotides CpXpG (Gruenbaum et al., 1981). Despite the lack of CpG suppression, it was still observed that recognition sites for restriction enzymes such as NotI provided landmarks for genes. The genes appeared to be located in undermethylated islands in wheat compared to the bulk of the repetitive sequences. Since these islands are rich in highly methylated CpG dinucleotides, they also provided sites for the restriction enzymes such as Not1, which cluster around genes. For this reason, the analysis of PstI or NotI restriction fragments of less than 0.5Kb derived from the wheat genome revealed that they are not only single copy sequences, but also when as probes for Southern blot analysis detect homoeologous sequences in rye, barley, maize and rice. This suggests that these fragments may well contain genes (Moore et al., 1993b).

Repeats

A number of repetitive sequences have been defined in the cereal genomes including several classes of retroelements, for example WIS-2 in wheat which are also present in rye, barley. and oats (Moore et al., 1991a). In fact BARE-1 recently isolated from the barley genome is a member of the WIS-2 family having a substantial level of homology at the sequence level (Manninen and Schulman, 1993). The hybridisation of BARE-1 and WIS-2 to barley DNA yields two markedly different figures for copy number of WIS-2 type elements in this genome. This probably reflects the divergence of these two copies with respect to the most common WIS-2 element in the genome. BIS-1 is another retroelement which was isolated from barley (Moore et al., 1991b). Analysis indicates that this element could represent more than 6% of the barley genome. When used as probes on Southern Blots, this element (and sets of primers) detect the similar size fragments in a range of Gramineae, from maize, millet, sorghum to *Lolium.* This analysis suggests that this family of elements may be an important component of these genomes. Studies will soon indicate whether in fact this family represents a major component of all these genomes. Sequence comparisons of a rye "specific" retroelement, 174, (Rogonsky et al., 1992) with BIS-1 not only indicates that 174 has regions of strong sequence homology to BIS-1 but that it has accumulated additional sequences with respect to BIS-1 (Alcaide and Moore, unpublished). It may well be that many cereal repeats apparently specific to particular species when assayed by Southern blot analysis, reveal homology when compared at the sequence level. Such elements may represent the "distant" members of the family being significantly more diverged than the bulk of the members and possessing additional sequences. It may be important to define these additional sequences which may well turn out to be genic in origin.

Other types of repetitive sequences have been found to be conserved across the Gramineae, for example the tourist elements, stem-loop structures, discovered in the maize genome (Bureau and Wessler, 1992). Also a Hi-10 element has been found in the genomes of rice, sorghum, maize, barley, wheat and *Lolium* (Abbo et al., 1995). This "element or gene" is also transcribed in rice (Alcaide and Moore, unpublished). Is this an example of a different type of "repeat", a gene which has become amplified during speciation of the cereals? Further studies should reveal the answer.

In contrast to the barley and wheat genomes, the small rice genome is undermethylated (Deshpande and Ranjekar, 1980). Analysis of sequences in short restriction fragments generated by either cleavage with PstI and NotI is not so clear cut as those derived from the wheat and barley genomes. A higher percentage of fragments contain repetitive sequences and far fewer cross-hybridise to DNA derived from other cereals revealing single copy hybridisation patterns. In the model of the evolution of methylated vertebrate genomes from undermethylated invertebrate genomes (Bird, 1987), rice would represent an "invertebrate" genome and the genomes of wheat and barley intermediate genomes between the two types. The CpG shortage in the bulk of the animal genomes is the result of methylated CpG undergoing a transition to TpG. The lack of CpG suppression in the bulk of the highly methylated large cereal genomes could suggest two possibilities, one that the majority of the additional bulk has arisen relatively recently or that there is not the same level or rate of CpG to TpG transitions in cereals. If the bulk of the genome has arisen relatively recently, comparative analysis might reveal how this occurred, the nature of the sequences involved and therefore provide some explanation for why it occurred.

SUB-MEGABASE LEVEL

Examination of cereal genomes at the megabase level also reveals non-random organisation (Moore et al., 1993b; Abbo et al., 1995). Although the most repetitive sequences in wheat and barley genomes are highly methylated, studies have indicated the presence of unmethylated recognition sites for the restriction enzymes MluI and NruI at non-random intervals. Cleavage of the genomes of rice, barley, rye and wheat with these enzymes yields discrete fragment size classes, or FSCs, ranging from hundreds of kilobases to several megabases in length.

The generation of FSCs suggests that unmethylated recognition sites for MluI are non-randomly distributed within these genomes. In addition, the elements BIS-1, WIS-2 and other dispersed repeats despite having different copy numbers within Triticeae genomes are distributed similarly between MluI FSCs. There are a number of possible explanations for this observation. One explanation is that the elements such as BIS 1 and WIS 2 are organised into units which then have undergone amplification.

MULTI-MEGABASE LEVEL

As indicated previously a cereal repeat sequence with homology to a rice cDNA has been reported (Hi-10) which is present in the genomes of many Gramineae studied (Abbo et al., 1995; Alcaide and Moore, unpublished). *In situ* hybridisation of Hi-10 to metaphase chromosomes of barley and rye reveals the compartmentalisation of this repeat to specific regions. The existence of such compartments within chromosomes would be explained by the divergence of repetitive sequences, their subsequent amplification, followed by further modification via mechanisms of moecular drive. Sequences homologous to Hi-10 may have diverged in particular chromosome regions so that they are no longer detectable by the specific copy of Hi-10 used for *in situ* studies. For example, although the retrotransposon (BIS 1), when used as a probe for *in situ* hybridisation studies. detects sequences dispersed along all the chromosome arms. Southern blot analysis reveals specific variants to be located to particular chromosomes.

CONSERVED LINKAGE SEGMENTS

A genetic (Restriction Fragment Length Polymorphism, RFLP) map of wheat has been constructed (Devos and Gale, 1992). This has a number of features. Firstly, markers are revealed which when used as probes detect sequences lying in the same genetic position on all three homoeologous chromosomes. This indicates that the genes are essentially colinear on all three homoeologous chromosomes within a chromosome group. Secondly, there is clustering on the genetic map of markers derived from the proximal/centromere regions.

The observed colinearity of markers on homoeologous chromosomes raises the issue of whether markers on barley and rye chromosomes are also colinear with those on wheat chromosomes. Although there are several large evolutionary translocations, it is now apparent that markers on homoeologous chromosomes of barley and rye are largely colinear with wheat (Devos et al. 1993a; Devos et al. 1993b). Rice and wheat evolved from a common ancestor some 60 million years ago (Martin et al. 1989; Wolfe et al. 1989). Rice now has 12 chromosomes and wheat three sets of 7 chromosomes. A wheat chromosome is some 25 fold larger than a rice chromosome. The question arises as to the relationship between the gene order on the chromosomes of these species which differ in size so dramatically. Reciprocal mapping of probes from the rice RFLP map (Kurata et al. 1994b) onto the wheat genetic map and vice versa indicated that strong colinearity has been retained by markers on wheat chromosome (group) 1 with those on rice chromosome 5, wheat chromosome 2 with rice chromosomes 4 and 7, wheat chromosome 3 with rice chromosome 1, wheat chromosome 4 with rice chromosome 3, wheat chromosome 5 with rice chromosome 9, wheat chromosome 6 with rice chromosome 2, and wheat chromosome 7 with rice chromosome 6 (Kurata et al. 1994a). Further studies now indicate that markers on rice chromosome 8 are colinear with the centromere region of wheat chromosome 7, rice chromosome 10 markers are syntenous with the centromere region of wheat chromosome 1, rice chromosome 11 markers are colinear with the centromere region of wheat group 5, and rice chromosome 12 markers are colinear with the short arm of wheat group 5 and with markers on the long arm of wheat chromosome 5. Nearly 400 markers from the rice and wheat genetic maps have analysed to date in this comparative study (Foote and Moore unpublished).

Comparison of the genetic maps of rice and wheat revealed that the maps of rice chromosomes could be dissected into a series of linkage segments (like lego blocks). These rice linkage segments could be used to construct wheat chromosomes. The availability of the data comparing markers/genes in the rice and maize genomes (and on duplicated loci in the maize genome) allowed the genetic maps of maize chromosomes to be also reconstituted from the same set of rice linkage segments (Moore et al., 1995a). This revealed that the maize genome was tetraploid and it indicated that the smaller 5 chromosomes may be derived from one parent and the larger from another parent. A comparison of the order of linkage segments in maize, wheat and rice indicated that a single stack of "lego" can construct an "ancestral" chromosome. This ancestral chromosome could be then cleaved in different places to give rise to all the wheat chromosomes, at other positions to give rise to one set of 5 maize chromosomes and different positions again for the other set of 5 maize chromosomes. This concept has been extended to include the chromosomes of sorghum, millet and sugarcane, thereby encompassing all the major cereal crops (Moore et al., 1995b). The cleavage of the same ancestral chromosome gives rise to the 56 chromosomes present in these species.

The implications of this observation are far reaching. All the information which has been generated by studying these species separately in the last 50 years can be collated and exploited with greater efficiency. By combining all this data with crops grown in different parts of the world, it will be possible to trace the threads of allelic variation during speciation to comprehend plant development within and between species. It will be possible to align the genes controlling stress, disease resistance and yield in these different genomes. Such a comparison makes rice, with its small genome, the model system for cereals.

Using markers already mapped on rice YACs, it has been shown that these are maintained in the same order on rice chromosomes and barley/wheat chromosomes (Dunford et al., 1995). Thus, gene-order is conserved at the micro-level. This implies that genes can be identified by map-based cloning of the homoeologue in rice, and then by homology back in the target species. The reconstruction of the rice genome in the form of ordered artificial chromosome (YACs) will reveal the general structure of most cereal genomes. This may provide clues as to how and why particular regions of some cereal genomes have undergone extensive expansion by non-random accumulation of repetitive sequences following speciation.

CHROMOSOMES

Although in cereals the rice linkage segments contain the homoeologous genes, the actual DNA content of the segments will vary depending on the extent of amplification of the repeats between the genes. Clearly the sum of the DNA in all the different combinations of linkage segments accounts for differences in DNA content of chromosomes.

The intense clustering of markers in the proximal/centromere regions on the genetic maps of wheat chromosomes begs the question of what is the relationship between the genetic and physical location of the genes/markers on the chromosome. Studies on the distribution of sites of recombination along a cereal chromosome using physical markers all show that recombination is predominantly confined to the distal regions (Linde-Laursen, 1982; Snape et al. 1985; Dvorak and Chen, 1984; Lawrence and Appels, 1986). As a result of reduced recombination towards the centromere, markers physically located in these regions are clustered compared to those located in the distal chromosome regions. A single rice chromosome contains thousands of genes. The colinearity of markers on several rice chromosomes (for instance 8,10,11) with the proximal/centromeric regions of various wheat chromosomes suggests that the proximal/centromeric regions of wheat which exhibit reduced recombination, also contain thousands of genes. This raises the important question as to what proportion of genes are being reassorted in wheat breeding programs.

REFERENCES

Abbo, S., Dunford, R.P. Foote, T.N., Reader, S.M., Flavell, R.B., and Moore, G., 1995, Organisation of retro-element and stem-loop repeat families in genomes and nuclei of cereals, *Chromosome Res.* 3: 5.

Ahn, S., and Tanksley,S.D., 1993, Comparative linkage maps of the rice and maize genomes, *Proc. Natl. Acad. Sci.* USA. 90:7980.

Bennett, M.D., and Smith, J.B., 1991, Nuclear DNA amount in angiosperms, *Phil. Trans. R. Soc. London* (Biol) 334: 309.

Bird, A.P., 1987, CpG islands as markers in the vertebrate nucleus, *Trends Genet.* 3: 342.

Bureau T.E. and Wessler S.R., 1992, Tourist, a large range of small-inverted repeat elements frequently associated with maize genes, *Plant Cell* 4:1283.

Deshpande. V.G., and Ranjekar, P.K., 1980, Repetitive DNA in three Gramineae species with low DNA content, *Hoppe-Seylers's Z. Physiol. Chem.* 361:1223.

Devos, K., and Gale, M.D., 1992, The genetic maps of wheat and their potential in plant breeding, *Outlook in Agriculture* 22: 93.

Devos, K., Millan, T., and Gale, M.D., 1993a, Comparative RFLP maps of homoeologous group 2 chromosomes of wheat, rye and barley, *Theor. Appl. Genet.* 85: 784.

Devos, K.M., Atkinson, M.D., Chinoy, C.N., Francis, H.A., Harcourt, R.L., Koebner, R.M.D., Liu, C.J., Masojc, P., Xie, D.X., and Gale, M.D., 1993b,

Chromosomal rearrangments in rye genome relative to that of wheat, *Theor. Appl. Genet. 85: 673.*

Devos, K.M., Chao, S., Li, Q.Y., Simonetti, C. and Gale, M.D. 1994, Collinearity of wheat and maize chromosomes, *Genetics* 138:1287.

Dvorak, J., and Chen, K-C., 1984, Distribution of nonstructural variation between wheat cultivars along chromosome arm 6Bp: evidence from the linkage map and physical map of the arm, *Genetics* 106:325.

Dunford, R., Kurata, N., Laurie, D, Money, T, Minobe, T., and Moore,G., 1995, Conservation of fine-scale marker order in the genomes of rice and the Triticeae, *Nucleic Acids Res.* 23:2724.

Gruenbaum, Y., Naveh-Many, T., Cedar, H., and Razin, A., 1981, Sequence specificity of methylation in higher plant DNA, *Nature* 292:860.

Kurata, N., Moore, G., Nagamura, Y., Foote, T., Yano, M., Minobe, Y., and Gale, M., 1994a, Conservation of genome structure between rice and wheat, *Bio/technology* 12: 276.

Lawrence, G.J., and Appels, R., 1986, Mapping the nucleolus organiser, seed protein loci and isozyme loci on chromosome 1R in rye, *Theor. Appl. Genet.* 71:742.

Linde-Laursen, I., 1982, Linkage map of the long arm of barley chromosome 3 using C-bands and marker genes, *Heredity* 49:27.

Martin, W., Gierl, A., and Saedler, H., 1989, Molecular evidence for pre-Cretaceous angiosperm origins, *Nature* 339:46.

Manninen, I., and Schulman, A.H., 1993, BARE-1, a copia-like retroelement in barley (Hordeum vulgare L.), *Plant Mol. Biol.* 22: 829.

Moore, G., Lucas, H., Batty, N., and Flavell, R.B., 1991a, A family of retrotransposons and associated genomic variation in wheat, *Genomics* 10: 461.

Moore, G., Cheung, W., Schwarzacher, T., and Flavell. R., 1991b, BIS 1,a major component of cereal genome and a tool for studying genomic organisation, *Genomics* 10:469.

Moore G., Gale., M.D., Kurata, N., and Flavell, R.B., 1993a, Molecular analysis of small grain cereal genomes, *Current Status and Prospects, Bio/technology* 11: 584.

Moore, G., Abbo, S., Cheung, W., Foote, T., Gale, M., Koebner, R., Leitch, A., Leitch, I., Money, T., Stancombe, P., Yano, M., and Flavell, R., 1993b, Key features of cereal genome organisation as revealed by the use of cytosine methylation-sensitive restriction endonucleases, *Genomics* 15: 472.

Moore, G., Foote, T., Helentjaris, T., Devos, K., and M Gale, 1995, Was there a single ancestral cereal chromosome, *Trends Genet.* 11: 81.

Moore, G., Devos, K., Wang, Z., and Gale, M.D., 1995b, Cereal genome evolution: Grasses, line up and form a circle, *Current Biol.* 5: 737.

Russell, G.J., Follett, E.A.C., Subak-Sharpe, J.H., and Harrison, B.D., 1971, The double stranded DNA of Cauliflower mosaic virus, *J. Gen. Virol.* 11:129.

Snape, J.W., Flavell, R.B., O'Dell,M., Hughes, W.G., and Payne, P.I., 1985, Intrachromosomal mapping of the nucleolar organiser region relative to three marker loci on chromosome 1B of wheat (Triticum aestivum), *Theor. Appl. Genet.* 69: 263.

Swartz, M.N., Trautner,.T.A., and Kornberg, A., 1962, Enzymatic synthesis of Deoxyribonucleic acid, *J. Biol. Chem.* 237: 1961.

Wolfe, K.H., Gouy, M., Yang, Y-W, Sharp,P.M., and Li, W.H., 1989, Date of the monocot-dicot divergence estimated from the chloroplast DNA sequence data, *Proc. Natl. Acad. Sci. USA.* 86:6201.

A RESPONSIVE REGULATORY SYSTEM IS REVEALED BY SENSE SUPPRESSION OF PIGMENT GENES IN *PETUNIA* FLOWERS

Richard A. Jorgensen and Carolyn A. Napoli

Department of Environmental Horticulture
University of California, Davis
Davis, California 95616-8587
USA

INTRODUCTION

Plant transgenes can suppress the expression of homologous genes in at least two ways. One is a phenomenon known as *epigene conversion*, whereby one copy of a gene causes the transcriptional silencing of another copy, and the altered expression state of the silenced gene is heritable after loss of the gene causing the silencing (as in paramutation). The other mode of homology-dependent gene silencing is best referred to as *sense suppression*, a post-transcriptional phenomenon usually resulting in the cosuppression of homologous transgenes and endogenes. The homology required for sense suppression can be limited to the transcribed regions of the genes. Sense suppression is observed frequently in transgenic plants, generally as a paradoxical outcome of attempts to increase expression of an endogenous gene by the introduction of a chimeric transgene comprised of a strong promoter fused to a copy of the endogene's coding sequence in the sense orientation. The varied observations and possible mechanisms for homology-dependent transgene silencing phenomena in plants have been much discussed in the literature, citations to which may be found in Matzke and Matzke (1995), de Carvahlo Niebel et al. (1995), and Jorgensen (1995).

The present paper exploits sense suppression of a flower pigment gene as an unusually sensitive reporter phenotype for investigation of higher order mechanisms that control gene expression in plant physiology and development. Sense suppression is a nonlinear, inverse response that appears to be threshold-dependent and so is able to amplify small, quantitative changes in gene dosage or transcription into large, qualitative changes in phenotypic expression. In the case of a flower pigment gene such as chalcone synthase (*Chs*), the effect is a visible loss of flower pigments. However, rather than uniform reduction in pigmentation across the corolla, an altered spatial distribution of pigments produces a wide variety of striking patterns. These patterns range from simple to complex, ordered to disordered, and stable to metastable. The white, pigmentless sectors that define the patterns are organized primarily according to petal morphology rather than clonal cell division patterns. A broad boundary zone, in which the pigment intensity declines in a steep, continuous gradient, separates fully pigmented sectors from white sectors (Napoli et al., 1990; Jorgensen, 1995). To explain these patterns it has been proposed (Jorgensen, 1995) that transcription factors responsible for *Chs* gene expression vary in concentration at different positions in the petal in a manner controlled by fields of morphological information. Such spatial variation in transcription factor concentration could then produce a gradient of *Chs* transcription rate across the petal, reflecting its morphology. Wherever the gradient of *Chs* transcript concentration crosses the threshold for induction of sense-suppression, it will be amplified into a steep, inverse gradient of transcript turnover rate, which will be visible as a steep gradient of anthocyanin pigment. Gene dosage experiments suggest that a two-fold change in gene expression can trigger sense suppression, and so a relatively shallow gradient of transcription rate might suffice to produce the very steep pigmentation gradients that are observed.

Plant transgenes often adopt multiple transcription states that are heritable and that change in a variety of ways: somatically or germinally, reversibly or progressively, and quantitatively or qualitatively (Matzke and Matzke, 1995). Sense suppression phenotypes can exhibit similar metastable behavior, the cause of which is probably no different than that of the metastability that is exhibited by plant transgenes in general. In fact, it has been demonstrated that sense suppression phenotypes can be just as stable as phenotypes controlled by naturally occurring genes (Courtney-Gutterson, 1994). Successful use of sense suppression to silence target genes should require simply that sufficient care be taken to select transformants that exhibit stable phenotypes, just as is necessary for any transgene-based trait in plants. What is potentially different about sense suppression phenotypes in transgenotes that are susceptible to epigenetic variations in transgene transcription is that these variations can be magnified by the nonlinearity of the sense suppression response. Thus, the threshold-dependence of sense suppression would seem to require that a sense transgene be expressed at levels reliably above the threshold in order for it to uniformly and stably silence homologous genes. Variations above the threshold will not have phenotypic consequences, while variations

Figure 1. Phenotypic diversity of plants with *T–Chs41*. F reprinted from Jorgensen (1993b), with permission of the Royal Society. H reprinted with permission from Jorgensen (1995); copyright 1995 American Association for the Advancement of Science.

Figure 2. Somatic events occurring on plants 2006.24 and 2006.1.

across the threshold will have, as is true for many kinds of transgenic traits.

The data presented here do not address sense suppression as a tool for genetic improvement of crop plants. Rather, this paper exploits the sensitivity of the sense suppression threshold as a tool for investigation of heritable gene expression states of a dynamic, responsive regulatory network. The most useful transgenic plants for these studies are not the stably sense-suppressed "winners" that are picked by genetic engineers for the purpose of crop improvement, but the metastable "losers" that are normally discarded because of the variability in their gene expression patterns. For experimental biologists, however, such "undesirable" abnormalities provide a valuable genetic window through which to view the operations of gene regulatory systems that could otherwise be investigated only by biochemical means.

Anthocyanin pigmentation is well known to be a useful phenotype for gene expression studies, especially for investigations of biological patterns, because it is visible, cell-autonomous, and non-essential. Sense suppression can produce an especially sensitive reporter phenotype because, as a threshold response, it can amplify even small changes in gene expression into qualitative phenotypic alterations as long as the expression level is sufficiently near the threshold and the direction of change crosses the threshold. For studies of epigenetic events, a herbaceous perennial like petunia can be advantageous: (1) The growth habit of petunia plants is easily manipulated by pruning to release multiple points of lateral branching. (2) Petunias can be vegetatively propagated from shoot tip or stem segment cuttings. (3) Petunias can be maintained in a continuously flowering state. (4) Despite being a long-lived perennial, petunia can flower within two months of seed germination. In sum, sense suppression of anthocyanin gene expression in petunia offers an extremely powerful reporter system for investigating the control of gene expression.

Napoli et al. (1990) reported substantial variations in the patterns of *Chs* co-suppression both within and among transgenotes and between them and their sexual progeny. Further study has shown that somatic changes in the pattern of *Chs* sense-suppression may be both heritable and reversible, and that such changes may arise either progressively or suddenly (Jorgensen, 1994, 1995). Furthermore, it was found that the physiological environment can cause heritable changes in patterns of sense suppression and that these changes appear to occur in a highly regular, meristem-based manner. Here, we confirm and extend our earlier observations on the diversity, complexity, and hierarchy of genetic controls exhibited in this system. The possible relevance of the behaviors of this system to understanding "normal" regulatory networks in plant physiology and development is noted. This paper largely completes the description of our observations on a particularly interesting sense *Chs* transgenote known as CHS41 (Napoli et al., 1990; Jorgensen, 1994; Jorgensen,

1995). (The latter is recommended to be read as a companion to the present paper.)

METASTABLE STATES OF *Chs* SENSE SUPPRESSION

After producing only white flowers during the first four months of flowering, CHS41 produced an axillary branch with fully pigmented flowers identical in appearance to flowers of its nontransgenic parent, the inbred V26. This purple "revertant" branch retained the *Chs* transgene and, unexpectedly, was found to "coexpress" the transgene and the homologous endogene *Chs A*. This coexpression state was the necessary point of reference for defining the opposite state of "cosuppression" of *Chs* gene expression exhibited by white corollas (Napoli et al., 1990).

In addition to this change in corolla phenotype, a concomitant change in anther phenotype also occurred: flowers with white corollas had white anthers, whereas flowers with revertant purple corollas had yellow anthers (Taylor and Jorgensen, 1992). The yellow pigment of petunia anthers is a chalcone that is produced in anther cells derived from the L2 layer of the shoot meristem. In contrast, the purple anthocyanin pigment of the corolla is produced in epidermal cells which are derived from the L1 layer of the shoot meristem. In petunia, cell division rarely results in L1 cells becoming part of the L2 layer or vice versa. Thus, somatic genetic events that occur in L1 are almost always restricted to L1-derived cells and are rarely transmitted to sexual offspring because gametes arise only from L2-derived cells (Cornu, 1970; R.J. and C.N., unpublished observations), as is typical in solanaceous species (Marcotrigiano and Bernatzky, 1995). The fact that corolla and anther phenotypes changed concomitantly during the reversion event seemed to suggest that it was a strictly developmental event that would not be sexually transmissible.

Germinal Inheritance Of White And Purple States

Progeny testing of both white- and purple-flowered branches of CHS41 was carried out by pollinating CHS41 flowers with V26 pollen. The resulting backcross progeny were tested in two ways. First, seeds were germinated on medium containing kanamycin and scored for resistance. It was found (Table 1) that kanamycin resistance segregated as a single Mendelian locus (designated as *T-Chs41*). Second, additional BC$_1$ seeds were sown in soil and grown to flowering. Flower color phenotypes of these are shown in Table 2. Progeny derived from white flowers inherited a co-suppression phenotype. (The fact that not only white-flowered plants were observed, but also plants with patterned flowers, will be discussed below in detail.) After scoring cosuppression phenotypes in population #2006, each plant was tested for kanamycin resistance by a leaf explant callusing assay. All plants exhibiting co-suppression were found to be resistant, whereas all

Table 1. Segregation of kanamycin resistance in BC_1 progeny of CHS41[1].

	Number of progeny		
Female parent	Sensitive	Resistant	Total
CHS41W	49	51	100[2]
CHS41P	45	47	92[3]

[1] CHS41 was used as the female parent and pollinated with V26 to produce BC_1 seed.

[2] X^2=0.04, P=0.81 for one degree of freedom

[3] X^2=0.04, P=0.83 for one degree of freedom

Table 2. Segregation of cosuppression phenotypes in BC_1 progeny of CHS41[1].

		Number of progeny			
Female parent	Population	Purple	Patterned	White	Total
CHS41W	#2006	15	7	2	24
CHS41W	#2314	20	2	21	43
CHS41P	#2007	24	0	0	24
CHS41P	#2313	38	0	0	38

[1]CHS41 was used as female parent and pollinated with V26 to produce BC_1 seed.

plants producing fully pigmented flowers were found to be sensitive. All progeny of the purple somatic derivative of CHS41 were also purple (Table 2), indicating that the purple state is germinally transmissible. This result seems contrary to expectation if the "reversion" to purple corollas and yellow anthers were strictly a developmental event. The reverse event has also been observed, i.e., re-reversion of the purple corolla, yellow anther phenotype to the white corolla, white anther state. This event too was sexually transmissible (Jorgensen, 1995).

One possibility explaining the germinal transmission of both corolla and anther phenotypes in these events is that the cosuppression state of the epidermis is controlled by the L2-derived cells underlying it (i.e., that a signal moves between cell layers in the corolla to determine the pigmentation pattern in the epidermis). Another possibility is that cells in both the L1 and L2 layers of the meristem have changed epigenetically, i.e., in nonclonal, "directed" fashion. While there is no direct evidence either supporting or excluding the former possibility, there does exist evidence that heritable changes in the state of cosuppression can arise nonclonally (Jorgensen, 1995), and so we favor the latter explanation.

Patterns Of Cosuppression Elicited By Locus *T-Chs41*

Long term observations have been made on backcross progeny hemizygous for locus *T-Chs41* by (a) monitoring flower color phenotypes over an extended period of time (4 - 12 months) in a relatively small number of plants in order to detect those somatic events that are more likely to arise in older plants and (b) observing only the first 3 to 10 flowers of larger numbers of plants derived from further backcrosses of both white flowers and patterned variants.

In this way, we have found that plants hemizygous for *T-Chs41* are capable of producing a diverse series of flower color patterns in which white sectors of *Chs* cosuppression can be determined not only by cells located at the junctions between petals (fused edges of adjacent petals), but also by cells located in or near the major veins. Examples of much of this diversity are shown in Figure 1 (see Jorgensen, 1995, for further examples). At least four different types of vein-based patterns can be readily distinguished; an example of each is shown in Figure 1D,E,F,O. Junction-based sectors are often superimposed on these vein-based patterns (e.g., Figure 1G-L,N). Sector size in vein-based patterns varies more widely than is shown here, and patterns may be almost totally white or almost totally purple. Predominantly white patterns generally possess a large junction contribution. Solid white flowers with white anthers and solid purple flowers with yellow anthers are extreme phenotypes flanking a continuum of intermediate patterns. These patterns vary with respect to both sector size and shape. Intermediate intensity of pigmentation is not observed in CHS41 and its progeny, except in boundary zones between white and fully pigmented sectors.

It is important to note that the petunia flower is not radially symmetric, but rather bilaterally symmetric about the vertical axis bisecting the uppermost petal and separating the two lowermost petals. Sense suppression patterns tend to be bilaterally symmetrical about this vertical axis. Sector shape is generally biased toward larger white sectors in lower petals and smaller sectors in upper petals. This is seen most clearly when the bias is extreme (e.g., Figure 1N). However, the coloration patterns produced by sense suppression are significantly more variable than is the morphology of the petunia flower (see also Figure 1 of Jorgensen, 1993a). In sense suppression then, pigment gene expression patterns follow morphological parameters in a more relaxed manner than do the cell division and growth patterns that determine corolla shape.

Somatic Changes In Sense Suppression Patterns

The origins and properties of the diverse variants of CHS41

Figure 1. Phenotypic diversity of plants with *T-Chs41*. F reprinted from Jorgensen (1993b), with permission of the Royal Society. H reprinted with permission from Jorgensen (1955); copyright 1995 American Association for the Advancement of Science. For a color representation of this figure, please see the color tip facing page 160.

have been investigated mainly by observations of a series of backcross generations, up to BC_5 in one instance. A single BC_1 population (#2006; Table 2) was the focus of these studies, and all further backcross generations discussed here were derived from it. BC_1 progeny of the original white-flowered state of transgenote CHS41 were not only plants with solid white flowers, but also plants with patterned flowers (Table 2). These patterned flowers had predominantly purple corollas, white anthers, and often a nonclonal white sector in the corolla tube. Boundaries between white and purple regions of the corolla were not coincident with the tube-limb boundary in the corolla. Rather, the white region varied in size within and among plants, the smallest examples being restricted to the base of the tube and the largest progressing from the tube into the limbs of the corolla, as in Figure 1A, rarely as extremely as in Figure 1B. White-flowered backcross progeny sometimes produced flowers with purple, wedge-shaped sectors located at petal tips (Figure 1C). Although the positions of sector boundaries were variable, inspection of many flowers indicated that petal junctions were the primary determinants of these patterns (Jorgensen, 1995). White sectors varied in size continuously between the extremes of fully purple and fully white flowers.

Because CHS41 had been found to be somatically variable, the nine BC_1 progeny exhibiting cosuppression in population 2006 were monitored over time to determine the somatic stability of their flower color phenotypes. Six of the seven patterned plants produced one or more branches that reverted to the original white state of the CHS41 parent. The most closely observed plant was 2006.24.

Upon flowering, the main shoot of 2006.24 produced patterned flowers. The main shoot was pruned just below the first flower to release axillary buds from apical dominance, and axillary shoots grew out of thirteen lower nodes and flowered. Ten of these produced patterned flowers (one of which is shown in Figure 1A), but three produced only white flowers. Flowers from all branches but one were backcrossed to inbred V26 for progeny testing. After seed harvest it was noted that one of the lateral shoots with patterned flowers had produced a secondary lateral branch with only white flowers. After cuttings were taken from most of the 13 branches, plant 2006.24 was allowed to flower again. Once again, a lateral with patterned flowers produced a new lateral with only white flowers. Altogether then, five axillary branches (three primary and two secondary) on plant 2006.24 produced white flowers.

The cuttings taken from 2006.24 were rooted and grown to flowering. Eight patterned branches produced a total of fifteen cuttings, all of which produced patterned flowers. However, three of these 15 cuttings also produced one or more white-flowered axillary branches, as in the plant shown in Figure 2. Thus, a total of eight such events were observed in plant 2006.24 and its rooted cuttings. As already mentioned, five similar events were observed in the six patterned plants in population 2006. Such events have been observed in many other *T-Chs41* lineages as well, as was noted elsewhere

(Jorgensen, 1995). In all these events, the first flower on the axillary branch was totally white, i.e., no flower was observed to be sectored along clonal lines, as would occur in a sectorial or mericlinal chimera. This absence of clonal sectors in such a large number of events indicates that the origin of the white-flowered phenotype in axillary branches is nonclonal. Furthermore, it appears to be a meristem-based phenomenon.

Nine cuttings were also taken from four white-flowered branches of 2006.24 (including a white axillary branch that was derived from a patterned branch) and grown to flowering. All branches on these plants produced only white flowers. We can conclude that both the white state and the patterned state are heritable somatically, although the patterned state is reversible to the white state in events occurring in some axillary meristems. As will be described below, changes from the white state to a patterned state are less frequent than the reverse, and the outcome is almost always a vein-based pattern or fully pigmented flowers.

The influence of petal junctions in patterned 2006.24 flowers was not always obvious, as can be seen in the patterned 2006.24 flower in Figure 2. This variation is due to some additional influence by veins, such that the sum of vein and junction influences, both being biased toward the base of the corolla, results in the "white tube, purple limb" phenotype seen in Figure 2. The junction-influenced flower in Figure 1A is also from plant 2006.24, i.e., there is some variation within patterned individuals. The "wedge" patterns (similar to Figure 1C) of three transgenotes reported by Napoli et al. (1990) are based on petal junctions, but with no vein influences observed at all (Jorgensen, 1993a). Interestingly, these transgenotes, unlike CHS41, are phenotypically stable. The 2006 "patterned" flowers (Table 2) are actually intermediate between the purely junction-based patterns of other transgenotes and the strongly vein-based patterns to be discussed later in this paper. Nonetheless, 2006 patterned flowers have only a minor vein influence and are biased toward the junction influence.

Phenotypic variation in backcross progeny

Considering the heritability and reversibility of the purple state that arose in plant CHS41, we performed progeny tests of white and patterned flowers in a series of backcross generations. It was found that both genetic and environmental factors could determine the frequency of white and patterned flowers in a population, though it was not until the BC_5 generation that this became clear. The experiments that led to this conclusion are described next.

Flower color phenotypes of kanamycin resistant (kan^r) progeny (i.e., all hemizygous for *T-Chs41*) of plant 2006.24 have been reported previously as the sum of 15 BC_2 populations derived from BC_1 plant 2006.24 (Jorgensen, 1995). Here these BC_2 data are shown in two parts, one comprised of six populations deriving from white branches and the other comprised of nine populations deriving from patterned branches (Table 3). Four phenotypic classes were arbitrarily defined among kan[r] progeny: (1) **P**, fully *purple* flowers with yellow anthers, (2) **PP**, *predominantly purple* flowers with white anthers, often with junction-based sectors extending upward from the base of the corolla tube but extending into no more than 20% of corolla limbs, as in the flower in Figure 1B, (3) **PW**, white or *predominantly white* flowers with junction pattern (>80% white limbs), (Fig 1C), and (4) **IM**, an *intermediate* class with white, junction-based sectors covering 20-80% of corolla limbs. In this scheme, the patterned flowers of plant 2006.24 are classified as PP, and the white flowers as PW. Among the hemizygous backcross progeny of plant 2006.24, IM class flowers and P class flowers

Table 3. Kanamycin resistant BC$_2$ progeny of BC$_1$ plant 2006.24[1].

| | Percent of progeny | | | | Number | |
Female Parent	P	PP	IM	PW	Progeny scored	Populations scored
PP (patterned)	1	45	1	54	166	9
PW (white)	0	45	0	55	128	6
Total	0.3	45	0.3	54	294	15

[1] Data in row "Total" have been previously published (Jorgensen, 1995).

occurred at low frequencies, whereas the PW class and the PP class were represented at high frequencies in similar proportions when summed over multiple populations. Regardless of whether the parent branch was classified as PP or PW, the result was the same when the data are summed over multiple populations. Because the phenotypes are exhibited in L1-derived cells and female gametes are derived from L2 cells, these data are not a test of whether the states are sexually heritable. However, the result is clearly different than was observed for the fully purple derivative of the white-flowered CHS41 transgenote (Table 2).

In order to determine whether the PP and PW states of the junction pattern might be sexually heritable, seven PW and ten PP BC$_2$ plants were backcrossed to V26. As was reported by Jorgensen (1995) and is reproduced in Table 4, the phenotype of the BC$_2$ parent had no influence on the frequencies of phenotypic classes in the BC$_3$ progeny. The phenotypic frequencies for BC$_3$ were similar to those for BC$_2$.

However, considerable variation was observed among individual BC$_2$ and BC$_3$ populations, as is presented in Table 5, which presents each of the 19 BC$_3$ populations that comprise the BC$_3$ data in Table 5.

Table 5 also presents eight BC$_3$ populations descended from plant 2006.4, a PP plant that produced a white axillary branch. These eight BC$_3$ populations were derived from four capsules, two populations ('a' and 'b') being from each capsule and grown at different times.

Table 4. Heritability of BC$_2$ phenotypes[1].

| Phenotypes of BC$_2$ Female Parents | Percent of progeny[2] | | | | Number | |
	P	PP	IM	PW	Progeny scored	Populations scored[3]
PP	0.4	36	4	60	242	12
PW	0	39	0.7	60	148	7
Total	0.3	37	3	60	390	19

[1] Previously published (Jorgensen, 1995).

[2] All progeny are kanamycin resistant BC$_3$ progeny.

[3] Number of populations were incorrectly reported by Jorgensen (1995), but are correct here.

Again, considerable variation can be seen among populations, although the summed data are similar to the summed data for the 2006.24-derived BC$_3$'s. More striking are the differences between populations grown from the same seed capsule, which indicate that the relative frequencies of the two pattern states, PP and PW, are determined after seed germination, rather than during sexual development in the parent or during seed development in the capsule.

In sum, the PP junction pattern arises after germination at variable frequencies that average about 40% in progeny of plants hemizygous for locus *T-Chs41*, regardless of the phenotype of the parent; however, the PP phenotype may arise at very high or very low frequencies in any given population. Factors that influence the frequency at which the PP phenotype arises are addressed next.

Environmental Induction Of Meristem States in a PP-biased line

In the BC$_2$ and BC$_3$ generations, the predominantly purple state of the junction pattern (PP-J) and the predominantly white state of the junction pattern (PW-J) were equally likely when averaged over multiple populations, however, individual populations deviated substantially from this average. However, selection for the PP-J state for several backcross generations produced, by BC$_5$, a bias for the PP-J state. As is shown in Table 6, PW-J plants appeared in each of five BC$_5$

populations at only low frequencies (average: 5%; range: 0-9%; spring planting). Sibling seed from each of the same seed capsules were

Table 5. Variation among populations of BC_3 plants.

Female parent[1]	Number of BC_3 progeny				
	P	PP	IM	PW	Total
2006.24-derived PP BC$_2$ plants					
2430.1	0	8	2	12	22
2430.3	1	6	0	15	22
2430.5	0	17	1	8	26
2431.1	0	9	0	4	13
2432.2	0	7	2	6	15
A110.2 a	0	11	2	7	20
A110.2 b	0	12	0	17	29
A110.3 a	0	0	0	19	19
A110.3 b	0	2	1	4	7
A111.2	0	8	0	10	18
A111.3 a	0	0	0	23	23
A111.3 b	0	8	1	19	28
Total	1	88	9	144	242
2006.24-derived PW BC$_2$ plants					
2430.2	0	0	0	25	25
2430.4	0	8	1	16	25
2431.2	0	3	0	11	14
A110.1 a	0	17	0	2	19
A110.1 b	0	20	0	5	25
A111.1 a	0	7	0	14	21
A111.1 b	0	3	0	16	19
Total	0	58	1	89	148
2006.4-derived PP BC$_2$ plants					
A112.2 a	0	11	0	5	16
A112.2 b	0	13	0	11	24
A112.3 a	0	12	3	5	20
A112.3 b	0	0	0	26	26
A112.4 a	0	16	0	1	17
A112.4 b	0	12	0	10	22
A112.6 a	0	15	1	5	21
A112.6 b	0	1	0	22	23
Total	0	80	4	85	169

[1]Pairs indicated by 'a' and 'b' are separate sowings of seed from the same seed capsule.

sown again and the five populations then produced a higher average frequency of 23% (range: 14-32%; summer planting) PW-J plants (Table 6). These data would seem to indicate that not only can lines be selected which favor the PP-J state, but also that seasonal environmental differences help determine which pattern will be expressed. Further study of these lines in appropriately controlled environments may allow us to determine what environmental factor is responsible, some possibilities being light intensity, light quality, and temperature.

Table 6. Phenotypes of kan^r BC$_5$ plants descended from BC$_1$ plant 2006.24: Seasonal influences.

Capsule		Number of progeny			
	P	PP	IM	PW	Total
Spring 1994					
#1	0	33	0	2	35
#2	0	32	0	3	35
#3	0	30	0	0	30
#4	0	29	0	3	32
#5	0	33	0	1	34
Total	0	157	0	9	163
Summer 1994					
#1	2	41	0	20	63
#2	1	49	6	9	65
#3	1	45	2	9	57
#4	1	44	0	21	66
#5	0	41	0	11	52
Total	5	220	8	70	303

Summary

Percent of progeny

	P	PP	IM	PW	Number of populations
Spring	0	95 range: 91-100	0	5 range: 0-9	5
Summer	2	73 range: 65-79	3	23 range: 14-32	5

Origins Of Vein-Based Patterns

Vein-based patterns first arose on a branch of the CHS41 transgenote, one of the flowers being shown in Figure 1K. Several such events have also been observed in white-flowered plants of population 2006 and these have been analyzed in some detail. Three events will be described here. Two occurred in the two white-flowered plants, 2006.2 and 2006.19. The third arose in a white revertant of a plant with patterned flowers called 2006.1. At flowering, this plant was comprised of a main shoot with patterned flowers and a single axillary branch with white flowers. After pruning, new branches were formed which produced only white flowers. Through several cycles of pruning and regrowth over a period of eight months, only white flowers were observed. After this period, one branch of plant 2006.1 produced flowers of a new pattern; Figure 2 shows this branch with its first three flowers from two vantage points, side and front. The first flower produced on the branch was of the PW-J type, almost completely white with a small purple wedge at the tip of one petal (as can seen in the side view but

171

is obscured in the front view in Figure 2). The next two flowers exhibited predominantly purple vein-based patterns, as can be seen in the front view. The flowers belonged to a class of vein pattern whose white sectors have sharp points protruding into petal limbs, as in Figure 1E,H,K (and in contrast to Figure 1D,G,J). None of these flowers appeared to be a mericlinal or sectorial chimera. This pattern continued to be produced on this variant branch for over 30 consecutive flowers, though with some variable contribution by junction-based sectors, Fig. 1N being the 26th flower on this branch.

Two months before this event, three white flowers on plant 2006.1 had been pollinated. After seed harvest, the plant was pruned to near its base and allowed to regrow. Upon flowering, the variant branch was noted. At that time, two additional white flowers were pollinated with V26, as well as the first three flowers and flowers 14, 18, and 22 on the variant branch. Pattern phenotypes of progeny of (a) the first three white flowers are presented in Table 7 as "White - September", (b) the next two white flowers as "White - November", and the six flowers of the new variant branch as "Variant branch". The September-pollinated white flowers gave progeny with PP and PW phenotypes as was observed for plant 2006.24. The November-pollinated white flowers produced also a large number of progeny with intermediately pigmented, vein-based patterns, like those in Figure 1H,K,M. The flowers of the variant branch produced PP and P progeny at similar frequencies, but no PW progeny. Such a high frequency of fully purple, yellow anther progeny had not been previously observed by us, with the exception of the progeny of the fully purple, yellow anther CHS41 somatic variant..

Table 7. Heritability of vein-based patterns.

| Parent | Percent of progeny[1] | | | | Number | |
	P	PP	IM	PW	progeny	populations
2006.1 x V26						
White - Sept.	0	7	0	93	58	3
White - Nov.	6	45	31	18	49	2
Variant branch	49	50	1	0	133	6
I1	50	36	14	3	72	3
Cutting 1B	84	16	0	0	135	6
1B axillaries	78	21	1	0	578	10
I5I'1-6	42	34	21	1	38	1
2006.2 x V26						
White	0	23	0	77	52	3
Patterned	11	37	23	29	238	10
2006.19 x V26						
White	0	26	2	72	98	4
Patterned	40	49	11	0	105	4

[1] All progeny are kanamycin resistant hemizygotes carrying *T-Chs41*.

An axillary branch that grew out immediately below the first flower of the variant branch of 2006.1 produced only flowers with purple corollas, most with yellow anthers, but some with anthers streaked white. (Often, the sharp-point vein pattern is accompanied by yellow anthers with white vertical streaks.) Three flowers on this branch were pollinated with V26. Progeny phenotypes were summed and are shown in Table 7 as "I1" (inflorescence branch 1). These progeny included few PW types, many P and PP types, and a number of vein-based IM types. A cutting taken from this axillary was rooted and grown to flowering ("Cutting 1B" in Table 7). Most progeny of its main flowering shoot were of the purple corolla, yellow anther P type, and none were IM or PW types. The data on progeny of flowers from plant 2006.1 suggest progressive increase in the heritability of the new purple state, approximately paralleling the progressive phenotypic change in 2006.1.

Photographs of the first six flowers of this shoot from Cutting 1B have been published (see Figure 1, f1-f6, of Jorgensen, 1994). These flowers had a pattern intermediate to the sharp-pointed vein pattern type (Figure 1H) and the round-tipped vein pattern type (Figure 1G) and exhibited progressive increase in the size of junction-based sectors as flowering progressed up the branch. No change in heritability was noted in progeny of the different flowers of this branch, and so the "Cutting 1B" data in Table 7 are summed over all six flowers. Some axillaries from this shoot exhibited the predominantly white version of this new vein-and-junction pattern, whereas some produced only fully purple flowers; changes between these two states were frequent and reversible (Jorgensen, 1994). The spectrum of phenotypes of sexual progeny, however, was the same for each type, and so these have been summed and are presented in Table 7 as "1B axillaries". The frequencies of phenotypes from these do not differ from the cutting 1B frequencies. There was, however, one exceptional seed capsule among these that is presented separately as "I5I'1-6". It produced a large number of IM flowers of a new pattern, shown in Figure 1 O, a vein-based "star" pattern similar to that known to petunia breeders; it also produced flowers with the related pattern of Figure 1F, as well as various intermediates between the two.

Plants 2006.2 and 2006.19 produced only white flowers for one year before producing branches with vein-based patterns (sharp-point type). Progeny testing of white-flowered branches and vein-patterned branches was carried out for both plants and is presented in Table 6. In each case, white-flowered branches produced progeny phenotypes at frequencies typical of 2006.24, whereas the patterned branches produced higher frequencies of P and IM plants, and IM plants were vein-based patterns.

An Unusual Variant On The Junction Pattern

Figure 1M shows a rare variant in which purple petal edges are superimposed onto the predominantly white state of the junction pattern; in addition, the corolla tube is purple. It arose on white-flowering CHS41 plants after four years of maintaining the original white state through vegetative propagation. Cuttings were rooted and grown to flowering and periodically plants were cut back to release new vegetative shoot growth, which then led to further flowering. The new flower color pattern of Figure 1M was not sexually transmissible, all progeny being predominantly purple. The purple petal edges of this variant appear as a negative image of certain irregular versions of the "Cossack Dancer" pattern (Jorgensen, 1994). It appears as if some morphologically controlled factor is reversing the suppression of *Chs* at petal edges.

A Hierarchy of Mechanisms

The following hierarchy of mechanisms has been proposed to explain the diversity and metastability of transgene-elicited patterns of *Chs* expression (Jorgensen, 1995).

(1) Because sense suppression appears to be threshold dependent, changes in the rate of transcription of any *Chs* gene could potentially have an impact on the cosuppression state; and so the cosuppression phenotype can act as a highly sensitive reporter of transgene transcription when transcript concentration is near a critical state or threshold.

(2) Transcription of the transgene is responsive to aspects of the nuclear environment (chromatin components, transcription factors) including perhaps ones that vary according to the morphology of the corolla. Boundaries between white and purple sectors will lie wherever in the corolla transgene transcript concentration reaches the threshold for sense suppression. Because of the nonlinearity of the response, shallow gradients are translated into steep gradients of *Chs* transcript and anthocyanin pigment concentrations. That the boundaries are diffuse would seem to indicate that cells may adopt intermediate states of transcript turnover at the threshold.

(3) The epigenetic state of the transgene can discriminate among transcription factors and hence determine which morphological fields will control expression. In other words, different epigenetic states of the transgene can elicit fundamentally different pigmentation patterns reflecting morphological fields.

(4) Physiological influences can induce new epigenetic states non-clonally in many or all the cells of a meristem. It is conceivable that nuclear factors can *cause* epigenetic alterations in susceptible genes, depending on the epigenetic state of those genes. Each epiallele of a gene may differ in lability to "epimutagenic" nuclear factors. In this manner, a series of imprints could be made onto an appropriately responsive gene as new nuclear factors are produced during the progression of plant development and as a result of changes in physiology. Under appropriate circumstances in a properly responsive system, the consequence would be a highly ordered sequence of epigenetic imprints that control gene expression, that are inherited

Figure 2. Somatic events occurring on plants 2006.24 and 2006.1. For a color representation of this figure, please see the color tip facing page 161.

somatically, and that are able to change in state reliably and functionally. Such a mechanism was originally proposed by R.A. Brink and B. McClintock some thirty years ago (reviewed by Jorgensen, 1994). The participation of threshold-responsive processes such as sense suppression in this dynamic, multistable regulatory scheme allows for great sensitivity to otherwise subtle shifts in transcription states.

CONCLUSIONS

Sense suppression of an anthocyanin gene can be a highly sensitive reporter of epigenetic states of gene expression, especially transcription states. This reporter phenotype is visible to the experimenter and dispensable to the plant. The morphology-based patterns of *Chs* sense suppression indicate that it is a reporter of fields of morphological information. The metastability and heritability of patterns suggests that transgene epigenetic states, perhaps transcriptional, can mediate pattern elaboration in response to these fields. The fact that some types of changes in these epigenetic states arise nonclonally in a population of cells comprising an axillary meristem and in response to environmental cues suggests a much greater degree of transgene participation in host regulatory processes than might have been anticipated. Although the transgene itself is only a small part of the pattern elaboration system, it is a centrally located component and so a useful tool for manipulating and exploring its operation.

ACKNOWLEDGEMENTS

This work was initiated at DNA Plant Technology Corporation. The authors greatly appreciate the efforts of John Bedbrook and Clinton Neagley to make available the CHS41 line for this study. We thank I. Sussex for helpful discussions on the nature of chimeras and meristems, J. Harding for greenhouse space, T. Robbins, X.-M. Sha, G. Shariat, and A.M. Napoli, for assistance with progeny tests.

REFERENCES

de Carvahlo Niebel, F., Frendo, PP., van Montagu, M., and Cornelissen, M., 1995, Post-transcriptional cosuppression of b-1,3-glucanase genes does not affect accumulation of transgene nuclear mRNA, *Pl. Cell* 7:347.

Cornu, A., 1970, Sur L'obtention de mutations somatique après traitements de graines de pétunias, *Ann. Amélior. Plantes* 20:189.

Courtney-Gutterson, N., 1994, The biologist's palette: genetic engineering of anthocyanin biosynthesis and flower color, *in*: "Genetic Engineering of Plant Secondary Metabolism," B.E. Ellis, G.W. Kuroki, H.A. Stafford, eds., Plenum, New York.

Jorgensen, R., 1993a, Elicitation of organized pigmentation patterns by a chalcone synthase transgene, *in*: "Cellular Communication in Plants," R.M. Amasino, ed., Plenum, New York.

Jorgensen, R., 1993b, The germinal inheritance of epigenetic information in plants, *Phil. Trans. R. Soc. Lond.* B 339:173.

Jorgensen, R., 1994, Developmental significance of epigenetic impositions on the plant genome: a paragenetic function for chromosomes, *Devel. Genet.* 15:523.

Jorgensen, R.A., 1995, Cosuppression, flower color patterns, and metastable gene expression states, *Science* 268:686.

Marcotrigiano, M., and Bernatzky, R., 1995, Arrangement of cell layers in the shoot apical meristems of periclinal chimeras influences cell fate, *Plant J.* 7:193.

Matzke, M.A., and Matzke, A.J.M., 1995, How and why do plants inactivate homologous genes? *Pl. Physiol.* 107:679.

Napoli, C., Lemieux, C., and Jorgensen, R., 1990, Introduction of a chimeric chalcone synthase gene into petunia results in reversible co-suppression of homologous genes *in trans*, *Plant Cell* 2:279.

Taylor, L.P., and Jorgensen, R., 1992, Conditional male fertility in chalcone-synthase-deficient petunia, *J. Hered.* 83:11.

ESCHERICHIA COLI — FUNCTIONAL AND EVOLUTIONARY IMPLICATIONS OF GENOME SCALE COMPUTER-AIDED PROTEIN SEQUENCE ANALYSIS

Eugene V. Koonin, Roman L. Tatusov, and Kenneth E. Rudd

National Center for Biotechnology Information, National Library of Medicine, National Institutes of Health, Bethesda, MD 20894

THE BACTERIAL GENOME AND GENES - PROBLEMS AND PROSPECTIVE

Complete sequencing of model genomes has recently become a reality. Hundreds of viral and more than 20 organellar genome sequences are currently available (Bork et al., 1994). The genomes of several bacteria, Archaea, and yeast are expected to be completed within 2-3 years. The ultimate value of genome projects is not to establish complete and accurate nucleotide sequences *per se*, but rather to use the sequence in order to deduce how the genome determines all cellular functions. Eventually, it should be possible to determine the whole pathway from the nucleotide sequence to the phenotype of an organism, which could be re-stated as the "first principles" of cellular structure and function. Numerous biochemical and genetic experiments will be indispensable for achieving this ambitious goal. Nonetheless, computer analysis of the amino acid sequences encoded in the genome is a necessary and complementary approach that allows one to systematically predict protein functions and derive possible evolutionary relationships. One cannot help but to note that computer-assisted sequences analysis, even though generally lacking the precision that is, at least in principle, achievable in laboratory experiments, is much less labor-consuming and costly. In fact, at this time, only computer methods allow one to analyze gene products encoded in a complete genome simultaneously and consistently and to obtain meaningful, readily comparable results for each of them in a relatively short time.

There is little doubt that among all the model organisms currently used in the laboratory, *Escherichia coli* has been studied in the greatest detail (Neidhardt et al., 1987, 1996; Riley, 1993). It has been estimated that over 80% of the metabolic pathways in E. coli are known, and functions have been determined for more than half of the projected total number of about 4,000 genes distributed along the 4.7 MB, circular *E. coli* chromosome (Riley, 1993). The *E. coli* genome sequencing project has been underway since 1991 (Daniels et al., 1992), and as of this writing (May 10, 1995), over 70% of the chromosome sequence has been determined. The analysis described in this chapter has been performed with an earlier version of the *E coli* genome sequence contained in the EcoSeq7 database (Rudd, 1993, and K. E. R., unpublished). EcoSeq7 includes 60% of the *E. coli* chromosome sequence that has been available by August, 1994. Translation of this database produced a set of 2,328 proteins and putative gene products. We choose this set of amino acid sequences for a pilot project on comparative analysis of the proteins encoded in a well characterized, model genome.

What do we hope to learn from a comparative sequence analysis of a large set of gene products that may be representative of the whole gene repertoire of the bacterium? It appears logical to consider five distinct, even though interrelated groups of problems that need to be addressed.

1.   Prediction of protein function.   Can we make experimentally testable, new functional predictions for a sizable proportion of proteins? Does application of increasingly sensitive methods for sequence analysis result in a significantly higher rate of prediction? Recent studies suggest positive answers to both of these questions. Careful evaluation of relatively weak sequence similarities using a combination of various computer methods indeed has resulted in a dramatic increase in the fraction of gene products whose function could be predicted, at least in general terms, on the basis of sequence comparison results (Bork et al., 1992; Koonin et al., 1994). These studies, however, concentrated on a relatively small number of genes and included a very detailed, case-by-case investigation of individual sequence alignments and motifs. Probably the principal challenge for genome scale protein sequence studies is to match the precision and completeness of functional prediction that has been achieved in studies on specific protein families, without dramatically compromising the speed of analysis.

2.   Sequence conservation.   How well are the sequence of various proteins conserved in evolution? In large scale genome projects, the average rate of sequence similarity detection is close to 50% of the analyzed protein sequences (Bork et al., 1994). In recent studies, when up to date computer methods have been used, this fraction has increased to 70% for yeast proteins (P. Bork, personal communication) and to 75% for *Mycoplasma capricolum* proteins (Bork et al., 1995). Since Mycoplasmas have the smallest genomes among bacteria (Maniloff et al., 1992), it appears reasonable to assume that they should have maintained only the basic set of house-keeping genes. Accordingly, one may expect that the fraction of conserved proteins found for these degenerate bacteria may be close to the upper limit for all organisms. This may be a useful reference point to compare the results obtained with a genome scale protein set.

Probably even more intriguing questions are - how many proteins, and which ones, contain ancient conserved regions or ACRs (Green et al., 1993; Green, 1994), that is sequences containing motifs that have been conserved through the time span of about 3.5 billion years separating *E. coli* from the point of the primary radiation of bacteria from the common ancestor of eukaryotes and Archaea? Recent estimates have strongly suggested that the overall number of distinct ACRs is probably relatively small - on the order of 1,000 (Chothia, 1992; Green et al., 1993). Therefore, given the majority of the bacterial gene product sequences, one may realistically hope to generate a nearly complete catalogue of ACRs. According to previous estimates, about 20% of the *E. coli* proteins contained ACRs (Green et al., 1993). It was of obvious interest whether this fraction will increase significantly as the result of the growth of sequence databases and application of new methods of analysis. The detection of each new ACR is important as these sequences are under extremely strong selective pressure and by inference, are the most functionally important parts of proteins.

3. Relationships among the genes and proteins within the *E. coli* genome. The existence of intraspecies homologous genes [known as paralogs (Fitch, 1970)] in *E. coli* has been recognized for some time; these genes should have evolved by duplication followed by diversification (Zipkas and Riley, 1975; Zipkas et al., 1978; Riley and Anilionis, 1978; Riley and Krawiec, 1987). However, now that the sequence of a major part of the genome has been determined, construction of a nearly complete catalogue of the paralogous clusters and evaluation of the extent of gene duplication in the *E. coli* genome has become feasible. With 60% of the *E. coli* protein sequences available, it is likely that the majority of the paralogous clusters are represented in this set, and the number of new clusters among the remaining 40% of the proteins will be relatively small. With the selection for compactness that is thought to be a major driving force of the bacterial genome evolution (Neidhardt et al., 1987), only those diverged gene duplicates that confer a significant selective advantage to the bacterium will be retained. Therefore correlations between the number of paralogs and protein function, if they exist and could be detected, should reveal important aspects of the bacterial cell physiology and evolution.

4. Bacterial genome evolution - the possibility of large scale duplications. In early studies, it has been noticed that certain paralogous genes, primarily those encoding enzymes of glucose catabolism, are separated by regular intervals on the *E. coli* chromosome. On the basis of this apparent periodicity, the intriguing hypothesis of two consecutive, whole chromosome duplications has been proposed (Zipkas and Riley, 1975; Zipkas et al., 1978). However, subsequent re-analysis using larger data sets has failed to fully confirm these findings (Riley, 1993; Labedan and Riley, 1995). Other authors who have analyzed a small set of paralogous genes have observed quasi-periodicity in their distribution on the *E. coli* chromosome, with the distances between the paralogs tending to be multiples of 7 min of the genetic map (Kunisawa and Otsuka, 1988). These findings have led to the hypothesis that the *E. coli* genome might have undergone four consecutive whole chromosome duplications. With the sequences and map positions of more than one half of the *E. coli* genes now available, the apparent periodicity in the paralog distribution can be assessed in a more definitive way.

5. *E. coli* **genome as a tool for exploring eukaryotic genomes.** The very notion of a model genome suggests that analysis of such genomes will be helpful in understanding the human genome and those of other eukaryotes. While genome organization in bacteria is dramatically different from that of eukaryotes, and direct inferences are unlikely, the amino acid sequences of many proteins are highly conserved. Therefore there is no doubt that the wealth of biochemical and genetic information that is available on *E. coli* genes can be used to predict functions of eukaryotic proteins. One of the goals of our work on *E. coli* genome is to assess the extent of such prediction, particular with respect to genes implicated in human diseases.

Each of these problems has been tackled previously from different perspectives and with different, relatively small data sets. Here we address them all systematically on a genomic scale.

STRATEGY AND METHODS OF COMPUTER ANALYSIS OF A
GENOME-SCALE PROTEIN SEQUENCE SET

There are two groups of methods for sequence conservation analysis: i) methods that directly compare sequences in search of pairwise similarity; and ii) methods that allow one to derive conserved motifs and then use them to identify similar segments in other sequences (Doolittle, 1990; Altschul et al., 1994). In our study of the *E. coli* proteins, we used both approaches. Database screening for pairwise similarity was performed using programs of the BLAST family (Altschul et al., 1990, 1994). The BLOSUM62 matrix for amino acid residue comparison, which generally appears to be the most sensitive of the currently available matrices, was used in all searches (Henikoff and Henikoff, 1992, 1993). The SEG program was used to mask low complexity (compositionally biased) regions in protein sequences that frequently produce spurious hits in database searches (Wootton and Federhen, 1993; Wootton, 1994a,b). For sequences that initially did not show similarity to any other sequences in the database, the search was repeated without SEG in order to rule out the possibility that a conserved region has been masked. Initial searches were performed using the BLASTP program to compare all the *E. coli* protein sequences to the non-redundant (NR) protein sequence database. Those sequences that did not show similarity to any sequences in the database were compared, using the TBLASTN program, to the nucleotide version of the NR database translated in all 6 reading frames, in order to detect possible similarity to putative proteins that have not been annotated in databases (Altschul et al., 1994). Those amino acid sequences alignments produced by BLAST that had a similarity score above 80, corresponding to the probability of matching by chance of about 0.01 (for an average-size protein), were accepted as an indication of an authentic relationship. The significance of the alignments in the "twilight zone" (scores between 60 and 80) was assessed through the analysis of conserved motifs using the CAP and MoST programs (Tatusov et al., 1994), multiple alignment analysis using the MACAW program (Schuler et al., 1991), as well as considerations of biological relevance. Recent analyses of large sets of yeast and bacterial proteins have shown that careful examination of relatively weak similarities detected in database searches using versatile, complementary computer methods is critical for increasing the rate of functional prediction and revealing evolutionary relationships (Bork et al., 1992, 1995; Koonin et al., 1994). We found that at least for *E. coli* proteins, a score greater than 80, computed using the BLASTP program, in all cases corresponded to a relationship that subsequently could be confirmed by multiple alignment analysis, search for conserved motifs, and functional assessment; furthermore, only a few spurious alignments with scores between 70 and 80 were detected. In contrast, among the alignments with scores between 60 and 70, the fraction of authentic ones was small.

We pursued two interrelated but distinct outlooks of the bacterial genome. The view "from the outside" pertains to the sequence conservation between *E. coli* proteins and proteins from other organisms. In order to evaluate this conservation, all *E. coli* protein sequences were compared to the NR database. The availability of 60% of the *E. coli* protein sequences makes it possible to derive reliable estimates of the number of highly conserved proteins containing ACRs, moderately conserved proteins similar to proteins from distantly related bacteria, and variable proteins that are conserved only in closely related bacteria, or are found in *E. coli* alone.

The view "from within" consists in delineating clusters of paralogs among the *E. coli* proteins sequence set. For clustering the proteins, we used a single-linkage, "greedy" algorithm. A cluster was defined as a group of protein sequences connected by similarity scores exceeding a chosen cut-off, but without the requirement that each pair of sequences within a cluster had such a score. Given the non-transitivity of BLAST, this method ensures the identification of a maximal number of paralogs belonging to a cluster. In the same vein, in order to make the clustering as complete as possible, we set a liberal cut-off, namely a BLASTP score of 70 or greater; among all the alignments between *E. coli* proteins in that score range, we found only very few that were clearly artifactual.

The further analysis included the delineation of conserved motifs typical of each of the paralogous clusters. These motifs were utilized to search first the NR databases and then the database of E. coli protein sequences using the MoST program (Tatusov et al., 1994), in order to delineate superclusters that may combine previously identified clusters and also may include some of the "loners", i. e. sequences not belonging to any of the original clusters.

The same motifs were used to resolve the difficulty encountered by the "greedy" clustering algorithm with proteins containing two or more conserved domains. Such proteins may artifactually bring together otherwise unrelated clusters. Therefore, in order to verify the validity of the delineated clusters, conserved motifs typical of each cluster were used to identify distinct domains of multidomain proteins which were accordingly included in different clusters.

Figure 1 schematically depicts the strategy that was employed in our analysis of the *E. coli* protein sequence set.

STATISTICS OF PROTEIN SEQUENCE CONSERVATION

View From the Outside — *E. coli* Proteins Are Highly Conserved

86% of the 2,328 *E. coli* proteins contained in the EcoSeq7 database are related to other proteins in the NR database (Figure. 2a). This is an unexpectedly high rate of sequence similarity detection as compared to the 70% recently reported for yeast proteins (P. Bork, personal communication) and the 75% observed for Mycoplasma capricolum (Bork et al., 1995). Thus, the fraction of *E. coli* proteins with no homologs in the current sequence databases is about half that for yeast.

Over 40% of the *E. coli* proteins contain ACRs, and about 60% contain regions that are conserved in proteins from distantly related bacteria [for the purpose of this analysis, bacteria "distantly related to *E. coli*" were defined as those that have been placed outside of the Proteobacteria in the latest 16S rRNA phylogenetic tree (Olsen et al., 1994)]. The observed fraction of the *E. coli* proteins containing ACRs

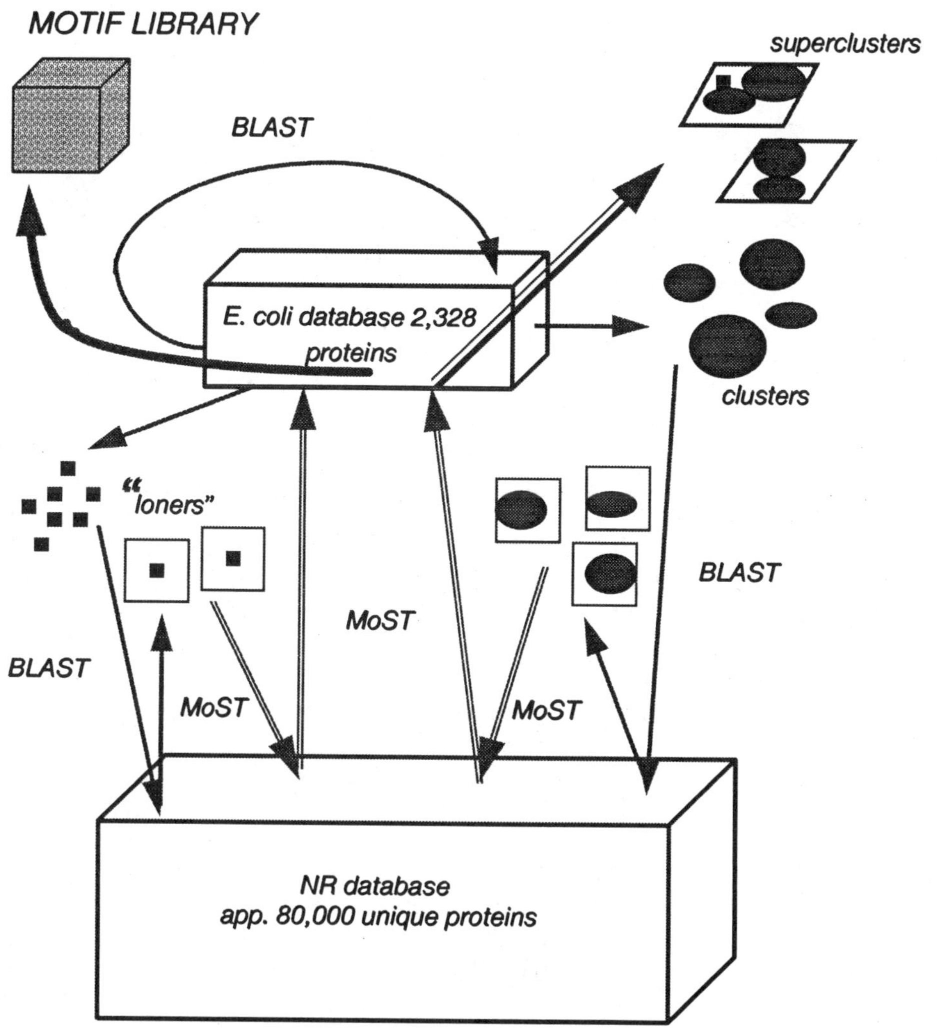

Figure 1. A general scheme of the computer analysis of the *E. coli* protein sequences.

was almost twice as high as the value reported previously (Green et al., 1993). Obviously, there are two components to this increase - inclusion of alignments from the "twilight zone" and database growth. A direct comparison with the previous study (Green et al., 1993) was complicated by the fact that different scoring matrices were used for BLAST searches. Approximately, the score of 75 with the PAM120 matrix used in (Green et al., 1993) corresponds to a score of 83 with the BLOSUM62 matrix. We found that 12.5% of the *E. coli* proteins contained ACRs that corresponded to scores between 60 and 82, but could be confirmed by additional analysis. As the increase in the fraction of the ACRs is about 20%, we estimate that the higher sensitivity accounted for approximately 60% of the additional ACRs. For the great majority of the ACRs that we detected in *E. coli* proteins - at least 97% - the related eukaryotic proteins have already been known to be conserved at least at the level of different animal phyla which is the original definition of an ACR (Green et al., 1993). Thus, only very few truly "new" ACRs have been detected, in accord with the notion that most of the ACR have already been discovered (Chothia, 1992; Green et al., 1993; Green, 1994).

Some of the genes coding for eukaryotic proteins with the highest similarity to their *E. coli* homologs by all likelihood, have evolved from endosymbiotic organellar (mitochondrial and chloroplast) genes that have been transferred to the nucleus (Palmer, 1985; Gray, 1989). Nevertheless, on average, *E. coli* proteins are less similar to their eukaryotic homologs than to the homologs from distantly related bacteria. Specifically, the mean similarity score, computed using BLASTP, between *E. coli* proteins and their eukaryotic or Archaeal homologs was 174 (median of the score distribution 109), and the respective value for homologs from distantly related bacteria was 230 (median 146). This suggests that most of the ACRs indeed are conserved from the time of the original bacteria-eukaryotic radiation which is thought to have occurred more than 3.5 billion years ago (Knoll, 1992).

Altogether, approximately 2/3 of the known *E. coli* proteins contain regions that have been conserved through more than a billion years of evolution that separate *E. coli* from the distantly related bacteria (Figure 2a). In other words, the high rate of protein sequence conservation in *E. coli* is not due to trivial similarity to proteins from closely related bacterial species. Rather, one has to conclude that the evolution of the majority of the bacterial proteins is strongly constrained, presumably due to the requirements to maintain a specific tertiary structure and to preserve functional sites. Contrary to expectations, the *E. coli* proteins are at least as conserved as the proteins encoded in the 8 times smaller, supposedly "minimal" *Mycoplasma capricolum* genome.

Given the wealth of functional data on *E. coli* genes and proteins, it is of interest to superimpose a synopsis of this information and the results of our sequence conservation analysis. As shown in Figure 2b, the majority of the *E. coli* proteins are characterized by both a known, at least in general terms, biological function and sequence similarity to other proteins. The second largest category - about 25% - includes *E coli* gene products that do not have a known function but have detectable homologs in sequence databases. For 2/3 of these proteins (approximately 400), function could be predicted based on sequence

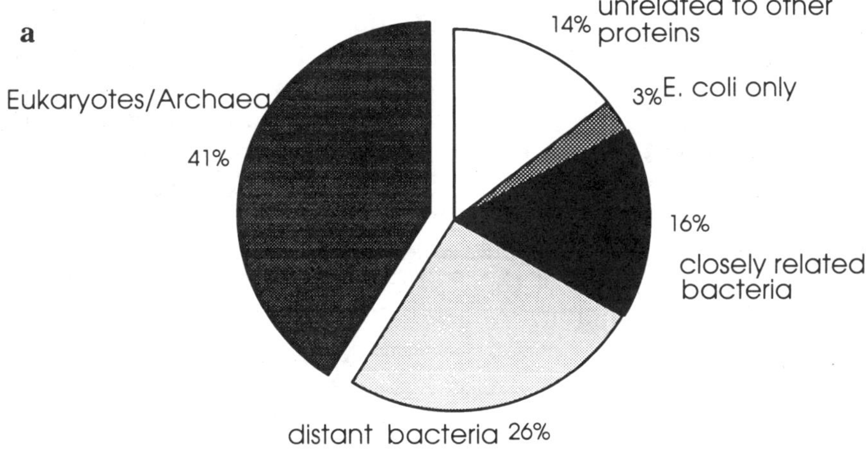

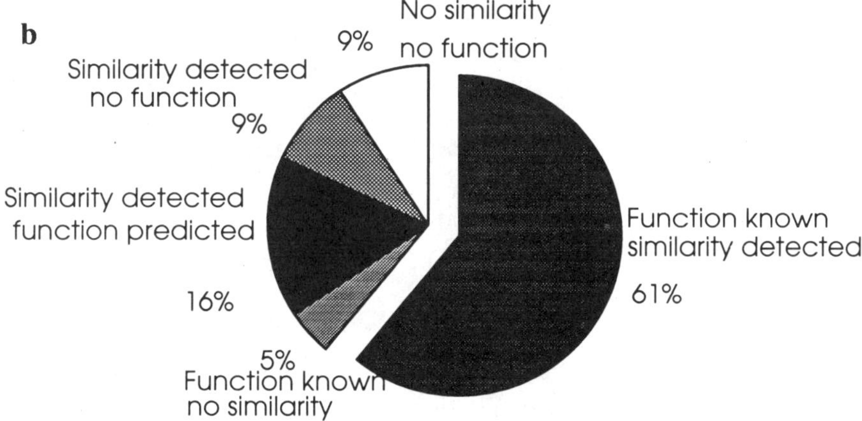

Figure 2. Sequence conservation in *E. coli* proteins. (a) Levels of sequence conservation. (b) Sequence conservation and functional information.

conservation. This group of proteins is of particular interest as sequence-based predictionof protein functions is one of the most important aspects of any genome project. In one of the subsequent sections, we briefly discuss an example of functional prediction for a group of *E. coli* proteins. About 200 *E. coli* gene products without a known function are related only to other uncharacterized proteins (Figure 1b). These observations are still significant as they show that these proteins have conserved functions and allow one to define new, functionally important motifs, but they cannot be immediately used for functional prediction.

Generally, the highest sequence conservation is observed among enzymes of intermediary metabolism. Proteins involved in genome replication and expression are somewhat less conserved, and regulatory and structural proteins are considerably more variable.

Among the 340 *E. coli* gene products (14% of the total) that do not show detectable similarity to any proteins in the current databases, 70% have no known function either. Those of these proteins whose functions are known, have a variety of activities (Table 1); conceivably, some of them have homologs in other organisms which have not yet been discovered, whereas others may be specific for *E coli* and closely related enterobacteria.

Only a small fraction of the *E. coli* gene products (less than 10%) remain completely uncharacterized, with neither an established

Table 1. Some *E. coli* proteins with known function in search of relatives.

Protein	Length (aa)	Function function in	Proteins with analogous other organisms
FucI	591	L-fucose isomerase	unknown
RecC	1122	Exonuclease V subunit, DNA repair and recombination	numerous groups of exonucleases
Dgt	505	dGTPase	unknown
PepD	485	b-Ala-His dipeptidase	unknown but other families of dipeptidases exist
SbcB	475	Exonuclease I, DNA repair	numerous groups of exonucleases
SelA	463	Selenocysteine synthase	unknown
CreD	450	Inner membrane protein	unknown
HipA	440	Resistance to inhibitors of peptidoglycane and DNA synthesis	unknown
ProX	330 periplasmic	glycine betaine-binding binding proteins protein	several families of solute-
HolA	343	DNA polymerase III δ subunit	unknown
Gsk	434	IMP-GMP kinase	numerous nucleotide kinases

function nor homologs among the proteins in the current sequence databases (Figure 2b).

View From the Inside — Clusters and Superclusters of Paralogs Among *E. coli* Proteins

About 46% of the *E. coli* proteins belong to 298 clusters of paralogs, defined on the basis of pairwise similarity. Another 10% could be included in 70 superclusters, which were defined on the basis of motif conservation and united different clusters, as well as "loners", i. e. sequences that did not show similarity to those of anyother *E. coli* proteins at the first stage of the analysis (Figure 3). For multidomain proteins, distinct conserved domain were included in different clusters, also on the basis of motif conservation that was analyzed using the MoST program. Our analysis revealed about 100 such proteins.

Remarkably, the fraction of proteins that have paralogs determined by our analysis is nearly identical to that reported in a recent study that has been performed with a smaller set of 1,264 *E. coli* protein sequences and used a different method for sequence comparison (Labedan and Riley, 1995). The virtual independence of the fraction of paralogs among *E. coli* proteins of the database size and the method of comparison strongly suggests that the observed value is objective and is unlikely to change significantly after the genome sequence is completed.

The majority of the clusters are small - 168 clusters (56%) contain only two members, and 247 (83%) contain from two to four members. In contrast, there are several clusters with more than 10 members (Figure 4 and Table 2). Even more dramatically, the four largest superclusters comprise nearly 25% of the *E. coli* proteins; these largest groups of paralogs include permeases; ATPases and GTPases with the conserved "Walker-type" motif; helix-turn-helix (HTH) regulatory proteins; and dinucleotide-binding proteins (Figure 3).

There is an obvious, strong correlation between protein function and the formation of paralogous clusters. Large clusters of paralogs in *E. coli* are mostly comprised by transport proteins and proteins involved in different types of gene expression regulation, whereas paralogous metabolic enzymes typically belong to small clusters (compare Table 2 and Table 3). Key enzymes of the genome replication and expression, notably catalytic subunits of DNA and RNA polymerases, and ribosomal proteins, do not have detectable paralogs. Physiologically, this correlation seems to make sense. Extensive diversification of transport and regulatory proteins provides a basis for the environmental adaptability and therefore, may confer a significant selective advantage to the bacterium. In contrast, the basic mechanisms of replication and expression apply uniformly to the whole genome - hence no driving force for paralogous evolution.

In the majority of the clusters, at least one protein contains an ACR, whereas only a small fraction of clusters consists of proteins with similarity only to other *E. coli* proteins (Figure. 5). Thus, most of the clusters include proteins with essential functions whose ancestor(s)

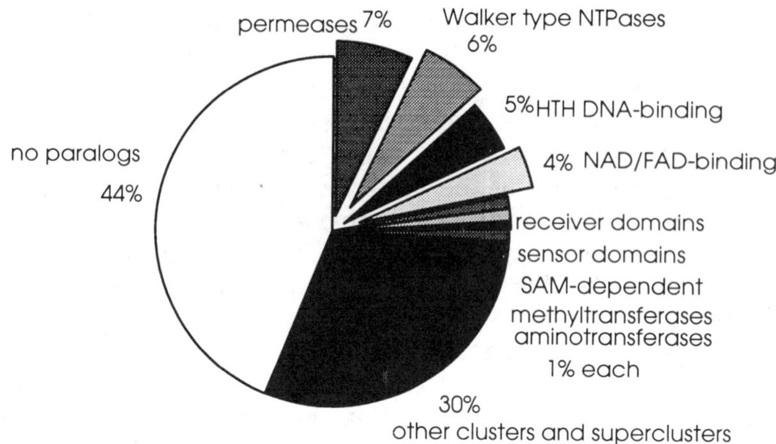

Figure 3. Clusters and superclusters of paralogous proteins in *E. coli.*

should have been encoded in the genome of the progenote - the common progenitor of bacteria, eukaryotes, and Archaea. Proteins that belong to clusters are, on average, more highly conserved than "loners" - 52% of the proteins that have paralogs contain ACRs, as opposed to 30% of the "loners". This probably should have been expected as the likelihood of detecting paralogs is higher for evolutionarily conserved proteins. Given this connection, it appears to be especially significant that in *E. coli*, such proteins as DNA and RNA polymerases which are highly conserved among bacteria, eukaryotes, and Archaea do not have detectable paralogs.

Given the overall high conservation of protein sequences in clusters, it is all the more remarkable that several clusters that are highly conserved between *E. coli* and distantly related bacteria do not seem to have homologs in eukaryotes or show only a very weak similarity to eukaryotic proteins. Examples include several clusters comprising the PTS system of sugar transport and phosphorylation (Saier, 1994; Saier and Reizer, 1994) that have no homologs among the known eukaryotic proteins, and the HTH transcription regulators whose similarity to homeodomains (Qian et al., 1989; Treisman et al., 1992) could be detected only using motif search methods. These unique transport and regulatory systems may define, at least in part, those features of the bacterial cell physiology that distinguish it from the eukaryotic cell.

The paralogous genes are non-randomly distributed along the *E coli* chromosome (Figure 6). The most prominent peak corresponds to closely spaced paralogs, many of which are actual tandem duplications.

The other peaks are due to a relatively weak, large-scale periodicity, with the distance between paralogs tending to be a multiple of 6-7 min of the E. coli chromosome (Figure 6). It is striking that the results of our analysis performed with 4,584 pairs of paralogous genes are in agreement with those obtained by Kunisawa and Otsuka (1988) with only 46 pairs of paralogs. Labedan and Riley (1995) reported a lack of periodicity in the distribution of paralogs along the *E. coli* chromosome. However, inspection of the actual graph published by these workers suggests that a regularity similar to that observed in our analysis may be detectable, albeit less convincingly. The discrepancy between the two studies may be attributed to the different methods used for sequence comparison, with the more restrictive measure employed by

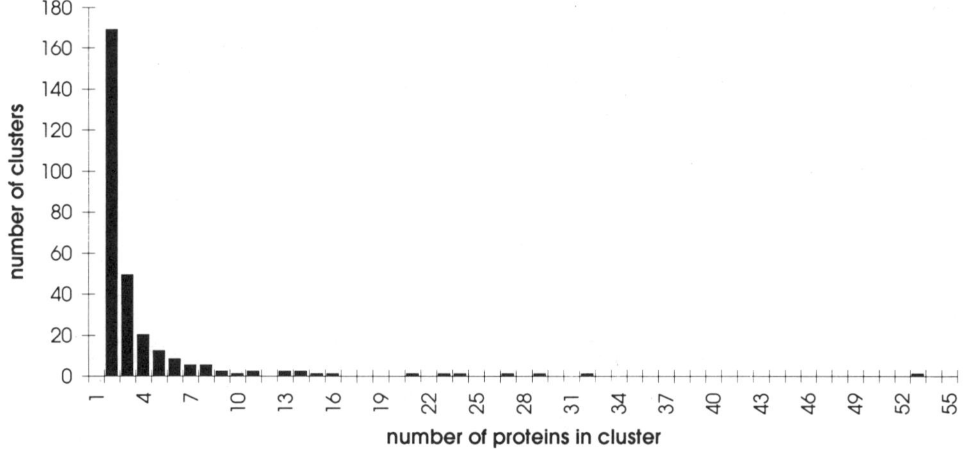

Figure 4. Distribution of the number of paralogs in *E. coli* protein clusters.

Labedan and Riley missing some of the paralogous pairs.

An evolutionary scenario with four consecutive duplications of an ancestral genome, originally proposed by Kunisawa and Otsuka (1988) is one possible explanation for the 6-7 minute periodicity. A difficulty with this theory is the aforementioned lack of duplication of genes coding for essential proteins involved in replication and expression which one would necessarily expect to be present even in a most primitive ancestor. On the other hand, if there is no selective advantage in maintaining duplicates of these genes, they could have been eliminated after each genome duplication. Alternatively, one may speculate that the observed periodicity might be related to the superstructure of the *E. coli* chromosome. If the size of a structural

Table 2. The 10 largest clusters of paralogs in *E. coli*—transport and regulatory proteins.

Cluster	Number of proteins	Function	Homologs in distantly related Bacteria	Homologs in Eukaryotes/Archaea
ABC transporter ATPases	54	ATP-dependent membrane transport; DNA repair	yes	yes
Permeases (AraE-related)	34	membrane transport	yes	yes
HTH proteins (Ada-related)	29	transcription regulation	yes	no
Receiver domains	28	membrane signal transduction	yes	yes
Sensor domains	24	membrane signal transduction	yes	yes
Permeases (ArtM-related)	24	membrane transport	yes	yes
HTH proteins (CynR-related)	21	transcription regulation	yes	no
Sugar-binding domains	15	metabolite transport	yes	
DEAD/H helicases	15	RNA/DNA duplex unwinding	yes	yes
GTPases	15	GTP-dependent processes	yes	yes

Table 3. Examples of small clusters of paralogous enzymes in *E. coli.*

Cluster	Proteins	Enzymatic activity	Homologs in distantly related bacteria	Homologs in eukaryotes/ Archaea
acetyltransferase	1. AccB	biotin carboxyl carrier protein	yes	yes
	2. AceF	dihydrolipoamide acetyltransferase	yes	yes
	3. SucB	dihydrolipoamide acetyltransferase	yes	yes
acetate kinase	1. AckA	acetate kinase	yes	yes
	2.YhaA'	??	yes	yes
acid phosphatase	1. Agp	glucose-1-phosphatase	no	no
	2. AppA	pH 2.5 acid phosphatase	no	no
alanine racemase	1. alr	alanine racemase	yes	no
	2. dadX	alanine racemase	yes	no
aminotransferase	1. AspC	aspartate aminotransferase	yes	yes
	2. TyrB	aromatic amino acid aminotransferase	no	yes

unit is close to 6-7 minutes and the units are juxtaposed in space, an increased likelihood of gene duplicates being transferred to new locations separated from the original ones by multiples of 6-7 minutes might be expected. It is well established that *E. coli* chromatin superstructure exists, but it has not been fully characterized (reviewed by Drlica, 1987). Further studies of the nucleoid organization may help explain the observed distribution of paralogous genes.

INSIGHTS FROM PROTEIN SEQUENCE COMPARISON

An Example of Functional Prediction for Uncharacterized *E. coli* Proteins — a New Family of Putative GTPases with Modified Conserved Motifs.

A detailed discussion of the functional prediction made in the course of the systematic analysis of the *E. coli* protein sequences is beyond the scope of this chapter. As an example, we discuss here the characterization of a new protein family that allows us to predict the enzymatic activity and function of 5 *E. coli* proteins at once. Inspection of the results of BLASTP searches indicated that 5 proteins showed a pattern of amino acid conservation that resembled that in GTPases,with the exception that a glutamic acid residue replaced the

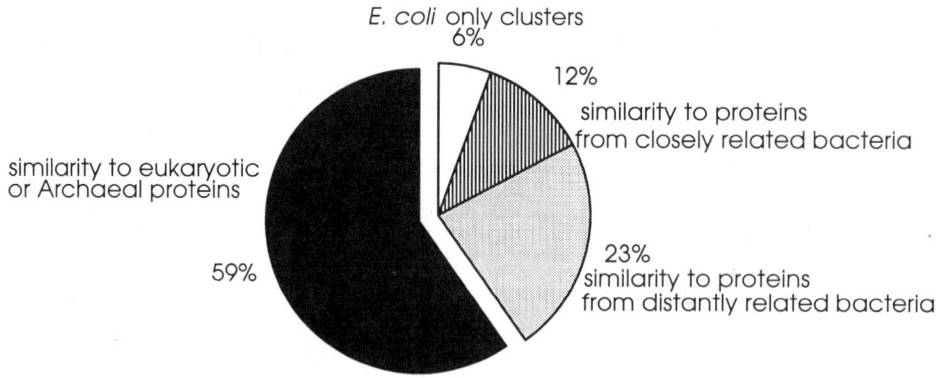

Figure 5. Sequence conservation in clusters of paralogous proteins in *E. coli*.

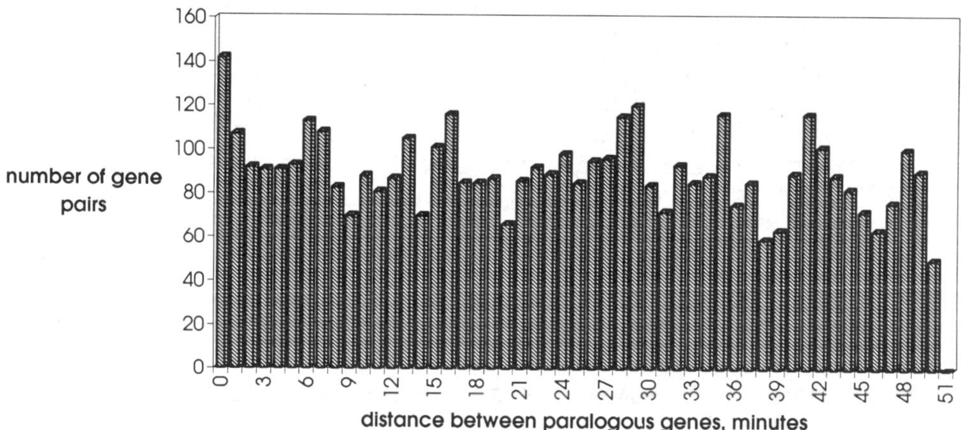

Figure 6. Distribution of paralogous genes on the *E. coli* chromosome.

conserved aspartic acid residue in the G3 motif which is implicated in Mg2+ binding (Dever et al., 1987; Bourne et al., 1990). These proteins included HypB (a hydrogenase subunit), UreG (an urease subunit), and 3 uncharacterized open reading frame (ORF) products - YeiR, YjiA, and YgfD. YeiR and YjiA showed up as a two-member cluster in our analysis of the *E. coli* paralogs, whereas the other three proteins were among the "loners". Additional, iterative BLASTP searches followed by a multiple alignment analysis using the MACAW program resulted in the delineation of a unique set of about 20 bacterial proteins with 3 conserved motifs corresponding to G1, G3, and G4 (Figure 7). We designate this set G3E proteins for the diagnostic substitution of a glutamic acid (E) for the conserved aspartic acid in the G3 motif. The G3E proteins belong to 2 distinct groups, one of which includes hydrogenase subunits (HypB) and urease subunits (UreG) from various bacteria, whereas the other one consists of the *Pseudomonas denitrificans* cobalamin biosynthesis protein CobW and several uncharacterized ORF products. The proteins within each of these families were highly similar to each other, with the probability of matching by chance below 10-9 (the CobW family) or below 10-4 (the HypB/UreG family). Among the *E. coli* G3E proteins, HypB, UreG and YgfD belonged to the HypB/UreG group, whereas YeiR and YjiA are in the CobW group.

The search among known protein sequences detected G3E proteins only in bacteria. However, when the translation of nucleotide version of the NR database was searched using TBLASTN, a highly significant similarity (P < 10-13) between the proteins of the CobW family and the putative products of two randomly cloned, overlapping human cDNAs (ESTs) was detected (Figure 7). Even though caution is due in the interpretation of high similarity to ESTs because of the possibility of bacterial contamination (Savakis and Doelz, 1993), in this particular case, this is very unlikely as the sequence of the putative human homolog of the CobW-related proteins was reconstructed from 2 independent clones (see legend to Figure 7). Therefore, even though complete sequences of eukaryotic G3E proteins so far have not been reported, it appears that similarly to other GTPase families, this group of proteins contains an ACR.

The similarity between the CobW and the HypB/UreG families was not statistically significant but concentrated in the G1, G3, and G4 motifs. The whole set of the G3E proteins could be retrieved from the NR database without any false positives using the following pattern of conserved amino acid residues - Ux4GxGK[ST]xnUUUEx2GxnUU[NTS]KxD (U indicates a bulky hydrophobic residue, i. e. I, L, V, M, F, Y or W). The proteins of the UreG/HypB group showed limited similarity to certain GTPases in the G1 and G4 motifs, and to ATPases in the G1 motif which corresponds to the P-loop, the wide-spread phosphate-binding site (Walker et al., 1982; Saraste et al., 1990). G3E proteins show some additional deviations from the general GTPase consensus. Specifically, nitride hydratase subunit from Rhodococcus erythropolis, which belongs to the CobW family, contains serine instead of asparagine or threonine in the G4 motif, whereas HypB proteins contain serine instead of the first glycine in the G1 motif (Figure 7).

The conservation of the G1, modified G3, and G4 motifs suggests that the G3E proteins comprise a new group within the GTPase superfamily. HypB binds GTP and has an intrinsic GTPase activity (Maier et al., 1993). This protein is required for nickel incorporation into the hydrogenase apoprotein (HycE), and on the basis on the analogy with other GTPases, it has been speculated that HypB complexed with GTP binds to HycE and facilitates Ni2+ incorporation; under this hypothesis, GTP hydrolysis is required for HypB release from the complex (Maier et al., 1993). UreG also is an accessory protein that is required for nickel incorporation into the urease apoenzyme (Lee et al., 1992; Cussac et al., 1992). CobW is the product of one of the genes of the cob operon that encodes proteins involved in cobalamin (coenzyme B12) biosynthesis including a cobaltincorporation step (Crouzet et al., 1991). Therefore, it seems likely that all the G3E proteins may be involved in the maturation of metal-containing enzymatic complexes.

We believe that the G3E family example illustrates the type of conclusions and predictions that are likely to result from protein sequence analysis, as well as the limitations of this analysis. Sequence comparison methods are very helpful in predicting the enzymatic activity and the general type of process in which an uncharacterized protein is likely to be involved.

```
                       G1
                 ▮▮▮▮▮▮▮▮▮▮▮▮▮▮▮▮▮
EFTU_ECOLI   13  NVGTIGHVDHGKTTLTAAI
consensus 1   1  ..UU$GULGOGKTT.U..UL.......JUOUUUNB      U..U..GCUC.........
YjiA         33  VTLLTGFLGAGKTTLLRHILNEQH-GYKIAVIENE  17  IKTLTNGCICCSRSNELED
YeiR          3  TNLITGFLGSGKTTSILHLLAHKDPNEKWAVLVNE  17  LKEIPGGCMCCVNGLPMQV
YTR2_SPIAU    9  VTVLTGFLGSGKTTLVNRLLKERP-GTRFGLVVNE   ?
COBW_PSEDE   12  ATVITGFLGAGKTTMIRNLLQNAD-GKRIGLIINE  23  IIELTNGCICCTVADDFIP
P47K_PSECL   11  VTVLSGFLGAGKTTLLNAILRNRQ-GLRVAVIVND  23  LIEMSNGCICCTLRADLLE
NIHY/RHOSP    7  VTVLSGFLGAGKTTLLNEILRNRE-GRRVAVIVND  24  LVEMTNGCICSTLRDLLSE
EST/HUMAN     ?  VTIITGYLGAGKTTLLNYILTEQH-SKRVAVILNE  32  WLELRNGCLCCSVKDNGLR
consensus 2      .U.U.O..GOGK$.UU...
UREG_BACSB    4  RIGIGGPVGAGKTMLVEKL
UreG          8  RIGVGGPVGSGKTALLEVL
HypB        105  VLNLVSSPGSGKTTLLTET
HUPM_RHOCA  148  ATNLVSSPGSGKTTLLVKT
YPLE_CAUCR   55  RIGITGVPGAGKSTTIERF
YgfD         30  RLGVTGTPGAGKSTFLEAF
consensus        ...U.O..GOGK$..U...
RASH_HUMAN    4  KLVVVGAGGVGKSALTIQL


                       G3                    G4
                 ▮▮▮▮▮▮▮▮▮▮▮▮          ▮▮▮▮▮▮▮▮▮▮▮▮
EFTU_ECOLI   41  TRHYAHVDCPGH  41  GVPYIIVFLNKCMDV  253  P02990
consensus 1      ..D.UUUE.$GU      U..A..UUU.K.DU.
YjiA         14  QFDRLVIECTGM  47  VGYADRILLTKTDVA  155  P24203
YeiR          9  KPDRLLIEPTGL   ?
COBW_PSEDE   10  RPDHIIIETSGL  68  LTAADLIVLNKTDLI  161  P33030
P47K_PSECL   10  RFDYLLIESTGI  69  VEYANVILVNKRDLI  227  P29937
NIHY/RHOSP    8  RFDYLLIESSGI  65  IEFADVILVSKADLI  212  P31521
EST_HUMAN    10  KFDYILLETTGL   ?
consensus 2      ...UUUUE..G.      U..O.LUUUNK.D..
UREG_BACSB   69  DLELIFIESGGD  33  MIKSVLFIINKIDLA   52  Q07403
UreG         69  NLDIVFVESGGD  33  ITHSDLLVINKIDLA   49  Q03287
HypB         59  DNGILFIENVGN  33  FAAASLMLLNKVDLL   47  P24190
HUPM_RHOCA   58  AGALLFIENVGN  33  FAAAGLAILNKVDLA   50  P26410
YPLE_CAUCR   71  GFDVVIIETVGV  33  IELADLLVINKADAD  121  P37895
YgfD         71  GYDVVIVETVGV  33  MEVADLIVINKDDGD   88  P27254
consensus        ....UUUE..G.      U..O..UUU.K.D..
RASH_HUMAN   26  TCLLDILDTAGQ  45  DDVPMVLVGNKCDLA   68  P01112
```

Figure 7. Conserved motifs in the G3E family of putative GTPases. The alignment was constructed using the MACAW program. The boundaries of the blocks were determined so as to ensure the maximal statistical significance (Schuler et al., 1991). The distances between the blocks are indicated by numbers. Only two representative sequences of UreG proteins and HypB proteins from distantly related bacteria are included. Consensus 1 is for the CobW family, and consensus 2 is for the HypB/UreG family. Consensus 2 was derived allowing one exception among the aligned sequences. The general consensus is the overlap between consensus 1 and consensus 2. B indicates an acidic residue (D or E); J indicates a basic residue (K or R); O indicates a small residue (G, A or S); $ indicates serine or threonine; U indicates a bulky hydrophobic residue (I, L, V, M, F, Y or W); and dot indicates any residue. The residues that conform to the consensus patterns are highlighted by bold type. The G1, G3, and G4 motifs are delineated by bars, and the respective sequences from GTPases with known tertiary structure (EF-Tu and p21ras) are shown for comparison. The sequences were from the SWISS-PROT (names with underline), PIR or GenBank databases; the accession numbers are indicated in the rightmost column. EST/HUMAN designates the combined products of 2 overlapping partial human cDNA clones (GenBank D20766 and T30462).

On the other hand, much caution is required when it comes to more specific biological interpretations.

E. coli Proteins Help Make Functional Predictions for Their Eukaryotic Homologs

An important goal of sequence studies on well characterized, model organism genomes is the prediction of eukaryotic, and particularly human gene functions. The importance of sequence similarity between human proteins and their homologs from model organisms has been dramatically highlighted by the recent discoveries of the human hereditary colon cancer genes based upon the identification of human homologs of bacterial and yeast DNA repair genes (Fishel et al., 1993; Leach et al., 1993; Service et al., 1994; Papadopoulos et al., 1994).

Our analysis of the *E. coli* protein sequences resulted in new functional predictions for a number of eukaryotic proteins. Some of the most interesting cases are presented elsewhere (Koonin, 1994; E. V. K., R. L. T., and K. E. R., submitted). Below we briefly describe another typical example.

We have shown previously that three functionally uncharacterized *E. coli* gene products - SpoU, YfifF, and YibK - belong to a distinct family of rRNA methyltransferases and are predicted to possess this activity (Koonin and Rudd, 1993). Our current analysis has identified two additional, recently sequenced *E. coli* gene products - LasT and YjfH - as members of the same family. Furthermore, we showed that the conserved motifs that are diagnostic of this family could be detected, at a statistically significant level, in the recently described yeast rRNA methylase (Figure 8).   The yeast rRNA methylase, the product of the PET56 gene, methylates the ribose of an invariant guanine nucleotide in the peptidyl transferase center of the mitochondrial 21S rRNA and is required for the formation of functionally active large subunits of the mitochondrial ribosomes (Sirum-Conolli and Mason, 1993). The detection of the conserved motifs allows one to make testable predictions as to the identity of the active center of this enzyme. In this example, comparative sequence analysis expands a family previously confined to bacteria into the eukaryotic domain.

Repetitive Binding Domains Come Across Kingdom Borders

Sequence conservation between bacterial and eukaryotic enzymes is well established. In contrast, domains implicated in macromolecular interactions and complex formation are much less conserved. In particular, a number of repetitive domains probably involved in different types of protein-protein interactions are thought to be unique for eukaryotic proteins or at least much more frequent in them. In the course of our analysis of the *E. coli* protein sequences, we showed that they contain at least four of such "eukaryotic" repeats. Typically, the identification of a repetitive domain was triggered by a limited similarity detected using BLASTP with an *E. coli* protein as a query and extended by database search with the derived alignment block using MoST. In each case, the results of the MoST searches were statistically highly significant (the details of this analysis will be presented elsewhere; EVK and KER, manuscript in preparation).

TPR repeats. TPR is a 34-amino acid repetitive domain that consists of two α-helices and is implicated in protein-protein interactions and assembly of macromolecular complexes (Sikorski et al., 1990, 1991; Goebl and Yanagida, 1991). Multiple (typically, about 10) TPR repeats have been found in a variety of eukaryotic proteins, primarily from yeast, that function in mitosis, transcription, splicing, protein transport and other processes. A single TPR unit also has been identified in *Bacillus subtilis* proteins GutR, putative ATPase involved in glucitol operon transcription regulation, and GsiA, glucose starvation-induced protein (Ye et al., 1994), as well as in an uncharacterized cyanobacterial protein (Sikorski et al., 1991). We identified the TPR domain in 9 *E. coli* gene products (Figure 9a). Among these (putative) proteins, NrfG, YejP, and yhjL belong to a cluster defined on the basis of pairwise similarity, whereas the other proteins were brought into the emerging supercluster based on the TPR motif conservation. Eight of the 9 *E. coli* TPR proteins contain only one or two readily detectable copies of the TPR domain but one protein, YhjL, contains at least five copies (Figure 9a). None of the *E coli* proteins containing the TPR motif has been characterized in detail. NrfG is required for nitrate reductase activity (Hussain et al., 1994), HemY appears to be involved in porphyrin biosynthesis (Alefounder et al., 1988), and NfrA is an outer membrane protein identified as the bacteriophage N4 receptor (Kiino et al., 1993). These data are insufficient for any strong functional predictions; it is tempting to speculate, nevertheless, that at least in NrfG and HemY, the TPR domains may be involved in the assembly of the respective multisubunit enzymatic complexes.

DHR repeats. The DHR domain has been identified as an approximately 90-amino acid repetitive element in a growing superfamily of eukaryotic peripheral membrane proteins that also contain a modified guanylate kinase domain (Woods and Bryant, 1991; Bryant and Woods, 1992; Koonin et al., 1992; Ruff et al., 1991; Lue et al., 1994). These proteins localize in cell junctions and are thought to be involved in cell-to-cell signal transduction; some of them are tumor suppressors (Woods and Bryant, 1991; Cho et al., 1992). In addition, the DHR domain has been found in the membrane-associated nitric oxide synthase (Bredt et al., 1991) and in *Drosophila* segment polarity protein Dishevelled (Klingensmith et al., 1994). The common property of these proteins is peripheral association with the plasma membrane, and it has been proposed that the DHR domain may mediate this binding (Klingensmith et al., 1994). We identified the bacterial homolog of the DHR domain in four periplasmic proteases from *E. coli*, namely HtrA, HhoA, HhoB that form a cluster of paralogs, and Prc that does not show significant similarity to the other three proteases, and in their homologs from other bacteria (Figure. 9b). We found that the most conserved motif of the DHR domain consists of 26 amino acid residues (Figure. 9b). HtrA and HhoA each contain two copies of the DHR domain, whereas HhoB and Prc only contain one copy each. The

```
consensus           U.UUU.....P.NUOOUUR$....G...UU       ..UUUG.E..
spoU          19    LTVCMEQVHKPHNVSAIIRTADAVGVHEVHA  63   TCILMGQEKT--
lasT           2    ITIILVAPARAENIGAAARAMKTMGFSDLRI  74   AALVFGREDS--
yibK           1    LNIVLYEPEIPPNTGNIIRLCANTGFRLHII  62   DYLMFGPETRgl
yfiF         197    CVLALENESNPHNLGGMMRSCAHFGVKGVVV  63   MVLVLGQEYE--
yjfH          95    FLLILDGVTDPHNLGACLRSADAAGVHAVIV  64   LALVMGAEGE--

SPOU/MYCCA    71    LVLVLDQIHDPYNFGAIIRSCSLLNVDGIII  65   SVVIIGNEQK--
YACO_BACSU    97    FFLILDELEDPHNLGSIMRTADAVGAHGIVI  64   LALVIGSEGK--
TSR_STRAZ    116    DVVVLDGVKIVGNIGAIVRTSLALGASGIIL  65   LALLFGSEKG--
YGL3_BACST     2    LHVVLFQPEIPANTGNIARTCAATNTALHLI  65   IFFIFGRETTg-

PET56/YEAST  241    LGLYLDEITDPHNIGAIIRSAYFLGVDFIVM  78   VVLVVGNESQ--


                    G.........D..U.UPU.              ..U.OUNUS.A.OOUU
spoU                GITQEALALADQDIIIPMI----------GMVQSLNVSVASALIL   P19396
lasT                GLTNEELALADVLTGVPMV----------ADYPSLNLGQAVMVYC   P37005
yibK                PASILDALPAEQKIRIPMV----------PDSRSMNLSNAVSVVV   P33899
yfiF                GLPDAARDPNDLRVKIDGT----------GNVAGLNISVATGVLL   P33635
yjfH                GMRRLTREHCDELISIPMA----------GSVSSLNVSVATGICL   P39290

SPOU/MYCCA          GVSELLTKNSDFNIYVPSN----------KNIDSFNSSVACSIIC   Z33076_2
YACO_BACSU          GMGRLVKEKCDFLIKLPMA----------GKVTSLNASVAAGLLM   Q06753
TSR_STRAZ           GPSDLFEEASSASVSIPMM----------SQTESLNVSVSLGIAL   P18644
YGL3_BACST          LPKELLAENEDRCLRIPMT----------ENVRALNLSNTAAILV   P32813

PET56/YEAST         GVRTNLKMRSDFFVEIPFGgiekgnrapePIVDSLNVSVATALLI   S48881
```

Figure 8. Conserved sequence motifs in the SpoU family of (putative) rRNA methylases. The alignment was generated using the MACAW program. The probability of obtaining each of the aligned blocks by chance was below 10-9. The designations are as in Figure 7. The accession numbers in the SWISS-PROT or PIR protein sequence databases are indicated in the rightmost column.

periplasmic localization of these proteases (Lipinska et al., 1990) is fully compatible with the proposed role of the DHR domain in peripheral association of proteins with membranes.

IRBP repeats. In addition to the DHR domain, the *E. coli* Prc protease was shown to contain another 35-amino acid motif that is a part of the repeated domain in mammalian interphotoreceptor retinol-binding proteins (Figure 9c; Silber et al., 1992). In this case, the statistically significant similarity between Prc (and its homologs from other bacteria) and IRBP has been detected by the original BLASTP search, whereas the subsequent motif analysis has not revealed additional occurrences of the motif. The actual function of the repetitive domain in IRBP is not known; the current hypothesis is that these domains may be involved in binding multiple retinoid and fatty acid ligands (Pepperberg et al., 1993).

Ankyrin repeats. The ankyrin (ANK) repeat, which consists of 33 amino acids residues, is one of the most wide-spread repetitive elements in eukaryotic proteins (Bennett, 1992; Michaely and Bennett, 1992; Bork, 1993). Like many other repetitive domains, ANK repeats are implicated in protein-protein interactions; in particular, it has been shown that they are responsible for coupling of various integral membrane proteins to spectrin (reviewed by Bennett, 1992).

```
consensus               U...U.........UU.UO..U...G....O...
HemY          317:  LRQQIKNVGDRPLLWSTLGQSLMKHGEWQEASLA  P09128   E. coli
NfrA          635:  LRAALELEPNNSNTQAALGYALWDSGDIAQSREM  P31600
NrfG           62:  LQDKIRANPQNSEQWALLGEYYLWQNDYSNSLLA  P32712
NrfG          133:  IDKALALDSNEITALMLLASDAFMQANYAQAIEL
YhbL          223:  LASRALTDDERAQLLYERGVLYDSLGLRALARND  P26428
YhbM          119:  FDSVLELDPTYnYAHLNRGIALYYGGRDKLAQDD  P39833
YhbM          223:  ATDNTSLAEHLSETNFYLGKYYLSLGDLDSATAL
YhjL          275:  LQQAVRANPKDSEALGALGQAYSQKGDRANAVAN  P37650
YhjL          357:  FQQARNVDNTDSYAVLGLGDVAMARKDYPAAERY
YhjL          467:  QRQRLALDPGSVWITYRLSQDLWQAGQRSQADTL
YhjL          574:  EAEVLRQQPPSTRIDLTLADWAQQRRDYTAARAA
YhjL          642:  LAKLPATDNASLNTQRRVALAQAQLGDTAAAQRT
YjcO           24:  SQYLKAAEAGDRRAQYFLADSWFSSGDLSKAEYW  P32713
YejP          211:  MRTQLQKNPGDIEGWIMLGRVGMALGNASIATDA  P33925
YejP          282:  LRQLVRTDHSNIRVLSMYAFNAFEQQRFGEAVAA
YchA'          81:  SEALLQFNPEDPYEIRDRGLIYAQLDCEHVALND  P20101


LCRH_YERPE     59:  FQALCVLDHYDSRFFLGLGACRQAMGQYDLAIHS  P21207   other bacteria
ACSC_ACEXY    314:  FQSALQINSHDADSLGGMGLVSMRQGDTAEARRY  P19450
ACSC_ACEXY    498:  LREAMAQAPRDPWVRINLANALQQQGDVAEAGRV  P19450
BCSC_ACEXY    513:  LRQAMTKAPDDPWLRINLANALQQQGDSAEAANV  P37718
YREC_SYNP2     66:  YRQALTLEANNARIHGALGYALSQLGNYSEAVTA  P19737
YREC_SYNP2    100:  YRRATELEDDNAEFFNALGFNLAQSGDNRSAINA  P19737
YREC_SYNP2    134:  YQRATQLQPNNLAYSLGLATVQFRAGDYDQALVA  P19737
YREC_SYNP2    202:  FPDLLRQRPNDAELRIKAAVTWFGLNDRDQAIAF  P19737
FRZF_MYXXA    446:  VQRLLADEPSDLDGLLTLGNLFSLTGRIPEAREA  P31759
FRZF_MYXXA    480:  FAQAIQREPLCVEARVFGGVAALQAGELSEARSE  P31759
FRZF_MYXXA    514:  LSKALFLEPTLAIGHYLLAQVHERTQDHEAARRS  P31759
YAD5_CLOAB     88:  LLQAIKLRPKTINDVYSFALSYHILGEPERALKY  P33746
YAD5_CLOAB    122:  FLRAVELQPNVGISYENLAWFYYLTGKYDKAIEN  P33746
YAD5_CLOAB    189:  LKKALDAEPEKPSTHIYFSYLKRKTNDIKLAKEY  P33746
GUTR_BACSU    725:  EISSTYGGEKTIEAYFNLGVAYVKCDQFEKAEEA  P39143
GUTR_BACSU    764:  YDKHNANQVELIYYHYGMAQLLYRKGEKTKAVES  P39143
GSIA_BACSU    211:  LAKKEGNPRLISSALYNLGNCYEKMGELQKAAEY  Q00828


YCF3_GALSU     61:  ALNLEEDPYDKSFILYNIGLIHASNGEYVKALDY  Q08814   chloroplast
YCF3_ORYSA     61:  ATRLESDPYDRSYILYNIGLIHTSNGEHTKALEY  P12203


CC16_YEAST    520:  SHRLAETFPKSAITWFSVATYYMSLDRISEAQKY  P09798   eukaryotic
CC16_YEAST    554:  YSKSSILDPSFAAAWLGFAHTYALEGEQDQALTA  P09798
CC16_YEAST    622:  FVLAYDICPNDPLVLNEMGVMYFKKNEFVKAKKY  P09798
CC16_YEAST    697:  FRCVLEKNDKNSEIHCSLGYLYLKTKKLQKAIDH  P09798
YCA1_PLAFA     30:  FSDAITNDPLDHVLYSNLSGAFASLGRFYEALES  P25407
YCA1_PLAFA     64:  ANKCISIKKDWPKGYIRKGCAEHGLRQLSNAEKT  P25407
INI6_HUMAN    206:  LRQAVRLNPDNGYIKVLLALKLQDEGQEAEGEKY  P09914
INI6_HUMAN    240:  IEEALANMSSQTYVFRYAAKFYRRKGSVDKALEL  P09914
INI6_HUMAN    329:  FESAVEKKPTFEVAHLDLARMYIEAGNHRKAEEN  P09914
INI6_HUMAN    426:  VLRKLRRKALDLESLSLLGFVYKLEGNMNEALEY  P09914
NCF2_HUMAN     60:  FTRSINRDKHLAVAYFQRGMLYYQTEKYDLAIKD  P19878
NCF2_HUMAN     26:  LDAFSAVQDPHSRICFNIGCMYTILKNMTEAEKA  P19878
```

Figure 9. "Eukaryotic" repetitive domains found in bacterial proteins. (a)
TPR repeats. The alignment includes all the bacterial and selected eukaryotic
sequence segments from the output of the MoST program. The number of the
starting amino acid of the TPR motif in the respective protein sequence is
indicated. The consensus includes amino acid residues that are conserved in the
majority of the TPR domain sequences. The sequences were from the SWISS-
PROT database; the accession numbers are indicated.

```
consensus              ...A....U..GDUUU.UN...U..U
HTRA_ECOLI       322:  NSSAAKAGIKAGDVITSLNGKPISSF   P09376    E. coli
HTRA_ECOLI       419:  GTPAAQIGLKKGDVIIGANQQAVKNI
HHOA_ECOLI       301:  GSGSAKAGVKAGDIITSLNGKPLNSF   P39099
HHOA_ECOLI       400:  GSPAAQAGLQKDDVIIGVNRDRVNSI
HHOB_ECOLI       290:  DGPAANAGIQVNDLIISVDNKPAISA   P31137
PRC_ECOLI        268:  GPAAKSKAISVGDKIVGVGQTGKPMV   P23865

HTRA/BRUAB       343:  DGPAAKAGIKAGDVITAVNGETVQDP   U07352    other bacteria
HTRA/BRUAB       450:  DSDAADRGIRSGDVIVSVNNQTVKTA   U07352
HHOA/BRUAB       307:  DGPAETAGLKVGDVVLSVQGVRVDNQ   U07351
HHOA/BRUAB       418:  GCPAARLGLRSGDIIRSINGNQIRTV   U07351
HTRA/MYCLE       470:  GSPAQKGGILENDVIVKVGNRKVADA   U15180
HTRA/ROCHE       440:  DSDAADKGIRPGDVIVTVNNKSVKKV   L20127
PRC/BARBA        117:  DTPAAKAGVLAGDFISKIDGKQISGQ   L37094
SP4B_BACSU       136:  KSPGETAGIEAGDIIIEMNGQKIEKM   P17896
YYXA_BACSU        18:  FSPAGKAGLKELDVITEFDGYKVNDI   P39668
ORF/SYNSP        135:  GSPAEAAGIEAKDQILAIDGIDTRNI   S18125

ENIG/HUMAN        36:  GGKAAQAGVAVGDWVLSIDGENAGSL   A55050    eukaryotic
CHEMA/HUMAN       68:  aASEQSETVQPGDEILQLGGTAMQGL   M90391
DLGA/RAT         198:  GAGHKDGRLQIGDKILAVNSVGLEDV   X66474
DAG3/HUMAN       121:  LAADQTEALFVGDAILSVNGEDLSSA   U01243
AF6/HUMAN       1028:  GAADVDGRLAAGDQLLSVDGRSLVGL   U02478
NOS1_HUMAN        50:  GAAEQSGLIQAGDIILAVNGRPLVDL   P29475
DLG1_DROME       518:  GPADLGSELKRGDQLLSVNNVNLTHA   P31007
PSD9_RAT         345:  GPADLSGELRKGDQILSVNGVDLRNA   P31016
ZO1_MOUSE         61:  KGGPAEGQLQENDRVAMVNGVSMDNV   P39447
ZO1_MOUSE        453:  DSPAAKEGLEEGDQILRVNNVDFTNI   P39447
YIA7_YEAST       143:  GSPSDKADIKVDDKLISIGNVHAANH   P40555
MPP1_HUMAN       103:  GMIHRQGSLHVGDEILEINGTNVTNH   Q00013
DVL-1/MOUSE      287:  GAVAADGRIEPGDMLLQVNDVNFENM   U10115
X104/HUMAN        71:  PGGPADGLLQENDRVVMVNGTPMEDV   L27476
X104/HUMAN       336:  GLATKDGNLHEGDIILKINGTVTENM   L27476
PTP/MOUSE        426:  QPAAESGKIDVGDVILKVNGAPLKGL   D28529
PTP/MOUSE        692:  DPAKGDGRLKAGDRLIKVNDTDVTNM   D28529
PTP/MOUSE        789:  SAAAVDGSLQLLDIIHYVNGVSTQGM   D28529
PICK1/MOUSE       56:  TPAALDGTVAAGDEITGVNGKSIKGK   Z46720
INAD/DROME        57:  SPAHLCGRLKVGDRILSLNGKDVRNS   U15803
INAD/DROME       615:  QYPEIDSKLQRGDIITKFNGDALEGL   U15803
INAD/DROME       283:  NGALGSVDIKPGDEIVEVNGNVLKNR   U15803
INAD/DROME       399:  GSNAELAGVKVGDMLLAVNQDVTLES   U15803
SAP97/RAT        262:  GAAAQDGRLRVNDCILRVNEADVRDV   U14950
PCF/RABIT         45:  GSPAEKAGLLAGDRLVEVNGENVEKE   U19815
PCF/RABIT        185:  DSPAEASGLREQDRIVEVNGVCVEGK   U19815
CNO/DROME        879:  GAADADGRLQAGDQLLRVDGQSLIGI   D49534
TIAM1/MOUSE      881:  TGLASKKGLKAGDEILEINNRAAGTL   A54146
MAST/MOUSE      1081:  GGPASEAGLRQGDLITHVNGEPVHGL   A54602
LIMPK/HUMAN      209:  MSPDVKNSIHVGDRILEINGTPIRNV   JP0078
ZIP/HUMAN         82:  DSPAHCAGLQAGDVLANINGVSTEGF   S43424
```

(b) DHR repeats. See legend to Figure 9(a). The sequences were from SWISS-PROT, PIR or GenBank.

```
consensus                UUUL....$A.A.E.UA..U....RAUUUGE.$.G
Prc              442:    LVVLVDRFSASASEIFAAAMQDYGRALVVGEPTFG    P23865    E. coli

PRC/BARBA        284:    IIVLINGGSASASEIVAGALQDHRRATIIGTQSFG    L37094    other
ORF1/NEIGO        15:    MAVLVNSGSASASEIVAGALQDHKRAVIVGTQSFG    U11547_1  bacteria
PRC/SYNSP        303:    LVVLVNQGTASASEILAGALQDNQRATLVGEKTFG    L25250
ORF/SYNSP        300:    LVVLVNQATASASEILAGALQDNGRAMLVGEKTFG    S18125

IRBP_HUMAN       227:    VVVLTSSQTRGVAEDIAHILKQMRRAIVVGERTGG    P10745    eukaryotic
IRBP_HUMAN       537:    VYLLTSHRTATAAEEFAFLMQSLGWATLVGEITAG    P10745
IRBP_HUMAN       840:    LYILMSHTSGSAAEAFAHTMQDLQRATVIGEPTAG    P10745
IRBP_HUMAN      1139:    MVILTSSVTAGTAEEFTYIMKRLGRALVIGEVTSG    P10745
IRBP_BOVIN       227:    VVVLTSSRTGGVAEDIAYILKQMRRAIVVGERTVG    P12661
IRBP_BOVIN       535:    VYLLTSHRTATAAEELAFLMQSLGWATLVGEITAG    P12661
IRBP_BOVIN       838:    LYVLVSHTSGSAAEAFAHTMQDLQRATIIGEPTAG    P12661
IRBP_BOVIN      1137:    MVILTSTLTAGAAEEFTYIMKRLGRALVIGEVTSG    P12661
```

(c) IRBP repeats. See legend to Figure 9(a). The sequences were from SWISS-PROT, PIR or GenBank.

Previously, ANK repeats havebeen found in the putative *E. coli* protein YjaC, as well as in two other bacterial proteins (Neuwald and Green, 1993; Bork, 1993). We confirmed these findings, and in addition, found a single copy of the ANK domain in another uncharacterized, putative E. coli protein, YifH (Figure 9d).

A "Prokaryotic" Repetitive Domain Found in Eukaryotes - the Isoleucine Patch Hexamer Repeats

The isoleucine patch (I-patch) domain is a hallmark of a large group of bacterial acyltransferases, as well as some other transferases whose substrates contain acyl groups (Vaara, 1992; Dicker and Seetharam, 1992; Vuorio et al., 1994). The domain consists of multiple, imperfect repeats with a unit length of 6 amino acid residues,with a bulky hydrophobic residue, mostly isoleucine, typically found in the first position of each unit and glycine found most frequently in the second position. Typically, acyltransferases contain 20-30 hexamer repeats, which in some of them (e. g. *E. coli* LpxA) comprise most of the protein length.

In addition to acyltransferases, we identified the I-patch domain in several (putative) nucleotidyltransferases of both bacterial and eukaryotic origin, and in two subunits of the eukaryotic translation initiation factor eIF-2B that also have been predicted to contain a nucleotidyltransferase-related domain (Koonin, 1995; Figure 10).

An all-β-sheet structure was strongly predicted for the I-patch domains, in agreement with the previous observations (Figure 10; Vuorio et al., 1994). The exact function of the I-patch domain is not known, but given that some acetyltransferases consist almost entirely of the hexamers, it seems likely that this domain may be involved in binding acyl groups. The function of the I-patch in the translation initiation factor remains enigmatic.   Thus, with the I-patch domain, the flow of information has been opposite to that with other repetitive domains - a repeat family that has been thought to be

```
consensus           U......O.$.U.UA...O...UU..UU....
YifH          139: LNWAYRRGKTTLRVATQMGNTAALKRYQSGANV   P27832   E.coli
YjaC          423: LEAKDKNGFSGLFLAISRKDKNVVTSILNALPK   P23325
YjaC          519: LKAKDFYGCPGLYLAMQNGHSDIVKVILEALPS   P23325
YjaC          567: LTAKSLARDTGLFMAMQRGHMNVINTIFNALPT   P23325

PHLB_SERLI    102: PAAAGLDGNSALHTAAMLQDAQYLRLLLAEGAQ   P18954   other
PHLB_SERLI    136: NVRNAVTGATPLAAAVLAGREEQLRLLLAAGAD   P18954   bacteria

LATA_LATMA    519: LNQPDKKGYTPIHVAADSGNAGIVNLLIQRGVS   P23631   eukaryotic
LATA_LATMA    553: NSKTYHFLQTPLHLAAQRGFVTTFQRLMESPEI   P23631
LATA_LATMA    587: INERDKDGFTPLHYAIRGGERILEAFLNQISID   P23631
LATA_LATMA    620: VNAKSNTGLTPFHLAIIKNDWPVASTLLGSKKV   P23631
LATA_LATMA    654: INAVDENNITALHYAAILGYLETTKQLINLKEI   P23631
LATA_LATMA    723: VNLKALGGITPLHLAVIQGRKQILSLMFDIGVN   P23631
LATA_LATMA    756: IEQKTDEKYTPLHLAAMSKYPELIQILLDQGSN   P23631
LATA_LATMA    789: FEAKTNSGATPLHLATFKGKSQAALILLNNEVN   P23631
LATA_LATMA    822: WRDTDENGQMPIHGAAMTGLLDVAQAIISIDAT   P23631
LATA_LATMA    856: VDIEDKNSDTPLNLAAQNSHIDVIKYFIDQGAD   P23631
LATA_LATMA    889: INTRNKKGLAPLLAFSKKGNLDMVKYLFDKNAN   P23631
LATA_LATMA    922: VYIADNDGMNFFYYAVQNGHLNIVKYAMSEKDK   P23631
LATA_LATMA    997: GTLGNFAICGPLHQAARYGHLDIVKYLVEEEFL   P23631
LATA_LATMA   1029: LSVDGSKTDTPLCYASENGHFTVVQYLVSNGAK   P23631
LATA_LATMA   1062: VNHDCGNGMTAIDKAITKNHLQVVQFLAANGVD   P23631
LATA_LATMA   1131: INEQNVDKDTALHLAVYYKNLQMIKLLIKYGID   P23631
PH81_YEAST    417: LNAQDIHSRVPLHYAAELGKLEFVHSLLITNLL   P17442
PH81_YEAST    452: VDPIDSDSKTPLVLAITNNHIDVVRDLLTIGGA   P17442
PH81_YEAST    500: VISSTKVQFDPLNVACKFNNHDAAKLLLEIRSK   P17442
PH81_YEAST    585: NEIDGFNKWTPIFYAVRSGHSEVITELLKHNAR   P17442
PH81_YEAST    618: LDIEDDNGHSPLFYALWESHVDVLNALLQRPLN   P17442
BCL3_HUMAN    120: ATRADEDGDTPLHIAVVQGNLPAVHRLVNLFQQ   P20749
BCL3_HUMAN    157: LDIYNNLRQTPLHLAVITTLPSVVRLLVTAGAS   P20749
BCL3_HUMAN    190: PMALDRHGQTAAHLACEHRSPTCLRALLDSAAP   P20749
BCL3_HUMAN    227: LEARNYDGLTALHVAVNTECQETVQLLLERGAD   P20749
BCL3_HUMAN    261: DAVDIKSGRSPLIHAVENNSLSMVQLLLQHGAN   P20749
BCL3_HUMAN    294: VNAQMYSGSSALHSASGRGLLPLVRTLVRSGAD   P20749
VB04_VARV     237: IDSVDFNGYTPLHYVSCRNKYDFVKSLISKGAN   P33823
VB04_VARV     333: INYRTIYNETSIYDAVSYNAYNMLVYLLNRNGD   P33823
```

(d) ANK repeats. See legend to Figure 9(a). All the sequences were from SWISS-PROT.

LpxA	20:	IGANAHIGPFCIVGPHVEIGEGTVL	P10440	E. coli
LpxD	224:	IGNGVIIDNQCQIAHNVVIGDNTAV	P21645	
ThgA	128:	YSFPITIGNNVWIGSHVVINPGVTI	P07464	
CaiE	98:	IGRDALVGMNSVIMDGAVIGEESIV	P39206	
YefH	135:	IGQRVWLGENVTVLPGTIIGNGVVV	P37750	
CysE	196:	IREGVMIGAGAKILGNIEVGRGAKI	P05796	
YefH	129:	ESSAVVIGQRVWLGENVTVLPGTII	P37750	
DapD	134:	VDEGTMVDTWATVGSCAQIGKNVHL	P03948	
GlmU	274:	IDTNVIIEGNVTLGHRVKIGTGCVI	P17114	
GlgC	373:	VGRSCRLRRC-VIDRACVIPEGMVI	P00584	
o256	155:	IGEDVTVGHKVMLH-GCTIGNRVLV	U18997	
CAT4_ECOLI	119:	IGSEAMIMPGIKIGHGAVIGSRALV	P26838	E.coli plasmid
E2BE_YEAST	333:	LAQSCKIGKCTAIGSGTKIGEGTKI	P32501	eukaryotic
eIFBe/RABIT	336:	LGHGSILEENVLLGSGTVIGSNCSI	NA	
E2BG_YEAST	423:	I-QSAAIGADAIVDPKCQISAHSNV	P09032	
PSA1_YEAST	263:	ISSTAKIGPDVVIGPNVTIGDGVRI	P41940	other bacteria
PPP1_CAEEL	331:	IGEKTKLKES-IIAKGVVIGNGCLH	P42004	
LPXD_RICRI	126:	IGKNCYIGHNVVIEDDVIIGDNSII	P32202	
THGA_ECOLI	134:	IGNNVWIGSHVVINPGVTIGDNSVI	P07464	
NODL_RHIME	134:	IGKHVWIGGGAIILPGVTIGDHAVV	P28266	
SATA/ENTFA	116:	IGNDVWIGKDVVIMPGVKIGDGAIV	L12033	
CAPG/STAAU	85:	IGNNVFIGINSIILPGVTIGNNVVV	U10927	
FIRA/PASMU	120:	IGANAVIEDGVTLGDHVVIGANCFV	S47342	
HRA196/YEAST	133:	IGDNVWIGGGVSIIPGVNIGKNSVI	S45154	
YR43_BACSU	133:	IGDQVWIGGRAVINPGVTIGDNAVI	P37515	
ORF/COXBU	120:	IGDHSVIAIGAIVMNNAIIGKNCII	S38220	
RFB/SALCH	152:	IGNCCWIGDNAVILAGSEICDGCVI	S22616	
CAT4_BACSH	74:	FKGDTVIGNDVWIGQNVTIMPGVII	P26840	
LPXD_YEREN	111:	LGENVSIGANAVIESGVVLGDNVVI	P32203	
Fbp/PSEAE	96:	VGINAVILNGAKIGKYCIIGANALI	M82832	
CAT/MORMO	113:	IGNDVWIGSEAMIMPGIKIGDGAVI	X82455	
SAT/ARATH	241:	IGAGSCILGNITIGEGAKIGSGSVV	L34076	
LPXA_RICRI	12:	IAEGAKLGKNVKIGPYCIIGPEVVL	P32199	
TMS_BACSU	326:	IGPFAHIRPDSVIGNEVKIGNFVEI	P14192	
LPXA_YEREN	2:	IDKTAVIHPSSIVEEGAVIGAGVHI	P32201	
NIFP_AZOCH	83:	HGACVVIGETAEIGRDVTLYHGVTL	P23145	
GLGC_ANASP	368:	IGPDTIIRRA-IIDKNARIGHDVKI	P30521	
GLGC_BACSU	323:	TVTSSVIMPDVTIGEHVVI-ENAIV	P39122	

Figure 10. "Isoleucine patch" repeats. The alignment includes segments from all the E. coli proteins and eukaryotic proteins, and selected proteins from other bacteria and was generated from the MoST output. The alignment block used for database screening consisted of 4 hexamer repeats, which are shown by arrows. Only one segment with the highest score is shown for each sequence. The bulky hydrophobic residues recurring in every 6th position and the adjacent glycines are highlighted by bold type. Gaps were introduced manually in some of the sequences in order to restore the periodicity. The sequences were from the SWISS-PROT database.

unique for bacteria was extended to include eukaryotic proteins.

Some general trends from the analysis of the repetitive domains. The five examples presented above show that ACRs can be found not only in enzymes, but also in domains implicated in macromolecular interactions. Predictably, these ACRs are less conserved than those that include active centers of enzymes, and motif searching methods are essential for their detection. When repeats formerly thought to be eukaryote-specific are detected in bacteria, they are found in a relatively small number of proteins, and these proteins typically contain a smaller number of repeat units than their eukaryotic counterparts. This is compatible with the genome compactness constraint that appears to be a major force in bacterial evolution (see above). With regard to the evolution of ANK repeats, it has been suggested that the genes coding for the respective proteins might have been horizontally transferred from eukaryotic to bacterial genomes (Bork, 1993). With some of the repeats found in multiple, not closely related bacterial proteins (e. g. TPR repeats), this seems unlikely to be the principal mechanism of introduction of these repeats into bacteria.

CONCLUDING REMARKS

Computer analysis of proteins sequences yields testable functional predictions for the majority of uncharacterized *E. coli* gene products. In addition, our understanding of proteins with known function is increased, for example via the prediction of active centers. Many of the predictions are made possible only by the application of methods for motif detection. An "extra benefit" of the *E. coli* protein sequence analysis is the consistent identification of eukaryotic homologs, including some that are implicated in human disease, and functional predictions for these proteins.

A large, greater than previously reported, fraction of bacterial amino acid sequences is highly conserved in evolution. The level of conservation appears to be higher than in yeastand even in *Mycoplasma* that have the supposedly "minimal" genome. Over 40% of the *E. coli* proteins contain ACRs. This is an approximately twofold increase as compared with the previous estimates, resulting in part from a detailed analysis of "weak" similarities, and in part to the database growth. While the fraction of proteins with ACRs may increase even further with the completion of eukaryotic and Archaeal genome sequences, the growth of the ACR list not expected to be dramatic since most of the ancient protein families are already represented in the database (Chothia, 1992; Green et al., 1993). Our finding that the great majority of the *E. coli* ACRs fall into protein families that have already been known include ACRs supports this notion. A realistic projection may be that about one half of bacterial proteins contain ACRs. Functional prediction and sequence conservation are closely linked. In fact, prediction is highly efficient mostly because many sequences are well conserved in evolution.

About one half of the bacterial proteins belong to clusters of paralogs. A strong correlation exists between protein function and gene duplication, or more precisely, retention of paralogs in evolution. All the large clusters include transport and regulatory proteins; metabolic enzymes form small clusters with less than 10 members; and no paralogs so far have been found for key proteins involved in genome replication and expression. The majority of the clusters correspond to essential, ancient functions and contain ACRs. There are, however, large clusters of transport and regulatory proteins that do not have counterparts in eukaryotes. These groups of proteins may define, in part, the uniqueness of the bacterial cell physiology.

Paralogous genes are non-randomly distributed along the chromosome. The conclusion on the periodicity of paralog distribution derived previously from a very small data set appears to have passed the genome scale test. Additional, more detailed analysis is necessary in order to validate or refute the whole genome duplication hypothesis.

The general conclusion from the study of *E. coli* protein sequences on a genomic scale is that even with the currently available computer methods, however imperfect, and databases, however incomplete, a wealth of useful functional and evolutionary information can be extracted from the bacterial genes. This information allows one to decipher the basic trends in the relationships within the *E. coli* protein set and in comparison with protein sequences from other organisms. As the majority of the proteins encoded in the bacterial genome have been included in the present study, these trends are not expected to change significantly when the genome sequence is complete.

We are only in the beginning of the genome analysis effort. The real game will start when the complete genome sequences of *E. coli* and other bacteria, Archaea, and simple eukaryotes become available for comparison. It is our hope that the work described here provides a part of the methodological framework and defines some of the questions that have to be addressed in the future genome studies.

SUMMARY

We describe the results of computer analysis of 2,328 protein sequences comprising about 60% of the *Escherichia coli* gene products. Arguably, the main trends revealed in a study of the majority of the genes can be extrapolated to the complete genome. An unexpectedly high fraction of *E. coli* proteins - 86% - show significant sequence similarity to other proteins in current databases; about 70% show conservation at least at the level of distantly related bacteria, and 40% contain ancient conserved regions (ACRs) shared with eukaryotic or Archaeal proteins. These values are a considerable increase over the percentage of conserved proteins observed in other genome projects as well as the previous estimates of the number of ACRs in *E. coli*. The apparent increase in sequence conservation is due mostly to the use of sensitive methods for amino acid sequence motif analysis, and to some extent, to the growth of sequence databases. For more than 90% of the *E. coli* proteins, either functional information, or significant sequence similarity, or both, are available. 46% of the *E. coli* proteins belong to 285 clusters of intraspecies homologs (paralogs) defined on the basis of significant pairwise similarity. Another 10% could be included in superclusters using methods for detection of conserved motifs. The majority of the clusters consist of only two to four proteins. On the other hand, almost 25% of all *E. coli* proteins belong to the four largest superclusters. These superclusters include permeases; ATPases and GTPases with the conserved "Walker-type" motif; DNA-binding proteins containing the helix-turn-helix motif; and dinucleotide-binding proteins. Genes encoding paralogous proteins are non-randomly distributed along the *E. coli* chromosome. In addition to the large number of tandem duplications, a weak but statistically significant periodicity was observed, with the preferred distance between paralogous genes being a multiple of 6-7 minutes of the chromosome. In addition to its intrinsic value for understanding the functions of the bacterial genes and the evolution of the bacterial genome, detailed analysis of the *E. coli* protein sequences is useful as a means for predicting the functions of eukaryotic genes on the basis of the amino acid sequence conservation. In the course of our systematic analysis of *E. coli* proteins such predictions were made for a number of important eukaryotic genes, including several whose products are implicated in human diseases. Using motif searching methods, we found that several widespread domains, apparently involved mostly in protein-protein interaction and previously thought to be confined to either eukaryotes or bacteria, actually are ACRs. Our general conclusion is that the majority of the bacterial protein sequences are highly conserved in evolution. Even with the currently available, incomplete databases and the still imperfect methods for sequence comparison, detailed computer analysis yields information on the functions and evolutionary relationships of most of the proteins encoded in a bacterial genome. Coupled with the wealth of biochemical and genetic information available for *E. coli*, this makes this bacterium, once again, the best current model for understanding the functions and evolution of a prokaryotic genome and a valuable tool for systematic prediction of eukaryotic gene functions.

ACKNOWLEDGMENTS

We are grateful to David Lipman for helpful discussions and to Peer Bork for communicating his results prior to publication.

REFERENCES

Alefounder, P. R., Abell, C., and Battersbym, A. R., 1988, The sequence of hemC, hemD and two additional *E. coli* genes, *Nucleic Acids Res.* 16:9871.

Altschul, S. F., Gish, W., Miller, W., Myers, E. W., and Lipman, D. J., 1990, Basic local alignment search tool, *J. Mol. Biol.* 215:403.

Altschul, S. F., Boguski, M. S., Gish, W., and Wootton, J. C., 1994, Issues in searching molecular sequence databases, *Nature Genetics* 6:119.

Bennett, V., 1992, Ankyrins. Adaptors between diverse plasma membrane proteins and the cytoplasm, *J. Biol. Chem.* 267:8703.

Bork, P., 1993, Hundreds of ankyrin-like repeats in functionally diverse proteins:mobile modules that cross phyla horizontally, *Proteins:Struct. Funct. Genet.* 17:363.

Bork, P., Ouzounis, C., Sander, C., Scharf, M., Schneider, R., and Sonnhammer, E., 1992, Comprehensive sequence analysis of the 182 ORFs of yeast chromosome III, *Prot. Sci.* 1:1677.

Bork, P., Ouzounis, C., and Sander, C., 1994, From genome sequences to protein function, *Curr. Opin. Struct. Biol.* 4:393.

Bork, P., Ouzounis, C., Casari, G., Schneider, R., Sander, C., Dolan, M., Gilbert, W., and Gillevet, P. M., 1995, Exploring the *Mycoplasma capricolum* genome:A small bacterium reveals its physiology, *Molec. Microbiol.*, in press.

Bourne, H. R., Sanders, D. A., and McCormick F., 1991, The GTPase superfamily:conserved structure and molecular mechanism, *Nature* 349:117.

Bredt, D. S., Hwang, P. M., Glatt, C. E., Lowenstein, C., Reed, R. R., and Snyder, S. H., 1991, Cloned and expressed nitric oxide synthase structurally resembles cytochrome P-450 reductase, *Nature* 351:714.

Bryant, P. J. and Woods, D. F., 1992, A major palmitoylated membrane protein of human erythrocytes shows homology to yeast guanylate kinase and to the product of a *Drosophila* tumor suppressor gene, *Cell* 68:621

Cho, K. O., Hunt, C. A., and Kennedy, M. B., 1992, The rat brain postsynaptic density fraction contains a homolog of the *Drosophila* discs-large tumor suppressor protein, *Neuron* 9:929.

Chothia, C., 1992, Proteins. One thousand families for the molecular biologist, *Nature* 357:543.

Crouzet, J., Levy-Schil, S., Cameron, B., Cauchois, L., Rigault, S., Rouyez, M. C., Blanche, F., Debussche L., and Thibaut, D., 1991, Nucleotide sequence and genetic analysis of a 13.1-kilobase-pair Pseudomonas denitirficans DNA fragment containing five cob genes and identification of structural genes encoding Cob(I)alanine adenosyltransferase, cobyric acid synthase, a nd bifunctional cobinamide kinase-cobinamide phosphate guanylyltransferase. *J. Bacteriol.* 173:6074.

Cussac, V., Ferrero, R. L., and Labigne, A., 1992, Expression of Helicobacter pylori urease genes in *Escherichia coli* grown under nitrogen-limiting conditions, *J. Bacteriol.* 174:2466.

Daniels, D., Plunkett, G., Burland, V., and Blattner, F. R., 1992, Analysis of the Escherichia coli genome:DNA sequence of the region from 84.5 to 86.5 minutes, *Science* 257:771.

Dever, T. E., Glynias, M. J., and Merrick, W. C., 1987, GTP-binding domain:three consensus sequence elements with distinct spacing, *Proc. Natl. Acad. Sci. USA* 84:1814.

Dicker, I. B. and Seetharam, S., 1992, What is known about the structure and function of the *Escherichia coli* protein FirA, *Mol. Microbiol.* 6:817.

Doolittle, R. F., ed., 1990, "Molecular Evolution", *Meth. Enzymol.*, 183.

Drlica, K., 1987, The nucleoid, in *"Escherichia coli* and *Salmonella typhimurium.* Cellular and Molecular Biology" Neidhardt, F., Ingraham, J. L., Low, K. B., Magasanik, B., Schaechter, M., and Umbarger, H. E., eds., *American Society for Microbiology*, Washington, DC, p. 91.

Fishel, R., Lescoe, M. K., Rao, M. R., Copeland, N. G., Jenkins N. A., Garber, J., Kane, M., and Kolodner, R., 1993, The human mutator gene homolog MSH2 and its association with hereditary nonpolyposis colon cancer, *Cell* 75:1027.

Fitch, W. M., 1970, Distinguishing homologous from analogous proteins, *Syst. Zool.* 19, 99.

Goebl, M. and Yanagida, M., 1991, The TPR snap helix:a novel protein repeat motif from mitosis to transcription, *Trends Biochem. Sci.* 16:173.

Gray, M. W., 1989, The evolutionary origin of organelles, *Trends Genet.* 5:294.

Green, P., 1994, Ancient conserved regions in gene sequences, *Curr. Opin. Struct. Biol.* 4:404.

Green, P., Lipman, D. J., Hillier, L., Waterston, R., States, D. J., and J. M. Claverie, 1993, Ancient conserved regions in new gene sequences and the protein databases, *Science* 259:1711.

Henikoff, S. and Henikoff, J., 1992, Amino acid substitution matrices from protein blocks, *Proc. Natl. Acad. Sci.* USA 89:10915.

Henikoff, S. and Henikoff, J., 1993, Performance evaluation of amino acid substitution matrices, *Proteins Struct. Funct. Genet.* 17:49.

Hussain, H., Grove J., Griffiths L., Busby S., and Cole J., 1994, A seven-gene operon essential for formate-dependent nitrite reduction to ammonia by enteric bacteria, *Mol. Microbiol.* 12:153

Kiino, D. R., Singer, M. S., and Rothman-Denes, L. B., 1993, Two overlapping genes encoding membrane proteins required for bacteriophage N4 adsorption, *J. Bacteriol.* 175:7081

Klingensmith, J., Nusse, R., and Perrimon, N., 1994, The *Drosophila* segment polarity gene dishevelled encodes a novel protein required for response to the wingless signal, *Genes Dev.* 8:118

Knoll, A. H., 1992, The early evolution of eukaryotes:a geological prospective, *Science* 256:622.

Koonin, E. V., 1994, Prediction of an rRNA methyltransferase domain in human tumor-specific nucleolar protein P120, *Nucleic Acids Res.* 22:2476.

Koonin, E. V., 1995, Multidomain organization of eukaryotic guanine nucleotide exchange translation initiation factor eIF-2B subunits revealed by analysis of conserved sequence motifs, *Protein Sci.* in press.

Koonin, E. V., Bork, P., and Sander, C., 1994, Yeast chromosome III:new gene functions, *EMBO J.* 13:493.

Koonin, E. V., Woods, D. F., Bryant, P. J., 1992, Dlg-R proteins: guanylate kinase homologues with an aberrant NTP-binding site, *Nature Genet.* 2:256.

Koonin, E. V. and Rudd, K. E., 1993, SpoU protein of *Escherichia coli* belongs to a new family of putative rRNA methylases, *Nucleic Acids Res.* 21:5519.

Kunisawa, T., and Otsuka, J., 1988, Periodic distribution of homologous genes or gene segments on the *Escherichia coli* K12 genome. *Protein Seq. Data Anal.* 1:263.

Labedan, B. and Riley, M., 1995, Widespread protein sequence similarities:origins of *Escherichia coli* genes, *J. Bacteriol.* 177:1585.

Leach, F. S., Nicolaides, N. C., Papadopoulos, N., Liu, B., Jen, J., Parsons, R., Peltomaki, P., Sistonen, P., Aaltonen, L. A., Nystrom-Lahti, M. et al, 1993, Mutations of a mutS homolog in hereditary nonpolyposis colorectal cancer, *Cell* 75:1215.

Lee, M. H., Mulrooney, B. S. B., Renner, M. J., Markowicz, Y., Hausinger, R. P., 1992, Klebsiella aerogenes urease gene cluster:sequence of ureD and demonstration that four accessory genes (ureD, ureE, ureI, and ureG) are involved in nickel metallocenter biosynthesis, *J. Bacteriol.* 174:4324.

Lipinska, B., Zylicz, M., and Georgopoulos, C., 1990, The HtrA (DegP) protein, essential for *Escherichia coli* survival at high temperatures, is an endopeptidase, *J. Bacteriol.* 172:1791

Lue, R. A., Marfatia, S. M., Branton, D., and Chishti, A. H., 1994, Cloning and characterization of hdlg: he human homologue of the Drosophila discs large tumor suppressor binds to protein 4.1, *Proc. Natl. Acad. Sci.* USA 91:9818

Maier, T., Jacobi, A., Sauter, M., and Bock, A., 1993, Purification of Rhizobium leguminosarum HypB, a nickel-binding protein required for hydrogenase synthesis, *J. Bacteriol.* 175:630.

Maniloff, J., McElhaney, R. N., Finch, L. R., and Baseman, J. B. (eds), 1992, "Mycoplasmas - molecular iology and pathogenesis", *American Society for Microbiology*, Washington, DC.

Michaely, P. and Bennett, V., 1992, The ANK repeat:A ubiquitous motif involved in macromolecular recognition, *Trends Cell Biol.* 2:127.

Neuwald, A. F. and Green, P., 1994, Detecting patterns in protein sequences, *J. Mol. Biol.* 239:698

Neidhardt, F., Ingraham, J. L., Low, K. B., Magasanik, B., Schaechter, M., and Umbarger, H. E. (eds.), *Escherichia coli* and *Salmonella typhimurium. Cellular and Molecular Biology, American Society for Microbiology*, Washington, DC.

Olsen, G. J., Woese, C. R., and Overbeek, R., 1994, The winds of (evolutionary) change:breathing new life into microbiology, *J. Bacteriol.* 176:1.

Palmer, J. D., 1985, Comparative organization of chloroplast genomes, *Annu. Rev. Genet.* 19:325.

Papadopoulos, N., Nicolaides, N. C., Wei, Y. F., Ruben, S. M., Carter, K. C., Rosen, C. A., Haseltine, W. A., Fleischmann, R. D., Fraser, C. M., Adams, M. D. et al., 1994, Mutation of a mutL homolog in hereditary colon cancer, *Science* 263:1625.

Pepperberg, D. R., Okajima, T. L., Wiggert, B., Ripps, H., Crouch, R. K., and Chader, G. J., 1993, Interphotoreceptor retinoid-binding protein (IRBP). Molecular biology and physiological role in the visual cycle of rhodopsin, *Mol. Neurobiol.* 7:61.

Qian, Y. Q., Billeter, M., Otting, G., Muller M., Gehring, W. J., and Wuthrich, K., 1989, The structure of the Antennapedia homeodomain determined by NMR spectroscopy in solution:comparison with prokaryotic repressors, *Cell* 59:573.

Riley, M., 1993, Functions of the gene products of *Escherichia coli, Microbiol. Rev.* 57:862.

Riley, M., and Anilionis, A., 1978, Evolution of the bacterial genome. *Annu. Rev. Microbiol.* 32:519.

Riley, M. and Krawiec, S., 1987, Genome organization, in:"*Escherichia coli* and *Salmonella typhimurium.* Cellular and Molecular Biology" F. Neidhardt, J. L. Ingraham, K. B. Low, B. Magasanik, M. Schaechter, and H. E. Umbarger, eds., *American Society for Microbiology*, Washington, DC, p. 967.

Rudd, K. E., 1993, Maps, genes, sequences, and computers:an Escherichia coli case study, *ASM News* 59:335.

Ruff, P., Speicher, D. W., Husain-Chishti, A., 1991, Molecular identification of a major palmitoylated erythrocyte membrane protein containing the src homology 3 motif, *Proc. Natl. Acad. Sci.* USA 88:6595.

Saier, M. H., Jr., 1994, Computer-aided analyses of transport protein sequences:gleaning evidence concerning function, structure, biogenesis, and evolution, *Microbiol. Rev.* 58:71.

Saier, M. H. Jr and Reizer, J., 1994, The bacterial phosphotransferase system:new frontiers 30 years later, *Mol. Microbiol.*13:755.

Saraste, M, Sibbald, P. R., and Wittinghofer, A., 1990, The P-loop - a common motif in ATP- and GTP-binding proteins, *Trends Biochem. Sci.* 15:430.

Savakis, C. and Doelz, R., 1993, Contamination of cDNA sequences in databases, *Science* 259:1677.

Schuler, G. D., Altschul, S. F., and Lipman, D. J., 1991, A workbench for multiple alignment construction and analysis, *Proteins:Struct. Funct. Genet.* 9:180.

Service, R. F., 1994, Stalking the start of colon cancer, *Science* 263:1559.

Sikorski, R. S., Boguski, M. S., Goebl, M., and Hieter, P., 1990, A repeating amino acid motif in CDC23 defines a family of proteins and a new relationship among genes required for mitosis and RNA synthesis, *Cell* 60:307.

Sikorski, R. S., Michaud, W. A., Wootton, J. C., Boguski, M. S., Connelly, C., and Hieter, P., 1991, TPR proteins as essential components of the yeast cell cycle, *Cold Spring Harb. Symp. Quant. Biol.* 56:663.

Silber, K. R., Keiler, K. C., and Sauer, R. T., 1992, Tsp:a tail-specific protease that selectively degrades proteins with nonpolar C termini. *Proc. Natl. Acad. Sci.* USA 89:295.

Sirum-Connolly, K. and Mason, T. L., 1993, Functional requirement of a site-specific ribose methylation in ribosomal RNA, *Science* 262:1886.

Tatusov, R. L., Altschul, S. F., and Koonin, E. V., 1994, Detection of conserved segments in proteins:iterative scanning of sequence databases with alignment blocks, *Proc. Natl. Acad. Sci. USA* 91:12091.

Treisman, J., Harris, E., Wilson, D., and Desplan, C., 1992, The homeodomain:a new face for the helix-turn-helix, *Bioessays* 14:145.

Vaara, M., 1992, Eight bacterial proteins, including UDP-N-acetylglucosamine acyltransferase (LpxA) and three other transferases of *Escherichia coli*, consist of a six-residue periodicity theme, *FEMS Microbiol. Lett.* 76:249.

Vuorio, R., Harkonen, T., Tolvanen, M., and Vaara, M., 1994, The novel hexapeptide motif found in the acyltransferases LpxA and LpxD of lipid A biosynthesis is conserved in various bacteria, *FEBS Microbiol. Lett.* 337:289.

Walker, J. E., Saraste, M., Runswick, M. J., and Gay, N. J.,1982, Distantly related sequences in the a- and b-subunits of ATP synthase, myosin, kinases and other ATP-requiring enzymes and a common nucleotide binding fold, *EMBO Journal* 1:945.

Woods, D. F. and Bryant, P. J., 1991, The discs-large tumor suppressor gene of Drosophila encodes a guanylate kinase homolog localized at septate junctions, *Cell* 66, 451.

Wootton, J. C. and Federhen, S., 1993, Statistics of local complexity in amino acid sequences and sequence databases, *Comput. Chem.* 17:149.

Wootton, J. C., 1994a, Non-globular domains in protein sequences: automated segmentation using complexity measures, *Comput. Chem.* 18:269.

Wootton, J. C., 1994b, Sequences with 'unusual' amino acid composition, *Curr. Opin. Struct. Biol.* 4:413.

Zipkas, D. and Riley, M., 1975, Proposal concerning mechanism of evolution of the genome of *Escherichia coli*, *Proc. Natl. Acad. Sci. USA* 72:4660.

Zipkas, D., Solomon, D., and Riley, M., 1978, Relationship between gene function and gene location, *J. Mol. Evol.* 11:47.

A PAIRING-LOOPING MODEL FOR POSITION-EFFECT VARIEGATION IN *DROSOPHILA*

Steven Henikoff

Howard Hughes Medical Institute and
Basic Sciences Division
Fred Hutchinson Cancer Research Center
Seattle, WA 98104

INTRODUCTION

L. J. Stadler's critical review, "The Gene" (Stadler, 1954), was published shortly after his death in 1954, when genes were defined abstractly by mutation, not molecularly by sequence, as they are today. In Stadler's time, position-effect variegation (PEV) represented a challenge to any simple view of the gene, because a PEV mutation could act to silence multiple linked genes, a feature commonly referred to as "spreading". One extreme view cited by Stadler was Goldschmidt's contention that PEV reveals that genes do not exist, at least not as discrete entities. Whereas such uncertainties concerning the existence and nature of the gene were soon cleared up during the molecular biology revolution, the problem of spreading seen with PEV remains: how can a single lesion affect multiple discrete genes in a region?

On the occasion of a symposium in honor of L. J. Stadler, I revisit the issue of spreading seen in PEV that so intrigued him and his generation of geneticists. Like the concept of the gene in 1954, the concept of spreading even today is mostly abstract; consequently, we must be mindful of Stadler's admonition that we stick to operational definitions, and not to confuse hypothesis with observation. Just such confusion is inherent in the use of the word "spreading" to describe a set of genetic observations: spreading implies continuity, an extrapolation of observation to hypothesis. Although this hypothesis dominates current opinion, I contend that the intuitive concept of

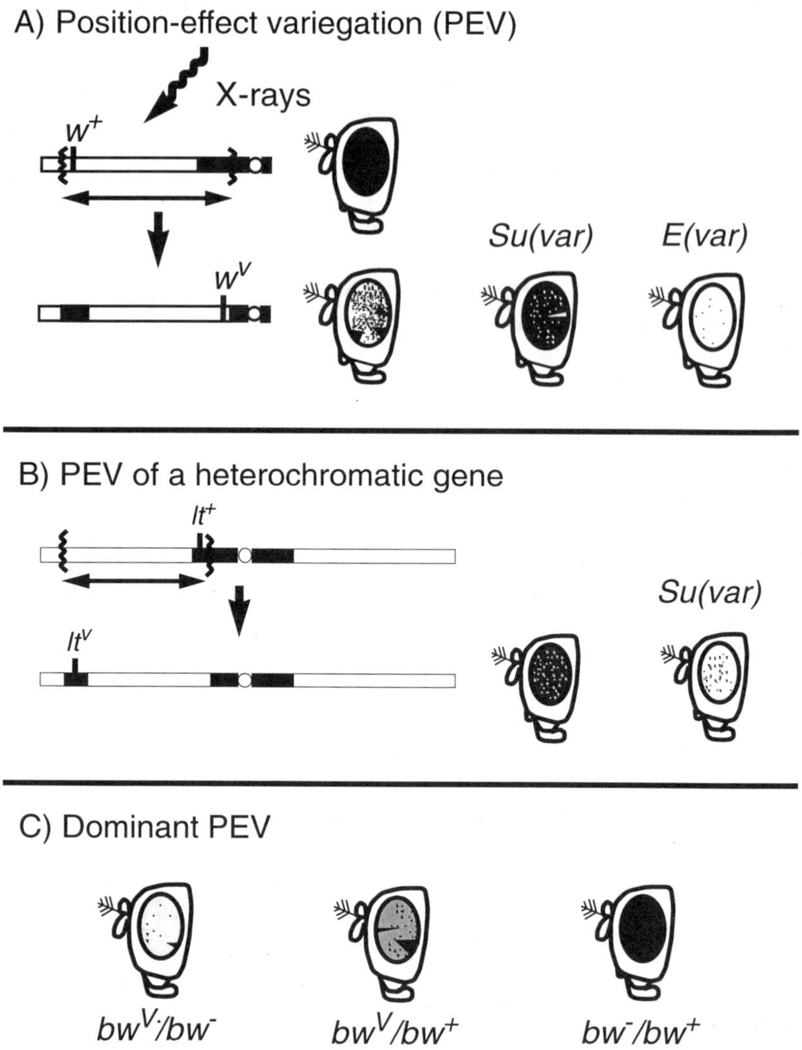

Figure 1. Three different categories of PEV. A) X-rays have induced an inversion, *white-mottled-4,* thus placing the wild-type *white (w+)* gene within about 25 kb of pericentric heterochromatin, giving rise to a mutant variegating allele, w^V (also see Figure 5). Here, the *white* gene is silenced in *cis. Su(var)*s suppress and *E(var)*s enhance the variegating phenotype. Filled boxes represent heterochromatin and open boxes euchromatin. B) Heterochromatic genes show PEV in chromosomal rearrangements. The *light (lt+)* gene displays variegation when moved to distal euchromatin. Variegation of *light (lt^V)* can be enhanced by *Su(var)*s and suppressed by *E(var)*s. C) Dominance is seen for a variegating allele of the *brown (bw)* gene. Heterozygotes between *brown-Variegated (bw^V)* and wild-type *brown, bw^V/bw +,* are variegated, although somewhat suppressed relative to bw^V/bw^-. Here, the *brown* gene is silenced both in *cis* and in *trans.*

continuous spreading is misleading. I will provide a different interpretation of PEV and spreading (Sabl and Henikoff, 1995) that can satisfactorily account for the observational evidence.

PEV was described by Muller (1930, 1932) as mosaic gene silencing that can occur when euchromatin and constitutive heterochromatin are juxtaposed (Figure 1A). Euchromatin is the chromosomal material that decondenses at interphase, whereas heterochromatin remains visibly condensed (Heitz, 1929) and forms a chromocenter or heterochromatic compartment associated with the nuclear envelope (Hochstrasser et al., 1986). The basic features of PEV have been known for more than half a century. However, fundamental questions remain: How are genes silenced from a distance? What differences between euchromatin and heterochromatin are responsible for silencing? Why is silencing somatically unstable? What accounts for allele-specific differences? How do proteins encoded by PEV modifiers act? Answering these questions is becoming increasingly important given the mounting evidence that some of the same questions arise for silencing phenomena in a wide variety of systems. For example, the uncertain basis for gene silencing caused by CGG repeat expansions in the fragile X syndrome (Bates and Lehrach, 1994; Willems, 1994) emphasizes how little we understand about any gene silencing mechanism in higher eukaryotes.

Our lack of understanding of PEV is not the result of a lack of models. Muller's original description was accompanied by several models (Muller, 1930), and each new observation generated more (*e.g.,* Schultz, 1936). However the failure of any model to explain the diverse observations led a frustrated Bill Baker in 1968 to suggest that further investigation would be unprofitable at that time (Baker, 1968). Subsequent studies only seemed to make matters worse. Eventually, most workers in the field accepted the view that PEV involves "spreading" of heterochromatic protein complexes along the chromosome from sites in heterochromatin into euchromatin (Tartof et al., 1989; Eissenberg, 1989; Grigliatti, 1991; Reuter and Spierer, 1992; Moehrle and Paro, 1994). Although the term "spreading" was originally coined to describe a set of genetic observations (Lewis, 1950), in recent years it has been used to describe an unproven molecular mechanism. To avoid perpetuating such confusion (to which I have contributed in the past), I will adopt the term "oozing", introduced by Ptashne (1986) to describe exactly this class of molecular models.

Not everyone in the field has embraced the oozing model: Spradling and his colleagues have criticized this model based on lack of evidence and have revived the old idea that somatic loss is responsible for the variegating phenotypes (Schultz, 1936; Karpen and Spradling, 1990; Spradling and Karpen, 1990; Glaser et al., 1992; Spradling et al., 1993). However, substantial contrary evidence exists (Henikoff, 1981; Rushlow et al., 1984; Hayashi et al., 1990; Umbetova et al., 1991; Dreesen et al., 1991) that is not easily explained away.

Here I consider a model (Sabl and Henikoff, 1995) that also incorporates an old idea, that the forces of somatic pairing underlie heterochromatin formation and PEV (Ephrussi and Sutton, 1944;

Henikoff, 1994; Henikoff, 1995). This model provides satisfactory explanations for both old and new observations and makes testable predictions. But first, let us consider the pros and cons of oozing, as this model dominates current thinking about PEV and similar gene silencing phenomena.

THE CASE FOR OOZING

The major motivation for the oozing model is that it can account for spreading, operationally defined as the silencing of closely-linked genes in a polar manner (reviewed in Lewis, 1950; Spofford, 1976). Spreading has been observed phenotypically for two genes adjacent to a heterochromatic breakpoint, in which case the gene closer to the breakpoint is more frequently inactive. It is notable that for the *white* and *roughest* genes, which affect the appearance of eye facets, no facets were found in which the gene closer to the breakpoint (*roughest*) was expressed and the gene farther away (*white*) was not (Figure 2, top). Spreading also has been observed cytologically in salivary gland polytene chromosomes as the variable loss of banding for regions in which affected genes lie (Hartmann-Goldstein, 1967; Hayashi et al., 1990; Belyaeva and Zhimulev, 1991).

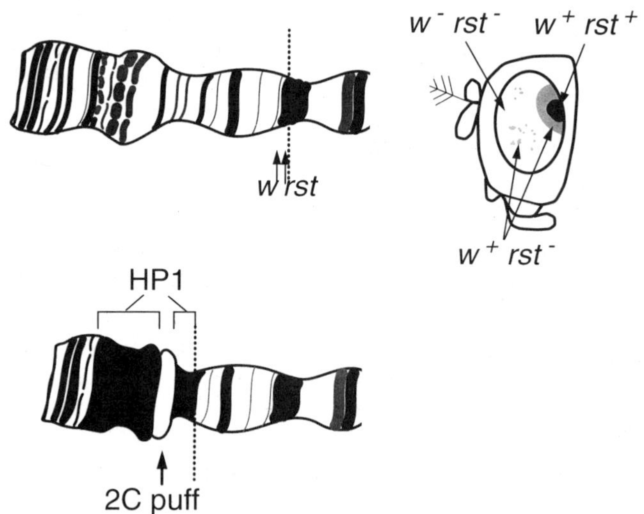

Figure 2. Depiction of the distal tip of the X chromosome (on the left) showing the location of the heterochromatic breakpoint (dotted line) causing PEV of *white (w)* and *roughest (rst)* genes (top). For simplicity, the locations of the euchromatic breakpoints are shown, not the resulting rearrangement chromosomes. The phenotype depicted on the right is adapted from a drawing by Demerec and Slizynska (1937). Discontinuous compaction of the region distal to a heterochromatic insertion into region 2E is depicted below, adapted from Belyaeva et al. (1993). Compaction beginning at the breakpoint (dotted line) causes shortening of the affected region (from 1C to 2E), a denser appearance and staining by anti-HP1 antibody. The 2C puff was not compacted or stained in the nucleus depicted, although it was in other nuclei.

These regions resemble heterochromatin in that they assume a diffuse
unbanded appearance. By both phenotypic and cytological criteria,
spreading decreases in extent when PEV is supressed and increases
when PEV is enhanced. Oozing models are straightforward extensions
of these phenotypic and cytological observations to the molecular
level.

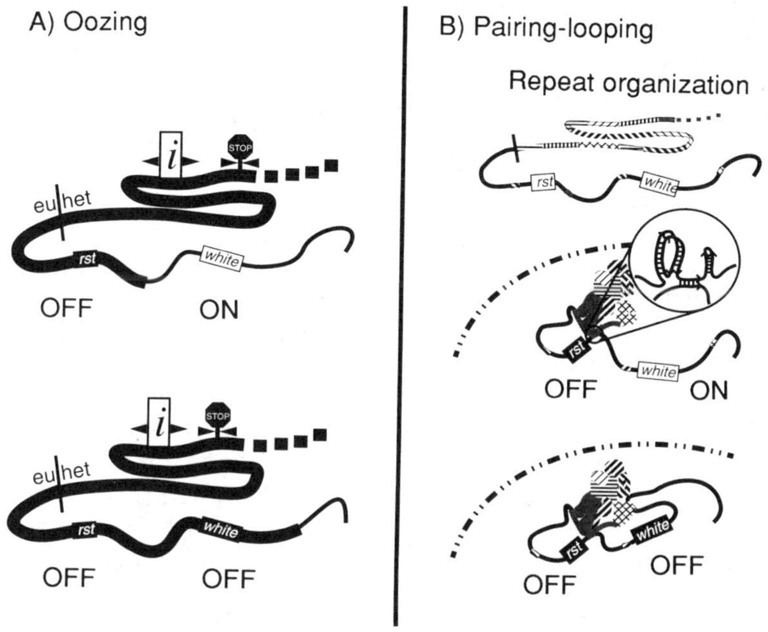

Figure 3. Models for PEV interpreting the observation illustrated in Figure 2
(top). A) Oozing. Heterochromatic protein complexes (thick bars) ooze from
initiation sites in heterochromatin (*i*) can pass through a
heterochromatin/euchromatin junction (eu/het) in the absence of a terminator
(stop) site, causing polar spreading into *rst* and *white*. B) Pairing-looping.
Looping of *white* and *rst* into the heterochromatic chromocenter (grey oval) on
the nuclear envelope (dashed and dotted line) results from pairing between
middle repetitive elements that are dispersed in euchromatin and concentrated
in heterochromatin [adapted from Sabl and Henikoff (1995)]. Different repeat
families are represented by different patterns, extended in the top cartoon and
folded into paired structures below. Paired associations between euchromatic
and heterochromatic repeats (shown in magnification) lead to inactivation of
rst or both *rst* and *white*, but not *white* alone, because compartmentalizing
white simultaneously draws in *rst*, which is held down on both sides.
Hypothetical pairing interactions are schematically illustrated as ladder rungs,
which are not meant to suggest any specific mechanism for association between
homologous sequences.

The idea that proteins could ooze along the DNA [or "creep"
(Zuckerkandl, 1974)] was conceived when our knowledge of chromatin
was at a rudimentary level and before it was known that the basic
unit of chromatin is two turns of DNA wrapped around a core histone
octamer separated by a histone H1 linker. As a result, it was possible
to envision a one-dimensional crystal consisting of chromosomal
proteins. The proteins in this crystal would behave according to the

law of mass action, and titration could account for PEV suppression (*i. e.* attenuation of the mutant effect) by the fully heterochromatic Y chromosome. Later, histones were implicated in the phenomenon based on genetic mapping criteria (Moore et al., 1983) and the finding that butyrate, which inhibits histone deacetylation, suppresses PEV (Mottus et al., 1980). In addition, the discovery of dosage-dependent suppressors of PEV, or *Su(var)*s, led to the notion that these encode non-histone chromosomal proteins, which form heterochromatin (Henikoff, 1979; Locke et al., 1988; Wustmann et al., 1989; Grigliatti, 1991). This possibility was confirmed by the evidence that one of these suppressors, *Su(var)205*, encodes HP1, a known component of heterochromatin (Eissenberg et

al., 1992). Eventually, a mass action model for modifiers of PEV was put forward to account for the sensitivity of PEV phenotypes to so many trans-acting factors (Locke et al., 1988). This model requires the formation of a multicomponent complex that causes heterochromatin to condense. If these complexes lie adjacent to one another along the DNA, then the genetic and cytological observations can be explained, because oozing would never skip a gene.

Polarity of the spreading effect requires that oozing of heterochromatic complexes starts at one place and ends at another. Oozing must start in constitutive heterochromatin, move through the heterochromatin-euchromatin junction, then end downstream of the farthest affected gene in that cell. To account for variegation, oozing must sometimes stop short of each gene. Specific models (Tartof et al., 1984; Tartof et al., 1989; Eissenberg, 1989; Grigliatti, 1991; Moehrle and Paro, 1994) hypothesize the existence of centers of ooze initiation, and (perhaps) terminators that limit oozing (see Figure 3A). Although models differ with respect to the frequency and location of initiators and terminators and whether they act uni- or bi-directionally, they share the crucial concept that modifiers of variegation act by changing the average distance along the chromosome from an oozing center to the site of termination as a result of the mass action law. The availability of more subunits of the heterochromatic complex is translated into more extensive oozing into neighboring euchromatin. If the specific site of termination differs from cell to cell, then oozing of heterochromatin complexes along the chromosome can account for the polar spreading effect.

THE CASE AGAINST OOZING

Molecular evidence is lacking

The original concept of a simple one-dimensional crystal is excluded by subsequent demonstations that heterochromatin, like euchromatin, is packaged similarly into nucleosomes consisting of an octamer core and H1 linkers (Wolffe, 1992). Rather, oozing must occur in the context of nucleosomes assuming a different higher order structure, seen as a 30 nm fiber. Evidence is lacking for structural differences between heterochromatin and quiescent euchromatin in the 30 nm fiber, so that one-dimensional crystallization of nucleosomes specific to heterochromatin remains purely speculative. The DNase I sensitivity of heterochromatin appears unremarkable, differing only slightly for the *white* gene between PEV-enhanced and PEV-suppressed states that are readily distinguished cytologically in polytene chromosomes (Hayashi et al., 1990). Restriction endonuclease probes (Schlossherr et al., 1994) or endogenous nucleases (Locke, 1994) also failed to detect differences at *white*. Although one might worry that nuclease treatments disrupt higher order structure, an *in vivo* assay using *E. coli dam* methylase in live flies likewise failed to detect consistent differences in accessibility between euchromatic and heterochromatic sequences (Wines et al., 1995). Thus, local effects of the postulated heterochromatic complex appear to be very subtle. This is hard to reconcile with the idea that oozing of protein complexes along the chromosome is responsible for the condensation of heterochromatin that can be seen on a cytological scale.

Allele-specific differences are not explained

The oozing model is difficult to reconcile with genetic evidence. For example, the effect on genes near a heterochromatic breakpoint is extremely variable from one PEV allele to the next. A comparison of PEV alleles reveals that the extent of spreading along the chromosome can vary from no spreading into the next gene to spreading over dozens of bands (*e. g.*, Figure 2). For example, *white-mottled-4 (w^{m4})*, which has become the prototypical example of PEV, has a strong silencing effect on *white* yet does not spread into the adjacent gene (Lindsley and Zimm, 1992). In contrast, $w^{m258-21}$ has a much weaker effect on *white* than w^{m4} and has been observed to spread for 67 bands (Hartmann-Goldstein, 1967; Reuter et al., 1982). If spreading is caused by heterochromatin complexes lining up along the chromosome, then why does spreading usually stop close to the *white* gene in w^{m4} but continue on in $w^{m258-21}$, even though w^{m4} silences *white* far more frequently? That is, how do the complexes know that they should terminate just past the *white* gene in w^{m4} but don't know in $w^{m258-21}$?

Another puzzling aspect of the phenomenon is that there seems to be a "fit" between the euchromatin and the heterochromatin comprising a PEV junction (Spofford, 1976). X-ray mutagenesis screens of the *brown* gene provide an illustration. In one study, none of the 21 PEV alleles were found to have translocated the chromosomal *brown* gene to X-heterochromatin and two were translocated to the Y chromosome (Slatis, 1955). However, in an X-ray mutagenesis screen of a P transposon carrying *brown*, three PEV alleles were recovered, of which two were translocated to X-heterochromatin and the other was to the Y chromosome (Dreesen et al., 1991). In the same study, another line of flies with a *brown*-bearing transposon at a different location yielded only a single PEV allele at a 10-fold lower frequency. Such examples indicate that heterochromatin is not uniform in its ability to induce PEV, a feature that depends upon the euchromatic location of the gene under study. The oozing model does not explain how a gene region discriminates between heterochromatic complexes oozing from different heterochromatic regions.

Prominent exceptions are not explained

The oozing model also fails to account for the two prominent "exceptions": PEV of heterochromatic genes and dominant PEV. These provide different vantage points for observing heterochromatin. Heterochromatic genes variegate when moved to distal euchromatin and show a reversed sensitivity to PEV modifiers. For example, the heterochromatic *light* gene variegates when breakpoints proximal to the gene move it far away from heterochromatin, and this effect can be enhanced by PEV suppressors (Hearn et al., 1991, see Figure 1B). If oozing of protein complexes is involved, then PEV should result from retreat of heterochromatin back past the junction into the *light* gene. However, breaks that cause *light* PEV can be extremely far away from the gene, translocating not only *light*, but also the large heterochromatic block in which *light* lies. If any ooze initiators lie within the translocated block between *light* and the breakpoint, then it is hard to understand how heterochromatin can retreat past these initiators to induce PEV.

Just as PEV of heterochromatic genes provides a view of heterochromatin from inside, dominant PEV provides a view from along side. We have focused on the genetic basis for dominant PEV of the *brown* gene (Figure 1C). Dominance is a consequence of somatic pairing of homologs: pairing is believed to bring the *cis*-inactivated copy of *brown* into contact with the *trans* copy, and *trans*-inactivation results (Henikoff and Dreesen, 1989). Susceptibility to *trans*-inactivation maps to the *brown* gene itself, most likely to a region of <1 kb in the 5' regulatory region (Martin-Morris et al., 1993). It appears that susceptibility to *trans*-inactivation and transcriptional regulation of the *brown* gene coincide, because both have been conserved despite 60 million years of divergence (Martin-Morris and Henikoff, 1995). Dominance of *brown* variegation is entirely dependent on the presence of heterochromatin, and is responsive to modifiers of PEV. In fact, various dominant *brown* PEV alleles have been used in screens for modifiers and are known to respond in a very consistent manner (Reuter and Wolff, 1981; Sinclair et al., 1992; Talbert et al., 1994). This situation is not explained by the oozing model: how is oozing along the chromosome in *cis* "felt" by a strictly localized region in *trans*?

Cytological compaction can be discontinuous

The aforementioned problems with the oozing model emphasize how little it actually explains. However, these failures do not constitute disproof, since new features might be added to the model to account for these observations. Rather, we would consider the model disproved if its key features were found not to hold.

One key feature is the continuity of oozing along the chromosome. However, discontinuous compaction has been reported for PEV alleles broken in the vicinity of the 2C salivary chromosome puff (Belyaeva and Zhimulev, 1991). This observation has been confirmed by staining using anti-HP1 antibody, which shows an active 2C puff flanked by HP1-associated chromatin (Belyaeva et al., 1993; Figure 2, bottom). How did the ooze skip the puff?

Artifical repeat arrays form heterochromatin

Another key feature of the oozing model is that initiator elements at oozing centers correspond to specific sequences in heterochromatin. How can this idea be tested critically? One way is to ask whether arbitrary sequences, not just sequences found specifically in heterochromatin, can nucleate heterochromatin formation. An intriguing clue came from the observation that an 11 Mb transgene array in the mouse genome appeared heterochromatic in cytological preparations (Manuelidis, 1991). Genetic evidence was provided by our recent results in *Drosophila,* where heterochromatin was found to form at repeats of a P transposon carrying the mini-*white* gene (Dorer and Henikoff, 1994). Repeat arrays of the transposon were found to nucleate heterochromatin formation, even when located far from heterochromatin. Here, the criterion for heterochromatin formation was based on PEV phenotypes and responsiveness to PEV modifiers that are identical to phenotypes seen for classical *white* PEV alleles. Since the sequences in the transposon are mostly of bacterial origin and are not derived from heterochromatin, it seems that an array of any sequence can nucleate heterochromatin formation. We proposed that paired structures form between neighboring repeats, resulting in condensation. This proposal was based on the observation that reversals of an element within a tandem array invariably strengthened mutant effects observed (see Figure 4), which paralleled earlier inferences based on differing genetic properties of a tandem and inverted duplication that could be observed cytologically (Gubb et al., 1990).

The strength of PEV is profoundly affected by changes at the gene

By the oozing model, distal sequence changes should have no effect, that is, it should not be possible to attract or repel the ooze from beyond. However, this notion is contradicted by our finding that

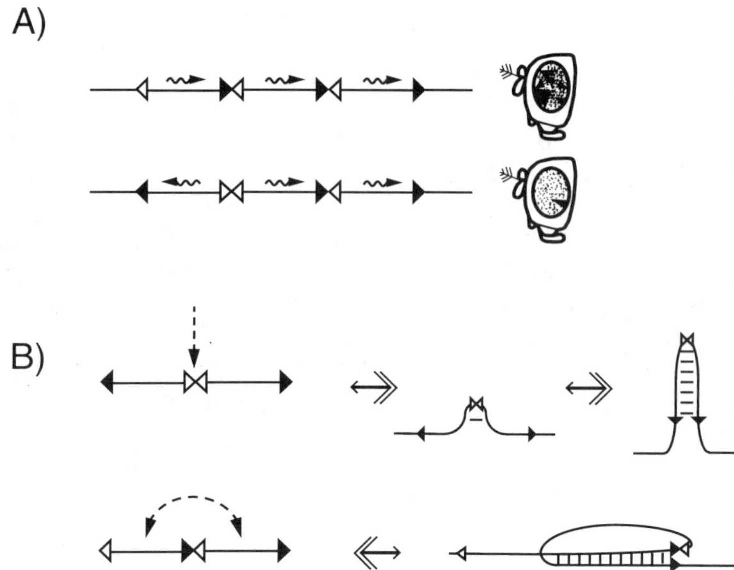

Figure 4. A) Experimental basis for proposing that pairing of repeats nucleates heterochromatin formation [adapted from Dorer and Henikoff (1994)]. Artificial repeat arrays of transposons carrying the *mini*-white reporter gene can form heterochromatin based on PEV phenotypes and sensitivity to PEV modifiers. Phenotypes strengthen with transposon reversals within an array. For example, a three-copy tandem array of a mini-*white* bearing transposon at the euchromatic 92E site is less strongly mutant than a three-copy array at the same site in which a single outside copy is reversed. Wavy line indicates the direction of mini-*white* transcription and arrowheads represent 5' (open) and 3' (filled) P-transposon ends. B) An explanation based on pairing between neighboring repeat units. When neighboring repeats are in head-to-head orientation, pairing begins at a single point in between (dashed arrow), then zippers up (ladder rungs). However, when in tandem orientation, pairing and zippering cannot begin until two points (curved dashed arrow) that lie a full repeat unit apart make contact.

alterations at a single site strongly regulate *brown* gene PEV (Sabl and Henikoff, 1995). Starting with a classical PEV allele consisting of an X-ray induced translocation that places pericentric heterochromatin adjacent to the *brown* gene on a P transposon, we obtained changes in the number and orientation of copies following transposase mutagenesis and phenotypic selection. Increasing transposon copy number enhanced PEV, and reversing the transposon suppressed PEV. If oozing from the heterochromatic breakpoint were responsible for the PEV phenotype, then more copies must inexplicably attract the ooze whereas a reversed orientation must repel.

Thus, key predictions of the oozing model are contradicted by recent evidence. Taken together with the mechanistic implausibility, the lack of molecular evidence in support and the failure to account for allele specificity, PEV of heterochromatic genes and dominant PEV, we are compelled to consider alternatives. One alternative is that PEV is a complex phenomenon (Reuter and Spierer, 1992), with multiple underlying causes (Karpen, 1994). However, the consistency of the phenomenon and its restriction to alleles that juxtapose euchromatin and heterochromatin reflect a basic simplicity. Furthermore, PEV suppressors are highly specific for PEV alleles; for example, PEV suppressors do not affect mosaic expression of the *white* gene caused by telomere position effect (Talbert et al., 1994), and at most a small subset of them might also be involved in analogous silencing phenomena (Grigliatti, 1991; Fauvarque and Dura, 1993). In general, similar modifications of unrelated mutations have been taken as evidence for a single underlying cause, a rationale that has served geneticists well (*e.g.* Modollel et al., 1983).

A PAIRING-LOOPING MODEL FOR PEV

The repetitive sequence organization of heterochromatin provides the basis for an alternative to the oozing model. Cytologically, heterochromatin is characterized by non-homologous associations, most easily observed in the salivary gland polytene chromosomes. Ever since these giant chromosomes were discovered, it was noticed that non-homologous associations in the unbanded chromocenter are similar to homologous associations that form the banded euchromatin and hold homologs together (Painter, 1934). Such observations led to the proposal that the same forces holding together chromatids and homologs cause non-homologous associations among heterochromatic regions both between different chromosomes and from repeats within a single chromosome (Ephrussi and Sutton, 1944). This basic idea led to the following prediction (Pontecorvo, 1944): "A heterochromatic segment should arise every time that a minute euchromatic region undergoes repeated reduplication in the genotype and the replicas remain adjacent to each other in the chromosome". Fifty years later this prediction was confirmed, with the demonstration that heterochromatin forms at repeat arrays consisting of as few as three copies of non-heterochromatic DNA (Dorer and Henikoff, 1994). Therefore, even limited stretches of repetitive sequences are sufficient to nucleate the formation of heterochromatin.

DNA in heterochromatin consists largely of blocks of simple sequences repeats (Lohe et al., 1993). Most of the remainder consists of interspersed blocks of middle repetitive sequences, often derived from transposable elements (Miklos et al., 1988; Devlin et al., 1990; Charlesworth et al., 1994; Zhang and Spradling, 1995). These middle repetitive sequences typically have counterparts scattered throughout euchromatin. Euchromatic repetitive elements might provide sites that are able to mediate homologous pairing associations between heterochromatin and euchromatin (Sabl and Henikoff, 1995). As a result, rearrangements that move repetitive elements present in heterochromatin close to homologous elements in euchromatin would form somatically paired structures, with the sequences in between looped out (Figure 3B). Homologous interactions would draw euchromatic elements to the surface of the heterochromatic chromocenter, and this would prevent access of diffusible factors needed for expression of affected genes. By the pairing-looping model, there is no need to invoke a novel force that causes one-dimensional protein crystallization, since the existence of somatic pairing is well established (Stevens, 1908; Metz, 1916), especially in *Drosophila* (Kopczynski and Muskavitch, 1992; Hiraoka et al., 1993).

ACCOUNTING FOR FEATURES OF PEV

Silencing by nuclear mis-localization

In addition to explaining our observations involving transgene manipulations (Dorer and Henikoff, 1994; Sabl and Henikoff, 1995), which motivated the pairing-looping model, the model can account for several features of PEV that the oozing model does not. Unlike predictions of the oozing model, no obvious differences between euchromatin and heterochromatin in nucleosomal or higher order structures are expected, because contact between homologous sequences is a feature of both types of chromatin. Rather, the exclusion of transcriptional components from euchromatic genes subject to PEV would arise from the somatic pairing-dependent compartmentalization of the nucleus, as envisioned by Wakimoto and co-workers in their model for PEV of heterochromatic genes (Devlin et al., 1990; Wakimoto and Hearn, 1990). In this way, the failure to detect differences between heterochromatized and quiescent chromatin is a result of similar local packaging, whereas the cytological differences arise from pairing-mediated networks present in heterochromatin but not in euchromatin. For example, cytological compaction in polytene chromosomes (Figure 2, bottom) would result from network formation stabilized by heterochromatic protein complexes that disrupt the regular band-interband pattern characteristic of the precisely paired euchromatic arms.

Silencing of multiple linked genes

Silencing of genes in a polar manner would result from the far greater likelihood for a proximal repetitive element to pair stably with its homolog(s) in heterochromatin than for a distal element. In the example of *white* and *roughest* genes (Figure 2, top), when the distal *white* gene is looped into the chromocentral enviroment, the proximal *roughest* gene is held down on both sides and unable to escape, so that $w^- rst^+$ cells are absent (Figure 3B). However, when the *roughest* gene is looped in, the *white* gene is not necessarily held down on the distal side, in which case it escapes, leading to the presence of $w^+ rst^-$ cells. Discontinuous compaction around the 2C puff (Figure 2, bottom) would result from the unusually large distance between the heterochromatic breakpoint and the compacted distal region. This would sometimes allow intervening chromatin to loop out far enough from the chromocenter to avoid being drawn in.

Epigenetic silencing

PEV phenotypes of cell-autonomous genes, such as the *white* and *brown* eye pigment genes, are of two basic types. Pepper-and-salt phenotypes suggest that the state of an affected gene in one cell is not correlated with that in a neighboring cell. Patchy or clonal phenotypes suggest that there is a high probability for an affected gene to be in the same state in neighboring cells that derive from a common progenitor. Such clonal phenotypes indicate that a decision made early in development is faithfully transmitted through multiple cell divisions, a form of cellular inheritance referred to as "epigenetic". DNA modification is highly unlikely to play a role in epigenetic silencing seen for PEV because of the absence of detectable methylation in *Drosophila* (Urieli-Shoval et al., 1982). Therefore, by the oozing model, one might hypothesize that ooze remains associated with position-affected genes throughout the cell cycle, a plausible possibility, but one that lacks experiment support. However, by the pairing-looping model, mislocalization of a position-affected gene to the chromocenter of the nucleus could persist through multiple rounds of cell division. This model is consistent with the observation of regularities from nucleus to nucleus in a salivary gland: the chromosome arms are similarly compartmentalized, and the Rabl orientation is seen, with the chromocenter always on the nuclear envelope (Hochstrasser et al., 1986). Perhaps this regular coalescence of the chromocenter indicates that at telophase, heterochromatic regions tend to fold into conformations that are similar to what they were in prior to the onset of mitosis.

Persistent instability

As its name implies, PEV is characterized by variegation, or as Muller first put it, the "eversporting" appearance of germinally stable mutations (Muller, 1930). Such persistent cell-to-cell variability is not specifically addressed by oozing models. However, if PEV is caused by homologous associations and the resulting nuclear mislocalization of a gene, then variegation could result from nucleus-to-nucleus differences in spatial arrangements of chromosomes (Hochstrasser et al., 1986). These differences might be due to physical disruption of order within the nucleus. Evidence that physical disruptions can affect spatial arrangements of chromosomes comes from examination of polytene gut nuclei in *Drosophila* larvae. Gut nuclei, unlike salivary gland nuclei, were highly disordered relative to one another (Hochstrasser and Sedat, 1987). This disorder has been attributed to peristalsis within the gut: the same forces that move food through the gut place stress on the nuclei and disrupt whatever order might have existed prior to hatching. Similarly, the persistent instability in gene activity characteristic of PEV would result from stresses on chromosomes that occur during mitosis and other cellular movements. These stresses would disrupt the associations between homologous elements responsible for gene mislocalization and silencing by the pairing-looping model. The absence of mitotic stresses during development of polytene nuclei might favor precise pairing of chromatids and homologs seen as transverse bands, in contrast to mitotic tissues, some of which appear to lack pairing of homologs (Kopczynski and Muskavitch, 1992).

Sequences found at PEV junctions

Allele-specific differences in the extent of spreading and in the dispersion of heterochromatic breakpoints can be accounted for by the distributions of middle repetitive elements in both euchromatin and heterochromatin. Gene-to-gene differences in the frequency with which PEV alleles are obtained would reflect differences in the distribution of elements nearby. Active transposable elements, such as *copia* and *hobo*, which show striking differences in euchromatic distributions between strains (Finnegan and Fawcett, 1986), might even lead to strain-specific differences in susceptibility of a gene to PEV.

The pairing-looping model can account for the observation that heterochromatic sequences that have been characterized at PEV junctions are not simple sequence repeats, but rather are middle repetitive sequences (Tartof et al., 1984; Tartof et al., 1989). It is particularly interesting that molecular mapping of the w^{m4} rearrangement revealed the presence of two types of repetitive elements: Type 1 insertion sequences are present on the heterochromatic side of the junction, and 3 copies of a 359 bp cognate sequence repeat are found just to the euchromatic side of the junction, about 20 kb from the *white* gene (Figure 5). The bulk of the 359 bp repeats are found in a large block that acounts for most of the proximal X-heterochromatin. As a result of the w^{m4} inversion, the cognate repeats are in a position to mediate a somatic pairing interaction with the homologous heterochromatic block. This study also mapped w^{mMc}, a PEV allele that is essentially like w^{m4}, except that the euchromatic breakpoint lies just to the other side of the 359 bp cognate, resulting in the separation of this cognate sequence from the *white* gene. Interestingly, this lack of a cognate sequence in w^{mMc} correlates with its weak non-clonal phenotype, while w^{m4}, which has the cognate, is strongly variegated and largely clonal. By the pairing-looping model, a mutant in which the cognate sequence is removed might result in a partial revertant phenotype like that of w^{mMc}, which presumably is caused by the presence of an unidentified repeat sequence nearby.

Alternatively, Type I repeats might be responsible for silencing as proposed (Reuter et al., 1985). By the pairing-looping model, there would be an undiscovered Type I cognate sequence not far from *white* that would mediate looping. To test these various possibilities and to explore the sequence requirements for the clonal w^{m4} phenotype, we have induced a large number of partial and full revertants of w^{m4}, and are currently examining those with lesions that map near *white*

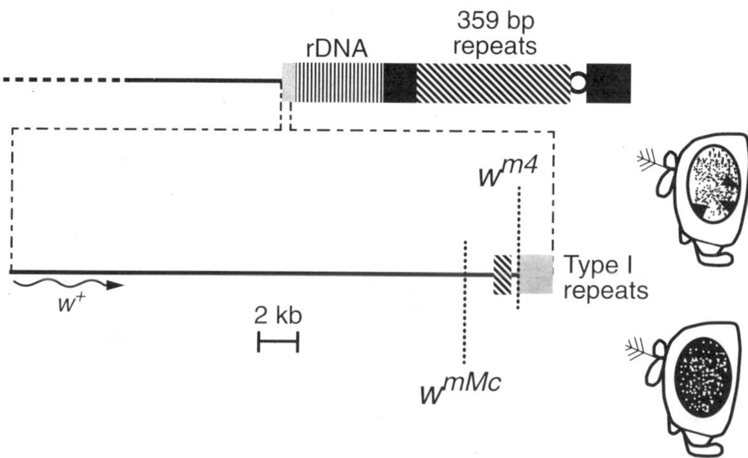

Figure 5. Map and phenotypes of w^{m4} and w^{mMc} [adapted from Tartof et al. (1989)]. The thin line represents euchromatin, rectangles represent heterochromatic sequences, the circle is the centromere and the wavy arrow is the *white* gene transcription unit. The region of the breakpoint is magnified about 1000X. The euchromatic breakpoint of w^{mMc} is shown distal to the 359 bp cognate sequence (small hatched rectangle) and the w^{m4} breakpoint fusing euchromatin to a block of Type I rDNA-associated repeats (grey rectangle).

(P. Talbert and S. H., unpublished results).

A converse prediction of the pairing-looping model is that simple sequence DNA might have little effect on neighboring genes juxtaposed nearby because of the lack of simple sequence DNA in euchromatin. This is the case for a hygromycin resistance gene, which is not silenced in human cells when sandwiched between a large block of X-centromeric alpha satellite DNA and a telomere (Bayne et al., 1994). Similarly, the *Drosophila brown*[Dominant] (*bw*[D]) heterochromatic element, which interrupts the *brown* gene coding region with extensive simple sequence DNA (Henikoff et al., 1995), appears to have no effect on nearby genes: *bw*[D] is homozygous viable, indicating a lack of effect in *cis*, even though it has the strongest silencing effect in *trans* of all known *brown* PEV alleles. A *cis* effect can be detected only by linkage enhancement, in which case *bw*[D] causes semi-lethality due to variegation of a nearby vital locus (Talbert et al., 1994). Thus very extensive simple sequence DNA that is unlikely to have cognate sequences near the *brown* gene is also weak in its silencing effect on nearby genes. More generally, the lack of effect of heterochromatic breaks near some genes, inferred from low frequencies with which

PEV alleles are obtained, would result from a scarcity of elements nearby that are cognates of elements concentrated in heterochromatin.

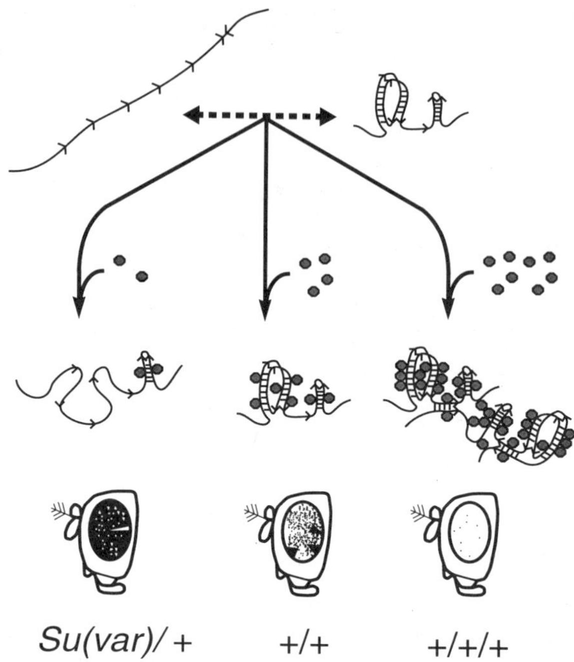

Figure 6. Dosage effects of *Su(var)* loci by the pairing-looping model. Fully dosage sensitive *Su(var)* loci encode proteins that are thought to be subunits of complexes (filled octagons) that package heterochromatin (Locke et al., 1988). These bind to paired structures that form transiently (dashed double arrow) at locally repetitive sequences (indicated by arrowheads on thin lines), causing stabilization of the structures and formation of networks between paired arrays nearby. If the concentration of each protein subunit is limiting for complex formation, then increasing the dose of a wild-type *Su(var)* gene (+) provides more complex for stabilization of paired repeats and formation of networks. Larger networks should become heterochromatic more readily, as is the case for mini-*white* and *P[bw+]* repeat arrays, which show a more strongly mutant phenotype as the numbers of copies increase (Dorer and Henikoff, 1994; Sabl and Henikoff, 1995). Larger networks would associate more readily with other blocks of heterochromatin (Eberl et al., 1993), increasing the frequency with which a gene is silenced by mislocalization to the chromocenter (Talbert et al., 1994; Henikoff et al., 1995).

In addition, differences in the extent of spreading (Figure 2) would reflect differences in density of shared repetitive elements on either side of the junction. Pairing between these elements would bring intervening euchromatic genes into a heterochromatic environment.

Exceptional PEV

PEV of heterochromatic genes has been accounted for by a model roughly similar to that described here, but where homologous interactions occur on a much larger scale (Devlin et al., 1990; Wakimoto and Hearn, 1990). This concept is supported by genetic observations on interstitial heterochromatin involving the *rolled* heterochromatic gene (Eberl et al., 1993) and the bw^D heterochromatic element (Talbert et al., 1994; Henikoff et al., 1995). In both cases, moving a block of heterochromatin closer to major heterochromatic blocks in *cis* leads to a stronger heterochromatic effect on a reporter gene, and *vice-versa*. In the case of the bw^D element, this heterochromatin-distance effect has a cytological counterpart, whereby phenotypic enhancement caused by movement closer to centric heterochromatin correlates with looping into the chromocenter of polytene chromosomes. This cytological observation represents a visible manifestation of pairing-looping.

The consistent effect of *trans*-acting PEV modifiers on dominant *brown-Variegated* alleles such as bw^D points to a quite different mechanism for heterochromatin formation than envisioned by the oozing model. If each association between members of a repeat pair is complexed by heterochromatic proteins, then overall condensation would be sensitive to complex concentration, which in turn is sensitive to the concentration of each individual component or to the overall dosage of heterochromatin in the cell (Figure 6). Heterochromatin complexes would be specific for paired hairpin or loop structures characteristic of closely-linked repeats and would stabilize them by binding (Dorer and Henikoff, 1994). More protein complexes available in the nucleus would lead to an increase in stability of homologous associations between linked repeats in heterochromatin and similar sequences present in adjoining euchromatin. The larger network of associated sequences in the vicinity of a position-affected gene would increase the probability that it will be sequestered from transcriptional components by being drawn to the chromocenter. In the case of the *brown* gene, which is thought to be inactivated in *trans* by contact with a component of the heterochromatic complex (Martin-Morris et al., 1993; Martin-Morris and Henikoff, 1995), the local increase in concentration of complex in the vicinity of *brown* will likewise increase inactivation in *trans*.

TOWARDS A MOLECULAR BIOLOGY OF HETEROCHROMATIN

What do *Su(var)*s bind?

The *trans*-acting modifiers of PEV provide an attractive route to understanding the molecular basis for properties of heterochromatin (Henikoff, 1979; Reuter and Wolff, 1981; Moore et al., 1983). Modifiers fall into two classes, commonly referred to as *Su(var)*s and *E(var)*s (*Enhancers-of-variegation*, discussed below). There may be hundreds of *Su(var)*s (Henikoff, 1979; Wustmann et al., 1989), of which only a much smaller subset show effects both in one dose and in three doses (Locke et al., 1988). It is these fully dosage-sensitive loci to which the law of mass action applies, and as such comprise the most likely candidates to encode structural constituents of chromatin (Locke et al., 1988). The role of *Su(var)* loci that show mutant effects only when in one dose (Garzino et al., 1992), or those that show effects only when homozygous (Henderson et al., 1994), might encode proteins that are less directly involved in heterochromatin formation. An example of the latter is the gene that encodes *Drosophila* PCNA, the homolog of a component of the replication apparatus (Henderson et al., 1994). A plausible role of this protein in PEV by the pairing-looping model is that replication defects would disrupt overall nuclear architecture. Indirect effects on PEV phenotypes are well-documented, including those caused by environmental changes such as temperature and pH during development (Spofford, 1976; Michailidis et al., 1988), and so mechanistic interpretations should be regarded with caution (Rushlow et al., 1984; Michailidis et al., 1988; Locke et al., 1988).

Full dosage sensitivity is seen for three *Su(var)* loci that have been cloned and sequenced (Reuter et al., 1990; Eissenberg et al., 1992; Tschiersch et al., 1994), of which HP1 is the best-studied example. Anti-HP1 antibody strikingly decorates the polytene chromocenter (James and Elgin, 1986), consistent with a role in packaging heterochromatin. But in addition, the antibody decorates the banded region of the fourth chromosome, Bridges' division 31, and at a much lower level, numerous dispersed regions in the euchromatic arms and telomeres (Powers and Eissenberg, 1993). It is intriguing that the Dr. D transposon and other middle repetitive elements show a roughly similar distribution of labeling by hybridization *in situ*, with heavy labeling at the chromocenter and the banded region of the fourth chromosome (James and Elgin, 1986). These observations are consistent with the possibility that HP1 is not specific for heterochromatin *per se*, but for repetitive sequences such as Dr. D (Miklos et al., 1988) and telomere-associated repeats (Karpen and Spradling, 1992; Levis et al., 1993). Because HP1 itself has no DNA binding activity (Powers and Eissenberg, 1993), its localization to chromosomes would reflect localization of a putative *Su(var)*-encoded complex to repetitive sequences. This evidence fits well with a model in which local repetitiveness alone can nucleate heterochromatin formation (Dorer and Henikoff, 1994): protein components of heterochromatin would bind to locally paired structures, increasing the probability that they will coalesce into a chromocenter.

How do E(var)s act?

If fully dosage-dependent *Su(var)*s encode structural components of a complex that packages heterochromatin (Locke et al., 1988), then what is the role of *E(var)*-encoded proteins? Recent work suggests that *E(var)*s constitute a diverse class of genes. Some *E(var)*-encoded proteins are also members of the *trithorax* group (trxG) (Farkas et al., 1994; Dorn et al., 1993), which act oppositely to *Polycomb* group (PcG) proteins in developmental silencing of homeotic gene complexes (Moehrle and Paro, 1994). The striking parallels between *Su(var)* and PcG proteins and between *E(var)* and trxG proteins has led to the idea that *E(var)*- and trxG-encoded complexes package active chromatin and so oppose gene silencing mediated by *Su(var)*- and PcG-encoded complexes (Locke et al., 1988; Moehrle and Paro, 1994). Indirect support for this idea comes from work in *S. cerevisiae*, in which gene activation can overcome silencing mediated by *SIR*-encoded proteins (Aparicio and Gottschling, 1994; Tartof, 1994). Indeed, cloned *E(var)*- and trxG-encoded proteins include a putative transcriptional activator protein (Dorn et al., 1993) and GAGA binding protein, a chromatin-remodeling protein that is thought to promote transcriptional activation (Farkas et al., 1994) and elongation (O'Brien et al., 1995). Interestingly, prior to blastoderm formation, GAGA protein binds strongly to GA-rich regions, which will later become heterochromatic (Raff et al., 1994). GAGA protein might be inhibiting the formation of heterochromatin at these sites, and its eventual removal (Tsukiyama et al., 1994) might provide the trigger for condensation into heterochromatin.

It should be noted that unlike the *E(var)*s and trxG genes, which have genes in common, the *Su(var)*s and PcG genes appear to be mostly non-overlapping sets of genes (Grigliatti, 1991; Fauvarque and Dura, 1993). The sharing of a short "chromodomain" between members of different groups, including 2 *Su(var)*-, one PcG- and one *E(var)*/trxG-encoded protein (Paro and Hogness, 1991; Dorn et al., 1993; Tschiersch et al., 1994), likely indicates parallel involvement in chromatin complex formation, but a more precise role for the chromodomain remains to be elucidated. Nevertheless, parallels between *Su(var)*- and PcG-encoded complexes in gene silencing suggest further similarities. Most notably, a recent model for PcG-silencing proposes local looping associations between PcG proteins that are bound to nearby but not to distant sites (Pirrotta and Rastelli, 1994). Similar protein-protein interactions might occur for *Su(var)*-encoded complexes, except that the enormous concentration of complexes within heterochromatic blocks could lead to interactions over much larger distances.

How do the recent molecular descriptions of some *E(var)*-encoded proteins (Dorn et al., 1993; Farkas et al., 1994) bear on models for PEV? By the oozing model, remodeling of chromatin to provide an active conformation might oppose the oozing of heterochromatic complexes. By the pairing-looping model, remodeling might oppose the binding of *Su(var)*-encoded complexes to paired structures formed by locally repetitive sequences. Current evidence does not distinguish these possibilities.

What are insulators?

If *E(var)*-encoded proteins oppose the formation of heterochromatin by promoting an open chromatin configuration, then removal of sites at which they act should cause a gene to become more susceptible to heterochromatic inactivation. This appears to be the case for the *white* mini-gene, which lacks eye-specific transcriptional enhancers (Steller and Pirrotta, 1985) and is exquisitely sensitive to both euchromatic and heterochromatic position effects (Kellum and Schedl, 1991; Roseman et al., 1993; Dorer and Henikoff, 1994). Sensitivity of mini-*white* to position effects has been the basis for a position-effect assay: the gene is flanked by elements that are thought to behave as boundaries and then is inserted into the genome at various sites to determine whether position effects occur (Kellum and Schedl, 1991). Although most insertions are into euchromatin, and so only test for insulation from euchromatic position effects, the binding sites for *Suppressor-of-Hairy-wing* protein appear to strongly insulate mini-*white* from heterochromatic position effects as well (Roseman et al., 1993).

It is important to realize that the mini-*white* assay is for insulation activity only; boundary function at the natural site is just an inference. An alternative possibility is that insulation is a general feature of promoters in an active or poised configuration. For example, the chicken lysozyme gene, once thought to be protected by boundary elements to either side based on assays in cell lines (Stief et al., 1989; Eissenberg and Elgin, 1991), is now known to instead require its full set of enhancer sequences to be insulated in a tissue-specific manner when the gene is present in tandem arrays (Bonifer et al., 1994). This alternative view is further encouraged by recent work on the scs and scs' insulators found on either side of the 87A heat shock genes. These were originally chosen for examination by this assay because they correspond to strong DNase I hypersensitive sites (Kellum and Schedl, 1991). Although both scs and scs' act similarly as insulators, there is no obvious sequence similarity between them or to other sequences identified as insulators by this assay (Roseman et al., 1993; Chung et al., 1993), so a multiplicity of insulators exist. Insulation activity of scs appears to map to individual hypersensitive sites, and the sites act additively (Vazquez and Schedl, 1994). Similarly, alterations in chromatin conformation that result in DNase I hypersensitivity of the chicken lysozyme gene appear to be responsible for insulation of transgenes in tandem arrays (Huber et al., 1994). Recently, it has been found that scs', which consists of two hypersensitive sites (Udvardy et al., 1985), coincides with divergent promoters for two genes adjacent to the 87A heat shock genes, and other hypersensitive sites mapped nearby correspond to promoter regions of other genes (Glover et al., 1995). Therefore, it seems unlikely that scs' has evolved to serve as a boundary element for heat shock genes, but rather exists as part of a shared promoter region for two unrelated genes nearby. It appears that the exquisite sensitivity of mini-*white* to position effects has made it possible to detect insulation activity for sequences that also cause hypersensitivity to DNase I, which is a common feature of sequences that regulate eukaryotic promoters (Weintraub and Groudine, 1976).

By this alternative view, *E(var)*-encoded proteins such as GAGA protein would act to remodel chromatin, form hypersensitive sites, and as a consequence, insulate genes from position effects. This view can account for differences that we have observed between mini-*white* and *brown* when they are present in repeat arrays at sites that are distant from heterochromatin. Mini-*white* shows PEV phenotypes even when 10 Mb from the nearest heterochromatic block (Dorer and Henikoff, 1994); however, *brown* appears to be naturally insulated from position effects (Dreesen et al., 1991; Martin-Morris et al., 1993), and is not comparably affected in repeat arrays (Sabl and Henikoff, 1995). The formation of an active or poised configuration at *brown* promoter regions might insulate the gene from heterochromatin formation by opposing the assembly of *Su(var)*-encoded complexes at somatically paired structures that form within the *brown* repeat array. In the same way, naturally repeated genes, such as those encoding the histones, are insulated and so would not form heterochromatin.

What is the molecular basis for pairing?

The pairing-looping model does not specify a mechanism whereby homologous sequences associate to form paired structures. The molecular basis for homologous pairing is a question that is relevant to other areas of chromosome research, such as meiosis, wherein pairing is crucial for regular segregation. Intriguing recent results in yeast argue that pairing precedes the initiation of meiosis (Weiner and Kleckner, 1994), providing a potential link between somatic and meiotic pairing. Homologous DNA sequences might interact directly, perhaps via multi-stranded structures (Kleckner and Weiner, 1993). It should be realized that these experiments in yeast address the initiation of pairing, leaving open the possibility that subsequent zippering occurs by a different mechanism.

One mechanism for zippering of homologs is suggested by current views on the structure of the 30 nm fiber, for many years thought to consist of regular folding of beaded nucleosomes into a solenoid. However, Woodcock et al. (1993) have argued that the 30 nm fiber follows a highly irregular trajectory because of differences in the spacings between nucleosomes, and recent studies of chromatin both in intact nuclei (Horowitz et al., 1994) and in purified preparations (Leuba et al., 1994) support this view. To the extent that nucleosomal spacing is ultimately a consequence of sequence specific interactions, then sequence-specific zippering of homologs might be a natural result. That is, homologous 30 nm fibers that have initiated pairing will zipper up because identical irregularities should follow identical and adjacent paths, just as adjacent strands of curly hair curl together rather than individually.

Pairing and polyteny

Formation of locally paired structures in heterochromain might have special consequences for polytene chromosomes, in which heterochromatin is severely under-represented. One model for under-representation is that replication forks are stalled at junctions between single-copy and repeat sequences (Laird et al., 1973). Another is that repeated sequences are replicated, but are then eliminated by transposase-like excision (Karpen and Spradling, 1990). Observations that are consistent with elimination (Glaser et al., 1992; Spradling et al., 1993; Spradling, 1993) can be understood in terms of pairing-looping. During the development of polytene chromosomes, paired structures would be substrates for cellular nucleases that excise the intervening loops, leading to under-representation of heterochromatic repeats. In diffusely endoreplicated ovarian nuclei, similar nuclease action on paired repeats might result in nested deletions and acentric fragments reaching into euchromatin, as has been observed (Spradling, 1993). The possible involvement of cellular nucleases specific for paired repeats rather than sequence-specific transposases (Karpen and Spradling, 1990) overcomes difficulties with the elimination model. One difficulty is that under-representation is most severe in heterochromatic regions, even though presumed transposon target sequences in heterochromatin are typically present in euchromatin as well. Another difficulty is that the sequences most subject to underrepresentation are simple sequence satellites, not middle repetitive sequences (Rudkin, 1969; Hammond and Laird, 1985a; Hammond and Laird, 1985b). However, if elimination occurs on any paired sequences, then scattered repeats in euchromatin would not be susceptible, middle-repetitive clusters would be moderately susceptible, and simple sequence blocks would be most susceptible, as observed (Hammond and Laird, 1985a; Hammond and Laird, 1985b).

An alternative explanation for under-representation is that polymerases are not permanently stalled, but rather switch templates. This copy-choice model is inspired in part by Ashburner's explanation for the regular presence of drawn-out fibers that connect specific sites in polytene chromosomes (Ashburner, 1980). There is a link between the occurrence of these ectopic fibers and under-representation of DNA sequences: ectopic fibers are found in numerous dipteran polytene nuclei, but only in tissues that show under-representation of satellite DNA (Ashburner, 1980). Moreover, ectopic fibers have been found to connect DNA sequences that are under-represented in polytene chromosomes (Lamb and Laird, 1987). In some cases, *in situ* hybridization has revealed the presence of middle-repetitive sequences in ectopic fibers, consistent with the possibility that pairing of homologous sequences underlies the ectopic interactions (Pardue, 1986). The similarities between these intercalary and chromocentral ectopic interactions suggest a similar causal basis.

Ashburner's model for ectopic fiber formation is that branch migration occurs at a stalled replication fork to displace a single-stranded whisker, and this can anneal with a similarly displaced whisker of opposite polarity to form a fiber (Ashburner, 1980). I suggest that strand-displacement and annealing can also result in under-representation by a copy-choice replication mechanism (Belling, 1933): replication begun on one template switches to a template strand that has been displaced by a converging polymerase on a paired duplex (see Figure 7 for details). In retroviruses, template switching between two RNA strands is thought to occur during first strand synthesis in the virion (Temin, 1993). A different template-switching mechanism has also been proposed to account for instability of inverted repeats in bacteria, whereby a hairpin is "skipped" by DNA polymerase (Leach, 1994). Similarly, template switching between paired repeats would result in underreplication of both the simple sequence DNA and any complex sequences in between, if it is assumed that no replication origins lie within the region that becomes under-

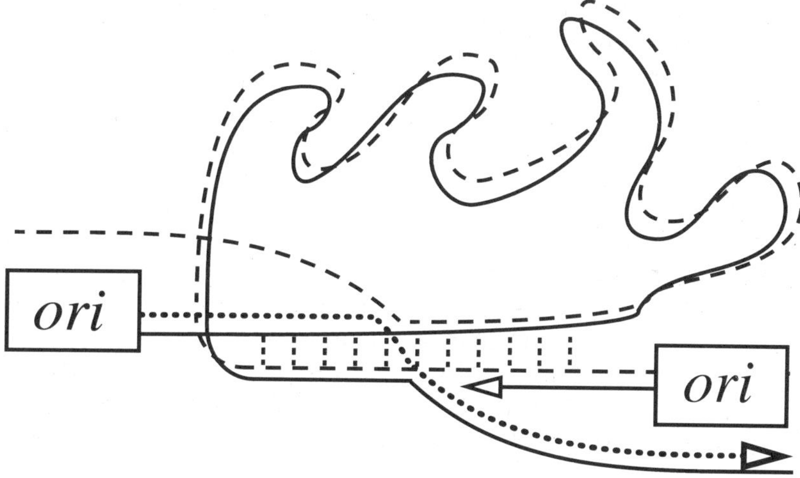

Figure 7. A copy-choice model for under-representation in polytene nuclei. When AT-rich satellite sequences [*e.g.*, $(AATAT)_n$] are paired, the inherent instability of AT-rich duplexes (solid line for Watson and dashed line for Crick) and the alignment of paired copies (at vertical dashes) allows a DNA polymerase in transit (at open arrowhead) to switch templates. If replication initiates outside of the tandem satellite block (at the boxed "*ori*" sites) but not inside, then a template switch originating from the left will allow the polymerase to exit right without replicating sequence in between, and vice-versa. Here, a polymerase using the Watson strand as template enters the paired region from the left (dotted line). Branch migration and displacement allow annealing to the Watson strand that has been displaced by a polymerase entering from the right using the Crick strand as template. If template switches occur frequently throughout the paired region, then an odd number of switches by a polymerase will lead to loss of the looped-out region, and an even number will have no effect. As a result, representation of the looped-out region will decrease relative to that of the genome as a whole with each round of replication.

represented. AT-rich satellites, which comprise the bulk of *Drosophila* simple sequence DNA (Lohe et al., 1993), might be especially prone to strand-displacement because of the formation of unstable duplexes.

Unlike the elimination model, copy-choice underreplication requires no special enzymology to accomplish under-representation, yet it could account for failure to detect replication forks (Glaser et al., 1992), for the observed spectrum of heterogeneous products (Spradling, 1993) and for the interspersion of different levels of polyteny (Zhang and Spradling, 1995). Furthermore, template switching might be occurring within euchromatic regions as well, knitting together aligned chromatids and homologues in polytene chromosomes, thus forcing them to remain precisely paired following acid treatment and squashing pressure that are applied in cytological preparations.

TESTS OF PAIRING-LOOPING

A useful model is one that makes testable predictions. In the absence of plausible alternatives to oozing, it was difficult to envision experimental tests that could establish that model to the exclusion of other models. As a result, oozing has been described more as an explanation of existing data, rather than as a model subject to contradiction (Tartof et al., 1984; Tartof et al., 1989; Eissenberg, 1989; Grigliatti, 1991; Moehrle and Paro, 1994). It is notable that evidence contradicting key features came only from subsequent unexpected observations of discontinuous compaction (Belyaeva and Zhimulev, 1991), heterochromatin formation involving artificial repeat arrays (Dorer and Henikoff, 1994), and dependence on copy number and orientation for PEV of a transgene (Sabl and Henikoff, 1995). Our current work is aimed at testing unique predictions of the pairing-looping model. One test has already been alluded to: generate partial revertants of w^{m4} to determine whether nearby middle repetitive sequences are required for the clonal phenotype.

Another unique prediction of pairing-looping model is that looping associations should correlate with gene silencing. We have shown that this is the case for PEV *trans*-inactivation by bw^D: increased looping of bw^D into the chromocenter occurs when the phenotype is enhanced by moving bw^D closer to heterochromatin, and both effects are simultaneously reverted when bw^D is moved farther away (Talbert et al., 1994; Henikoff et al., 1995). Recently, this correlation has been extended to diploid cells (A. Dernburg, J. Sedat, A. Csink and S. H., unpublished results). It will be interesting to determine whether the *cis*-inactivation phenotypes seen for mini-*white* repeat arrays also correlate with looping into the chromocenter of diploid cells.

The pairing-looping model also predicts that similarities might be detected between DNA sequences on either side of euchromatic-heterochromatic junctions in PEV rearrangements, such as has been noted for w^{m4} (Tartof et al., 1984). Just how similar is difficult to predict, although the inverse copy number and orientation dependences of PEV at repeat arrays that lack known heterochromatic sequences suggest that very limited similarities might be effective. It is also difficult to predict how close to one another the similar sequences must be. A more direct approach would be to use cloned middle repetitive DNA from a particular heterochromatic region, insert it next to a reporter gene and ask whether PEV affecting the reporter is induced by DNA from that region. Although there are several candidate sequences that might be used, the 359 bp repeat is especially attractive, both because of its possible involvement in w^{m4} variegation and the fact that large blocks are located exclusively in the proximal heterochromatin of the X chromosome (Lohe et al., 1993).

CONCLUSIONS

I hope that the foregoing critique of the heterochromatin oozing model and the presentation of evidence in favor of the pairing-looping model will help clarify thinking about gene silencing phenomena in organisms that lack the powerful genetic tools available in *Drosophila*. For example, numerous homology dependent phenomena in plants have been discovered in recent years (Matzke et al., 1994; Rossignol and Faugeron, 1994; Flavell, 1994; Jorgensen, 1995), and these might well involve mechanisms similar to PEV. Other examples include the many cases in which insertion of transgenes into mammalian cells is accompanied by a lack of copy number dependence on expression (Palmiter and Brinster, 1986; Bonifer et al., 1991). These observations are typically attributed to effects of neighboring sequences on transgene expression, although the fact that these effects are always negative relative to expectation is consistent with gene silencing by heterochromatin formation. In such cases, the frequency of gene silencing might depend on the distance of an affected insertion to blocks of heterochromatin along the chromosome, so that copy number would not correlate with overall expression levels. In this way, gene inactivation would occur by mislocalization to a heterochromatic compartment of the nucleus. Other examples of local heterochromatin formation might be trinucleotide (CGG) repeat expansions that are responsible for fragile sites in diseases such as the fragile X syndrome (Laird, 1987; Hansen et al., 1993). Perhaps locally paired structures form that are recognized by the cellular machinery and targeted to the heterochromatic compartment of the nucleus. Although details are certainly likely to differ, it is attractive to think that the forces of somatic pairing might underlie diverse gene silencing phenomena in higher eukaryotes. A mechanistic understanding of these forces is a major challenge for future research.

ACKNOWLEDGEMENTS

The rich history of *Drosophila* heterochromatin research and the interactive group of current practitioners have contributed heavily to thoughts presented in this essay. I thank Joy Sabl for sharing unpublished results and ideas (and the curly hair metaphor), and also Amy Csink, Doug Dorer, Linda Martin-Morris and Paul Talbert for their intellectual and experimental contributions to the model. Work in my laboratory is supported by the Howard Hughes Medical Institute.

REFERENCES

Aparicio, O.M., and Gottschling, D.E., 1994, Overcoming telomeric silencing: a trans-activator competes to establish gene expression in a cell cycle-dependent way, *Genes and Dev.* 8:1133.

Ashburner, M., 1980, Some aspects of the structure and function of the polytene chromosomes of the Diptera. *in*: "Insect cytogenetics. Tenth Symposium of the Royal Entomological Society," R.L. Blackman, G.M. Hewitt and M. Ashburner eds., pp. 65-84.

Baker, W.K., 1968, Position-effect variegation, *Adv. Genet.* 14:133.

Bates, G., and Lehrach, H., 1994, Trinucleotide repeat expansions and human genetic disease, *Bioessays* 16:277.

Bayne, R.A.L, Broccoli, D., Taggart, M.H., Thomson, E.J., Farr, C.J., and Cooke, H.J., 1994, Sandwiching of a gene within 12 kb of a functional telomere and alpha satellite does not result in silencing, *Hum. Mol. Genet.* 3:539.

Belling, J., 1933, Crossing-over and gene rearrangement in flowering plants, *Genetics* 18:388.

Belyaeva, E.S., and Zhimulev, I.F., 1991, Cytogenetic and molecular aspects of position effect variegation in *Drosophila* III. Continuous and discontinuous compaction of chromosomal material as a result of position effect variegation, *Chromosoma* 100:453.

Belyaeva, E.S., Demakova, O.V., Umbetova, G.H., and Zhimulev, I.F., 1993, Cytogenetic and molecular aspects of position-effect variegation in *Drosophila melanogaster*. V. Heterochromatin-associated protein HP1 appears in euchromatic chromosomal regions that are inactivated as a result of position-effect variegation, *Chromosoma* 102:583.

Bonifer, C., Hecht, A., Saueressig, H., Winter, D.M., and Sippel, A.E., 1991, Dynamic chromatin: the regulatory domain organization of eukaryotic gene loci, *J. Cell. Biochem.* 47:99.

Bonifer, C., Yannoutsos, N., Kruger, G., Grosveld, F., and Sippel, A.E., 1994, Dissection of the locus control function located on the chicken lysozyme gene domain in transgenic mice, *Nucleic Acids Res.* 22:4202.

Charlesworth, B., Jarne, P., and Assimacopoulous, S., 1994, The distribution of transposable elements within and between chromosomes in a population of *Drosophila melanogaster*. 3. Element abundances in heterochromatin, *Genet. Res.* 64:183.

Chung, J.H., Whiteley, M., and Felsenfeld, G., 1993, A 5' element of the chicken beta-globin domain serves as an insulator in human erythroid cells and protects against position effect in Drosophila, *Cell* 74:505.

Demerec, M., and Slizynska, H., 1937, Mottled white 258-18 of *Drosophila melanogaster*, *Genetics* 22:641.

Devlin, R.H., Bingham, B., and Wakimoto, B., 1990, The organization and expression of the *light* gene, a heterochromatic gene of *Drosophila melanogaster*, *Genetics* 125:129.

Dorer, D.R., and Henikoff, S., 1994, Expansions of transgene repeats cause heterochromatin formation and gene silencing in Drosophila, *Cell* 77:993.

Dorn, R., Krauss, V., Reuter, G., and Saumweber, H., 1993, The enhancer of position-effect variegation of *Drosophila, E(var)3-93D,* codes for a chromatin protein containing a conserved domain common to several transcriptional regulators, *Proc. Natl. Acad. Sci. USA* 90:11376.

Dreesen, T.D., Henikoff, S., and Loughney, K., 1991, A pairing-sensitive element that mediates *trans*-inactivation is associated with the *Drosophila brown* gene, *Genes and Dev.* 5:331.

Eberl, D.F., Duyf, B.J., and Hilliker, A.J., 1993, The role of heterochromatin in the expression of a heterochromatic gene, the *rolled* locus of *Drosophila melanogaster, Genetics* 134:277.

Eissenberg, J.C., 1989, Position effect variegation in *Drosophila*: Towards a genetics of chromatin assembly, *Bioessays* 11:14.

Eissenberg, J.C., and Elgin, S.C.R., 1991, Boundary functions in the control of gene expression, *Trends Genet.* 7:335.

Eissenberg, J.C., Morris, G.D., Reuter, G., and Hartnett, T., 1992, The heterochromatin-associated protein HP-1 is an essential protein in Drosophila with dosage-dependent effects on position-effect variegation, *Genetics* 131:345.

Ephrussi, B., and Sutton, E., 1944, A reconsideration of the mechanism of position effect, *Proc. Natl. Acad. Sci. USA* 30:183.

Farkas, G., Gausz, J., Galloni, M., Reuter, G., Gyurkovics, H., and Karch, F., 1994, The *Trithorax-like* gene encodes the *Drosophila* GAGA factor, *Nature* 371:806.

Fauvarque, M.-O., and Dura, J.-M., 1993, *polyhomeotic* regulatory sequences induce developmental regulator-dependent variegation and targeted *P*-element insertions in *Drosophila, Genes and Dev.* 7:1508.

Finnegan, D.J., and Fawcett, D.H., 1986, Transposable elements in Drosophila melanogaster, *Oxfd. Surv. Eukaryot. Genes* 3:1.

Flavell, R.B., 1994, Inactivation of gene expression in plants as a consequence of specific sequence duplication, *Proc. Natl. Acad. Sci. USA* 91:3490.

Garzino, V., Pereira, A., Laurenti, P., Graba, Y., Levis, R.W., Le Parco, Y., and Pradel, J., 1992, Cell lineage-specific expression of *modulo*, a dose-dependent modifier of variegation in *Drosophila, EMBO J.* 11:4471.

Glaser, R.L., Karpen, G.H., and Spradling, A.C., 1992, Replication forks are not found in a *Drosophila* minichromosome demonstrating a gradient of polytenization, *Chromosoma* 102:15.

Glover, D.M., Leibowitz, M.H., McLean, D.A., and Parry, H., 1995, Mutations in *aurora* prevent centrosome separation leading to the formation of monopolar spindles, *Cell* 81:95.

Grigliatti, T., 1991, Position-effect variegation-An assay for nonhistone chromosomal proteins and chromatin assembly and modifying factors. *in*: "Functional organization of the nucleus: A laboratory guide," B.A. Hamkalo and S.C.R. Elgin eds., Academic Press, San Diego, pp. 587-627.

Gubb, D., Ashburner, M., Roote, J., and Davis, T., 1990, A novel transvection phenomenon affecting the *white* gene of *Drosophila melanogaster, Genetics* 126:167.

Hammond, M.P., and Laird, C.D., 1985a, Chromosome structure and DNA replication in nurse and follicle cells of *Drosophila melanogaster, Chromosoma* 91:267.

Hammond, M.P., and Laird, C.D., 1985b, Control of DNA replication and spatial distribution of defined DNA sequences in salivary gland cells of *Drosophila melanogaster, Chromosoma* 91:279.

Hansen, R.S., Canfield, T.K., Lamb, M.M., Gartler, S.M., and Laird, C.D., 1993, Association of fragile X syndrome with delayed replication of the FMR1 gene, *Cell* 73:1403.

Hartmann-Goldstein, I.J., 1967, On the relationship between heterochromatization and variegation in *Drosophila* with special reference to temperature-sensitive periods, *Genet. Res.* 10:143.

Hayashi, S., Ruddell, A., Sinclair, D., and Grigliatti, T., 1990, Chromosomal structure is altered by mutations that suppress or enhance position effect variegation, *Chromosoma* 99:391.

Hearn, M.G., Hedrick, A., Grigliatti, T.A., and Wakimoto, B.T., 1991, The effect of modifiers of position-effect variegation on the variegation of heterochromatic genes of *Drosophila melanogaster*, *Genetics* 128:785.

Heitz, E., 1929, Heterochromatin, Chromocentren, Chromomeren, *Ber. Dtsch. bot. Ges.* 47:274.

Henderson, D.S., Banga, S.S., Grigliatti, T.A., and Boyd, J.B., 1994, Mutagen sensitivity and suppression of position-effect variegation result from mutations in *mus209*, the *Drosophila* gene encoding PCNA, *EMBO J.* 13:1450.

Henikoff, S., 1979, Position effects and variegation enhancers in an autosomal region of *Drosophila melanogaster*, *Genetics* 93:105.

Henikoff, S., 1981, Position-effect variegation and chromosome structure of a heat shock puff in *Drosophila*, *Chromosoma* 83:381.

Henikoff, S. 1994, A reconsideration of the mechanism of position effect, *Genetics* 138:1.

Henikoff, S., 1995, Gene silencing in *Drosophila*. *in*: "Gene Silencing in Higher Plants and Related Phenomena in Other Eukaryotes," P. Meyer eds., *Current Topics in Microbiology and Immunology:*, vol. 197. Springer-Verlag, Berlin, pp. 193-208.

Henikoff, S., and Dreesen, T.D., 1989, Trans-inactivation of the *Drosophila brown* gene: evidence for transcriptional repression and somatic pairing dependence, *Proc. Natl. Acad. Sci. USA* 86:6704.

Henikoff, S., Jackson, J.M., and Talbert, P.B., 1995, Distance and pairing effects on the *brownDominant* heterochromatic element in Drosophila, *Genetics* 140:1007.

Hiraoka, Y., Dernburg, A.S., Parmelee,S.J., Rykowski,M.C., Agard,D.A., and Sedat,J.W., 1993, The onset of homologous chromosome pairing during *Drosophila melanogaster* embryogenesis, *J. Cell Biol.* 120:591.

Hochstrasser, M., and Sedat, J.W., 1987, Three-dimensional organization of Drosophila melanogaster interphase nuclei. I Tissue-specific aspects of polytene nuclear architecture, *J. Cell Biol.* 104:1455.

Hochstrasser, M., Mathog, D., Gruenbaum, Y., Saumweber, H., and Sedat, J.W., 1986, Spatial organization of chromosomes in the salivary gland nuclei of *Drosophila melanogaster*, *J. Cell Biol.* 102:112.

Horowitz, R.A., Agard, D.A., Sedat, J.W., and Woodcock, C.L., 1994, The three-dimensional architecture of chromatin in situ: electron tomography reveals fibers composed of a continuously variable zig-zag nucleosomal ribbon, *J. Cell Biol.* 125:1.

Huber, M.C., Bosch, F.X., Sippel, A.E., and Bonifer, C., 1994, Chromosomal position effects in chicken lysozyme gene transgenic mice are correlated with suppression of DNase I hypersensitive site formation, *Nucleic Acids Res.* 22:4195.

James, T.C., and Elgin, S.C.R., 1986, Identification of a nonhistone chromosomal protein associated with heterochromatin in *Drosophila* and its gene, *Mol. Cell. Biology* 6:3862.

Jorgensen, R.A., 1995, Cosuppression, flower color patterns, and metastable gene expression states, *Science* 268:686.

Karpen, G.H., 1994, Position-effect variegation and the new biology of heterochromatin, *Curr. Op. Genet. Dev.* 4:281.

Karpen, G., and Spradling, A.C., 1990, Reduced DNA polytenization of a minichromosome region undergoing position-effect variegation in *Drosophila*, *Cell* 63:97.

Karpen, G.H., and Spradling, A.C., 1992, Analysis of subtelomeric heterochromatin in the Drosophila minichromosome *Dp1187* by single *P* element insertional mutagenesis, *Genetics* 132:737.

Kellum, R., and Schedl, P., 1991, A position-effect assay for boundaries of higher order chromosomal domains, *Cell* 64:941.

Kleckner, N., and Weiner, B.M., 1993, Potential advantages of unstable interactions for pairing of chromosomes in meiotic, somatic, and premeiotic cells, *Cold Spring Harbor Symp. Quant. Biol.* 58:553.

Kopczynski, C.C., and Muskavitch, M.A.T., 1992, Introns excised from the *Delta* primary transcript are localized near sites of *Delta* transcription, *J. Cell Biol.* 119:503.

Laird, C.D., 1987, Proposed mechanism of inheritance and expression of the human fragile-X syndrome of mental retardation, *Genetics* 117:587.

Laird, C.D., Chooi, W.Y., Cohen, E.H., Dickson, E., Hutchinson, N., and Turner, S.H., 1973, Organization and transcription of DNA in chromosomes and mitochondria of Drosophila, *Cold Spring Harbor Symp. Quant. Biol.* 38:311.

Lamb, M.M., and Laird, C.D., 1987, Three euchromatic DNA sequences under-replicated in polytene chromosomes of *Drosophila* are localized in constrictions and ectopic fibers, *Chromosoma* 95:227.

Leach, D.R.F., 1994, Long DNA palindromes, cruciform structures, genetic instability and secondary structure repair, *Bioessays* 16:893.

Leuba, S.H., Yang, G., Robert, C., Samori, B., van Holde, K., Zlatanova, J., and Bustamante, C., 1994, Three-dimensional structure of extended chromatin fibers as revealed by tapping-mode scanning force microscopy, *Proc. Natl. Acad. Sci. USA* 91:11621.

Levis, R.W., Ganesan, R., Houtchens, K., Tolar, L.A., and Sheen, F.-m, 1993, Transposons in place of telomeric repeats at a Drosophila telomere, *Cell* 75:1083.

Lewis, E.B., 1950, The phenomenon of position effect, *Adv. Genet.* 3:73.

Lindsley, D.L., and Zimm, G.G., 1992, "The genome of *Drosophila melanogaster*". Academic Press, San Diego.

Locke, J., 1994, Examination of DNA sequences undergoing chromatin conformation changes at a variegating breakpoint in *Drosophila melanogaster*, *Genetica* 92:33.

Locke, J., Kotarski, M.A., and Tartof, K.D., 1988, Dosage-dependent modifiers of position effect variegation in *Drosophila* and a mass action model that explains their effect, *Genetics* 120:181.

Lohe, A., Hilliker, A.J., and Roberts, P.A., 1993, Mapping simple repeated DNA sequences in heterochromatin of *Drosophila melanogaster*, *Genetics* 134:1149.

Manuelidis, L., 1991, Heterochromatic features of an 11-megabase transgene in brain cells, *Proc. Natl. Acad. Sci. USA* 88:1049.

Martin-Morris, L.E., and Henikoff, S., 1995, Conservation of *brown* gene *trans*-inactivation in Drosophila, *Genetics* 140:193.

Martin-Morris, L.E., Loughney, K., Kershisnik, E.O., Poortinga, G., and Henikoff, S., 1993, Characterization of sequences responsible for *trans*-inactivation of the *Drosophila brown* gene, *Cold Spring Harbor Symp. Quant. Biol.* 58:577.

Matzke, M.A., Matzke, A.J.M., and Mittelsten-Scheid, O., 1994, Inactivation of repeated genes - DNA-DNA interaction? *in*: "Homologous recombination in plants," J. Paszkowski ed., Kluwer, Amsterdam, pp. 271-307.

Metz, C.W., 1916, Chromosome studies on the Diptera II. The paired association of chromosomes in the diptera, and its significance, *J. Exp. Zool.* 21:213.

Michailidis, J., Murray, N.D., and Graves, J.A.M., 1988, A correlation between development time and variegated position effect in *Drosophila melanogaster*, *Genet. Res.* 52:119.

Miklos, G.L.G., Yamamoto, M.-T., Davies, J., and Pirrotta, V., 1988, Microscloning reveals a high frequency of repetitive sequences characteristic of chromosome 4 and the beta-heterochromatin of *Drosophila melanogaster*, *Proc. Natl. Acad. Sci. USA* 85:2051.

Modollel, J., Bender, W., and Meselson, M., 1983, *Drosophila melanogaster* mutations suppressible by the suppressor of Hairy-wing are insertions of a 7.3-kilobase mobile element, *Proc. Natl. Acad. Sci. USA* 80:1678.

Moehrle, A., and Paro, R., 1994, Spreading the silence: epigenetic transcriptional regulation during *Drosophila* development, *Dev. Genet.* 15:478.

Moore, G.D., Sinclair, D.A., and Grigliatti, T.A., 1983, Histone gene multiplicity and position effect variegation in *Drosophila melanogaster*, *Genetics* 105:327.

Mottus, R., Reeves, R., and Grigliatti, T.A., 1980, Butyrate suppression of position-effect variegation in *Drosophila melanogaster*, *Mol. Gen. Genet.* 178:465.

Muller, H.J., 1930, Types of visible variations induced by X-rays in *Drosophila, J. Genet.* 22:299.

Muller, H.J., 1932, Further studies on the nature and causes of gene mutations, *Proc. Intl. Congr. of Genet.* 1:213.

O'Brien, T., Wilkins, R.C., Giardina, C., and Lis, J.T., 1995, Distribution of GAGA protein on *Drosophila* genes in vivo, *Genes and Dev.* 9:1098.

Painter, T.S., 1934, Salivary chromosomes and the attack on the gene, *J. Hered.* 25:464.

Palmiter, R.D., and Brinster, R.L., 1986, Germ-line transformation of mice, *Ann. Rev. Genet.* 20:465.

Pardue, M.L., 1986, *In situ* hybridization to DNA of chromosomes and nuclei. *in:* "Drosophila, a practical approach," D.B. Roberts ed, IRL Press, Oxford, pp. 111-137.

Paro, R., and Hogness, D.S., 1991, The Polycomb protein shares a homologous domain with a heterochromatin-associated protein of *Drosophila, Proc. Natl. Acad. Sci. USA* 88:263.

Pirrotta, V., and Rastelli, L., 1994, *white* gene expression, repressive chromatin domains and homeotic gene regulation in *Drosophila, Bioessays* 16:549.

Pontecorvo, G., 1944, Structure of heterochromatin, *Nature* 153:365.

Powers, J.A., and Eissenberg, J.C., 1993, Overlapping domains of the heterochromatin-associated protein HP1 mediate nuclear localization and heterochromatin binding, *J. Cell Biol.* 120:291.

Ptashne, M., 1986, Gene regulation by proteins acting nearby and at a distance, *Nature* 322:697.

Raff, J.W., Kellum, R., and Alberts, B., 1994, The *Drosophila* GAGA transcription factor is associated with specific regions of heterochromatin thoughout the cell cycle, *EMBO J.* 13:5977.

Reuter, G., and Spierer, P., 1992, Position effect variegation and chromatin proteins, *Bioessays* 14:605.

Reuter, G., and Wolff, I., 1981, Isolation of dominant suppressor mutations for position-effect variegation in *Drosophila melanogaster, Mol. Gen. Genet.* 182:516.

Reuter, G., Werner, W., and Hoffmann, H.J., 1982, Mutants affecting position-effect heterochromatinization in *Drosophila melanogaster, Chromosoma* 85:539.

Reuter, G., Wolff, I., and Friede, B., 1985, Functional properties of the heterochromatic sequences inducing *wm4* position-effect variegation in *Drosophila melanogaster, Chromosoma* 93:132.

Reuter, G., Giarre, M., Farah, J., Gausz, J., Spierer, A., and Spierer, P., 1990, Dependence of position-effect variegation in *Drosophila* on dose of a gene encoding an unusual zinc-finger protein, *Nature* 344:219.

Roseman, R.R., Pirrotta, V., and Geyer, P.K., 1993, The *su(HW)* protein insulates expression of the *Drosophila melanogaster white* gene from chromosomal position-effects, *EMBO J.* 12:435.

Rossignol, J.-L., and Faugeron, G., 1994, Gene inactivation triggered by recognition between DNA repeats, *Experientia* 50:307.

Rudkin, G.T., 1969, Non-replicating DNA in Drosophila, *Genetics (suppl.)* 61:227.

Rushlow, C.A., Bender, W., and Chovnick, A., 1984, Studies on the mechanism of heterochromatic position effect at the *rosy* locus of *Drosophila melanogaster, Genetics* 108:603.

Sabl, J.F., and Henikoff, S., 1995, Copy number and orientation determine the susceptibility of a gene to silencing by nearby heterochromatin in Drosophila, *Genetics* (accepted for publication).

Schlossherr, J., Eggert, H., Paro, R., Cremer, S., and Jack, R.S., 1994, Gene inactivation in *Drosophila* mediated by the *Polycomb* gene product or by position-effect variegation does not involve major changes in the accessiblity of the chromatin fibre, *Mol. Gen. Genet.* 243:453.

Schultz, J., 1936, Variegation in Drosophila and the inert heterochromatic regions, *Proc. Natl. Acad. Sci. USA* 22:27.

Sinclair, D.A., Ruddell, A.A., Brock, J.K., Clegg, N.J., Lloyd, V.K., and Grigliatti, T.A., 1992, A cytogenetic and genetic characterization of a group of closely linked second chromosome mutations that suppress position-effect variegation in *Drosophila melanogaster*, *Genetics* 130:333.

Slatis, H.M., 1955, Position effects at the brown locus in *Drosophila melanogaster*, *Genetics* 40:5.

Spofford, J.B., 1976, Position-effect variegation in *Drosophila*. *in*: "Genetics and Biology of Drosophila," M. Ashburner, and E. Novitski eds, vol. 1c. Academic Press, London, pp. 955-1019.

Spradling, A.C., 1993, Position effect variegation and genomic instability, *Cold Spring Harbor Symp. Quant. Biol.* 58:585.

Spradling, A.C., and Karpen, G.H., 1990, Sixty years of mystery, *Genetics* 126:779.

Spradling, A.C., Karpen, G., Glaser, R., and Zhang, P., 1993, Evolutionary conservation of developmental mechanisms: DNA elimination in *Drosophila*. *in*: "Evolutionary conservation of developmental mechanisms," A.C. Spradling ed., Wiley-Liss, New York, pp. 39-53.

Stadler, L.J., 1954, The gene, *Science.* 120:811.

Steller, H., and Pirrotta, V., 1985, A transposable P vector that confers selectable G418 resistance to *Drosophila* larvae, *EMBO J.* 4:167.

Stevens, N.M., 1908, A study of the germ cells of certain Diptera with reference to the heterochromosomes and the phenomena of synapsis, *J. Exp. Zool.* 5:359.

Stief, A., Winter, D.M., Stratling, W.H., and Sippel, A.E., 1989, A nuclear DNA attachment element mediates elevated and position-independent gene activity, *Nature* 341:343.

Talbert, P.B., LeCiel, C.D.S., and Henikoff, S., 1994, Modification of the Drosophila heterochromatic mutation *brownDominant* by linkage alterations, *Genetics* 136:559.

Tartof, K.D., 1994, Position effect variegation in yeast, *Bioessays* 16:713.

Tartof, K.D., Hobbs, C., and Jones, M., 1984, A structural basis for variegating position effects, *Cell* 37:869.

Tartof, K.D., Bishop, C., Jones, M., Hobbs, C.A., and Locke, J., 1989, Towards an understanding of position effect variegation, *Dev. Genet.* 10:162.

Temin, H., 1993, Retrovirus variation and reverse transcription: Abnormal strand transfers result in retrovirus genetic variation, *Proc. Natl. Acad. Sci. USA* 90:6900.

Tschiersch, B., Hofmann, A., Krauss, V., Dorn, R., Korge, G., and Reuter, G., 1994, The protein encoded by the *Drosophila* position-effect variegation suppressor gene *Su(var)3-9* combines domains of antagonistic regulators of homeotic gene complexes, *EMBO J.* 13:3822.

Tsukiyama, T., Becker, P.B., and Wu, C., 1994, ATP-dependent nucleosome disruption at a heat-shock promoter mediated by binding of GAGA transcription factor, *Nature* 367:525.

Udvardy, A., Maine, E., and Schedl, P., 1985, The 87A7 chromomere: identification of novel chromatin structures flanking the heat shock locus that may define the boundaries of higher order domains, *J. Mol. Biol.* 185:341.

Umbetova, G.H., Belyaeva, E.S., Baricheva, E.M., and Zhimulev, I.F., 1991, Cytogenetic and molecular aspects of position effect variegation in *Drosophila melanogaster* IV. Underreplication of chromosomal material as a result of gene inactivation, *Chromosoma* 101:55.

Urieli-Shoval, S., Gruenbaum, Y., Sedat, J., and Razin, A., 1982, The absence of detectable methylated bases in *Drosophila melanogaster* DNA, *FEBS Lett.* 146:148.

Vazquez, J., and Schedl, P., 1994, Sequences required for enhancer blocking activity of scs are located within two nuclease-hypersensitive regions, *EMBO J.* 13:5984.

Wakimoto, B., and Hearn, M., 1990, The effects of chromosome rearrangements on the expression of heterochromatic genes in Chromosome *2L* of *Drosophila melanogaster*, *Genetics* 125:141.

Weiner, B.M., and Kleckner, N., 1994, Chromosome pairing via multiple interstitial interactions before and during meiosis in yeast, *Cell* 77:977.

Weintraub, H., and Groudine, M., 1976, Chromosomal subunits in active genes have an altered conformation, *Science* 193:848.

Willems, P.J., 1994, Dynamic mutations hit double figures, *Nature Gen.* 8:213.

Wines, D.R., Talbert, P.B., Clark, D.V., and Henikoff, S., 1995, Introduction of a DNA methyltransferase in *Drosophila* to probe chromatin structure *in vivo*, *Chromosoma* (in press).

Wolffe, A.P., 1992, *Chromatin: Structure and function.* Academic Press, San Diego.

Woodcock, C., Grigoryev, S.A., Horowitz, R.A., and Whitaker, N., 1993, A folding model for chromatin that incorporates linker DNA variability produces fibers that mimic the native structures, *Proc. Natl. Acad. Sci. USA* 87:7603.

Wustmann, G., Szidonya, J., Taubert, H., and Reuter, G., 1989, The genetics of position-effect variegation modifying loci in *Drosophila melanogaster*, *Mol. Gen. Genet.* 217:520.

Zhang, P., and Spradling, A.C., 1995, The Drosophila salivary gland chromocenter contains highly polytenized subdomains of mitotic heterochromatin, *Genetics* 139:659.

Zuckerkandl, E., 1974, A possible role of "inert" heterochromatin in cell differentiation. Action of and competition for "locking" molecules, *Biochimie* 56:937.

MATRIX ATTACHMENT REGIONS AND TRANSGENE EXPRESSION

William F. Thompson [1,2], George C. Allen[1], Gerald Hall, Jr.[2,3], and Steven Spiker[2]

1. Department of Botany, and 2. Department of Genetics, North Carolina State University, Raleigh, North Carolina 27695, and 3. Mycogen Plant Genetics, Madison, Wisconsin 53716

INTRODUCTION

Many of the questions we have about how biological systems work are ultimately questions about the regulation of gene expression. For this reason, the control of transcription is fundamental and has received well-deserved attention. From a simplistic point of view, transcription can be though of as being regulated at two levels. The first level (coarse control) involves access of RNA polymerase and transacting factors to the specific DNA sequences with which they interact. Access is a function of chromatin structure. In chromatin fibers of both condensed metaphase chromosomes and interphase chromatin, DNA is highly compacted and essentially inaccessible to RNA polymerase and trans-acting factors. In order to make the DNA accessible, chromatin fibers must in some way take on a more open, less compact structure. Once an open (transcriptionally poised) chromatin structure is obtained, further regulation of transcription involving availability and interactions of transacting factors (fine control), come into play. For reviews see Paranjape et al. (1994) and Reeves (1984).

One of the hallmarks of transcriptionally poised chromatin is heightened general sensitivity to the endonuclease DNase I (Weintraub and Groudine, 1976). it has long been known that the DNase I sensitivity extends far beyond the immediate region of the transcribed gene and that "domains" of transcriptionally poised chromatin exist (Stalder et al., 1980). In at least some cases the domains of transcriptionally active chromatin (as assayed by DNase I sensitivity) correspond to structural domains in chromatin that have been described as "loop domains" (Bode and Maass, 1988; Bonifer et al., 1991; Farache et al., 1990; Gasser and Laemmle, 1986; Levy-Wilson and Fortier, 1989; Mirkovitch et al., 1984; Phi-van and Strätling, 1988).

Over the last several years abundant evidence has appeared in support of the loop domain model of chromatin structure (reviewed in Laemmli et al., 1992). According to this model, chromatin fibers are organized into a series of loops attached at their bases to a proteinaceous network called the nuclear matrix. The loops range in size form about 5 kb to more than 100 kb and attachment is through a specific interaction between proteins of the nuclear matrix and DNA sequences (approximately 1 kb) called Matrix Attachment Regions (MARs[1]) (see Figure 1). The loop domains have been hypothesized to form topologically isolated units of transcription and replication (Amati and Gasser, 1988; Bonifer et al., 1991; Goldman, 1988; Jack and Eggart, 1992).

Some early experiments with transfected animal cells in culture indicated that MAR sequences can overcome the problems of low and variable expression of introduced genes that have traditionally been attributed to "position effects" - i.e., to the site of insertion of transgenes into host chromatin in transfected cell lines. In these experiments, introduced genes were flanked with MAR sequences with the result that the average level of gene expression increased approximately tenfold and the formerly variable expression became copy-number dependent (Bonifer et al., 1990; Phi-Van et al., 1990; Stief et al., 1989). An early model to explain these results is depicted in Figure 1. According to this model, an introduced gene can become incorporated into the host chromatin in either an open, transcriptionally poised domain (Figure 1A, left) or a highly coiled, transcriptionally inactive domain (Figure 1A, right). An introduced gene without flanking MARs takes on the chromatin structure of the domain into which it becomes incorporated. Because most domains in multicellular eukaryotes would be expected to be inactive domains, most introduced genes would have a transcriptionally inactive chromatin structure. Only genes that become incorporated into transcriptionally active domains would be expected to have a high potential for transcription. Conversely, when an introduced gene is flanked by cloned MARs, it can form its own independent loop domain insulated from the structure of the domain into which it becomes incorporated (Figure 1B). There is no *à priori* reason to believe that the artificial domains created by the cloned MARs should be

---

[1]The nuclear matrix and MARs have also been called the nuclear scaffold and SARs. Although there is a formal distinction between the two sets of terms, based on method of preparation, the biological concept is the same. We will use the terms Matrix and MARs in this work.

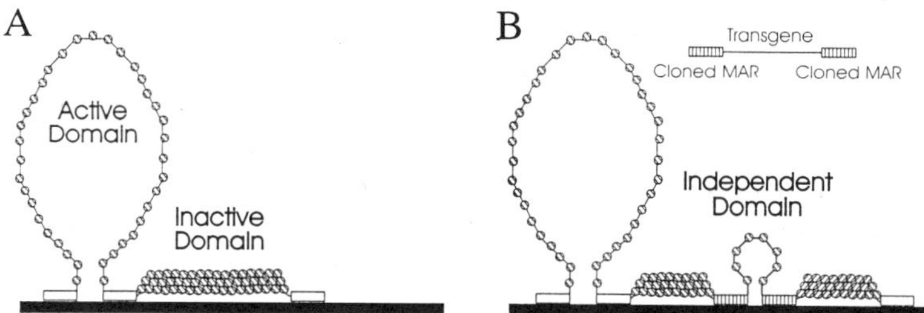

Figure 1. Models depicting the organization of chromatin into active and inactive loop domains and the formation of independent transgenic loop domains. **A.** MAR sequences (open boxes) interact with nuclear matrix fiber (filled bar) to form two loop domains. The active domain is depicted as a 11-nm nucleosome fiber and the inactive domain as a 30-nm fiber formed by supercoiling of the 11-nm fiber. **B.** An independent domain formed by the integration of MAR-flanked transgene into the inactive domain.

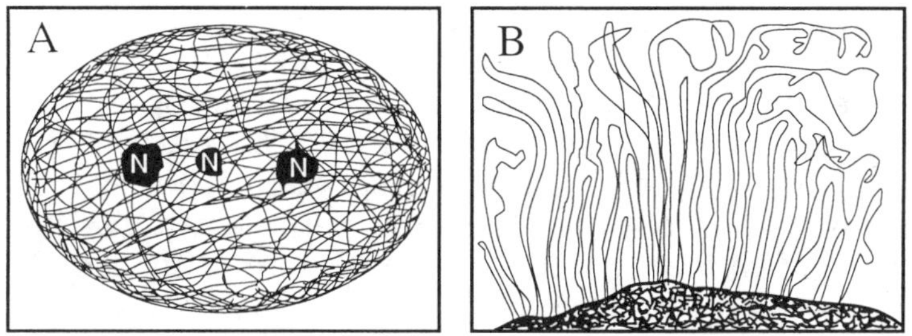

Figure 2. Nuclear matrix and halo structures. **A.** Drawing based on an electron micrograph (Capo et al., 1982) of a nucleus from which the nuclear membrane has been removed by detergents and chromatin removed by high salt extraction leaving the fibers of the nuclear matrix. N=nucleoli. **B.** Drawing based on electron micrograph of a portion of a metaphase chromosome (Paulson and Laemmli, 1977) from which chromatin proteins have been extracted. Extraction of histones removes coiling restraints allowing DNA to spill out and form a "halo." DNA loops are presumed to be attached by MARs to the chromosome scaffold (below).

transcriptionally poised domains as shown in the figure, but a tenfold increase in average gene expression and copy-number dependence would indicate that they are.

The prospect of overcoming "position effect" problems has created widespread interest in the use of MARs in transgenic organisms. Work in this field is still in the early stages but it is clear that the effects of MARs in transgenic organisms cannot be accounted for solely by the model shown in Figure 1. The early observations of MAR-mediated increases in gene expression have in general been corroborated. Sometimes the increases have been much greater than tenfold, but in a few instances they have been negligible or nonexistent. MAR-mediated reduction in variability of transgene expression has not been consistently demonstrated. Because several experiments have demonstrated lack of a simple correlation between copy number and gene expression, the idea that MARs increase gene expression by overcoming "position effects" has been questioned. In the following pages we will discuss some of the properties of MARs and examine the evidence concerning their effects on the expression of transgenes. We will also discuss some of the models that have been put forth to explain the biological activity of MARs and how our interpretation of data may be complicated by the effects of homology-dependent gene silencing. We will use data from all biological systems in our attempts to describe the current state of knowledge in the field, but we will place emphasis in examining the data from plant systems.

THE NUCLEAR MATRIX AND MATRIX ATTACHMENT REGIONS

After years of controversy, the existence of a nuclear matrix seems to be well established (Jack and Eggert, 1992). Figure 2A is a drawing representing a transmission electron micrograph of a portion of a nucleus (including nucleoli) from which chromatin has been removed leaving the residual, proteinaceous fibers of the nuclear matrix. Figure 2B is another drawing representing a portion of a chromosome scaffold with loops of DNA emanating from it. In the original micrograph upon which this figure is based, histones have been extracted from a metaphase chromosome, thus removing the soiling restraints and allowing the DNA to spill out and form a "cloud" or "halo" around the chromosome scaffold. The DNA sequences at the point of attachment of the loops to the scaffold or matrix fibers are the MARs. There is a great deal of evidence indicating that the MARs are not random-sequence DNAs, but rather that the MAR-nuclear matrix interaction is specific. As will be discussed below, the sequences required to bestow matrix-binding activity are imprecisely known. MARs in general are very AT-rich (typically>70%), but high AT-content alone does not insure matrix binding. A number of MAR consensus sequences have been proposed, but the relevance of the consensus sequences is questionable because they all are very AT-rich and might be expected to occur frequently by chance in such high AT-content DNA.

MARs can be identified by two assays. The endogenous assay identifies DNA sequences as MARs by their co-isolation with purified nuclear matrices. The exogenous assay identifies MARs by their ability to bind isolated nuclear matrices in vitro. The isolation of the nuclear matrix and the endogenous and exogenous MAR assays are outlined in Figure 3. The procedure begins with isolated nuclei. Histones and other chromatin proteins are removed form isolated nuclei by a variety of treatments. We use the procedure involving extraction with lithium diiodosalicylate (LIS) (Hall et al., 1991; Hall and Spiker, 1994; Mirkovitch et al., 1984). Removal of histones results in formation of nuclear halos (compare to Figure 2B) The DNA in the loops is then solubilized with endonucleases (usually restriction enzymes). The solubilized DNA can then be separated form the insoluble nuclear matrices by centrifugation. In the endogenous assay the solubilized DNA (DNA in the loops) and the DNA associated with the insoluble matrices (MARs by operational definition) can be assayed in Southern blots probed with any DNA sequence present originally in the nuclei. The exogenous assay is not so limited. In this assay end-labeled DNA from any source is incubated with the isolated nuclear matrices along with competitor DNA. MAR sequences will bind to the insoluble nuclear matrix and thus be found in the pellet after centrifugation. Non-MAR sequences remain in the supernatant fraction.

Figure 4 shows examples of "endogenous" assays (4A) and "exogenous" assays (4C) with tobacco (NT1 cells). In Figure 4B is a map of a portion of a genomic clone harboring a tobacco root-specific gene that has been called RB7 (Conkling et al., 1990). The coding portion of the gene is represented by a 1.8 kb HindIII fragment marked with an "a" and with an arrow to show the direction of transcription. Several kb downstream from the coding region is an XbaI fragment, marked with a "b," in a region that contains a MAR. In the endogenous assay (4A), DNA form the matrix (pellet;P) and supernatant (S) fractions are isolated and equal quantities are separated on agarose gels and blotted as outlined on the right side of Figure 3. A lane of restriction-enzyme digested total DNA (T) is included for comparison. The blots are probed with the "a" fragment, representing the coding region (left three lanes) or the "b" fragment, representing the region of the MAR DNA (right three lanes). Note that when the coding region ("a") is used as a probe, HindIII is used in the nuclear matrix preparation and to digest the total DNA used for comparison. In this case the DNA hybridizing to the probe is essentially all in the supernatant lane, indicating that the DNA in this region does not contain a MAR. When the "b" fragment is used as a probe, EcoRI is used in the nuclear matrix preparation and to digest the total DNA. Here, essentially all the DNA hybridizing to the probe is found in the pellet lane, indicating that fragments containing the "b" DNA contain MARs. Note that a prominent 4.3 kb hybridizing fragment is found in the pellet lane. This may indicate that one of the EcoRI sites is partially protected by interaction with the nuclear matrix.

In Figure 4C, two "exogenous" assays are shown. Purified nuclear matrices are made and incubated with end-labeled fragments as shown on the left side of Figure 3. Competitor DNA in both assays is tobacco genomic DNA. After incubation, DNA bound to the insoluble nuclear matrix (MARs by operational definition) is separated form non-bound DNA by centrifugation. Both DNA populations, along with a lane (total, T) of labeled fragments that were not incubated with the nuclear matrix preparations, are subjected to agarose gel electrophoresis and analyzed directly by autoradiography. In the left three lanes a plasmid containing an insert of the coding region in a HindIII site, was digested with HindIII and end-labeled. The vector (V) and the insert (H-H) are both found in the supernatant. Thus neither contains MAR DNA. In the right three lanes a plasmid containing an insert at the HindIII/SalI sites is cut with HindIII, SalI and XbaI. The fragments are end-labeled, incubated with the nuclear matrix and analyzed in the same way. Again the vector fragment (V) partitions with the supernatant as it does not contain a MAR. The XbaI-SalI (X-S) fragment also does not contain a MAR. The XbaI-XbaI (X-X) fragment partitions almost entirely with the pellet.

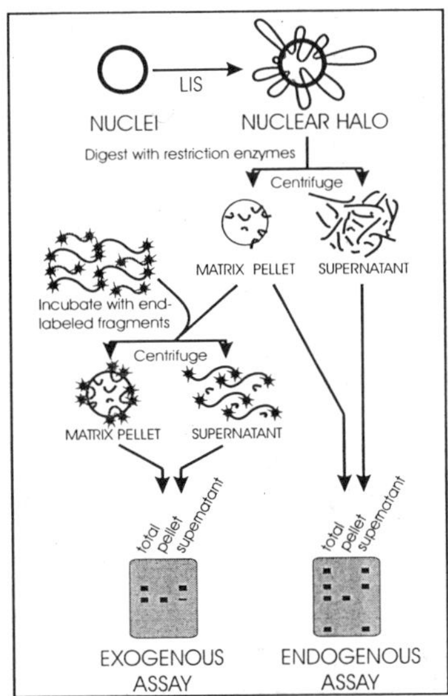

Figure 3. Isolation of the nuclear matrix and characterization of MARs by exogenous and endogenous assays. Chromatin proteins are extracted form isolated nuclei (we use lithium diiodosalicylate, LIS) thus removing coiling restraints and allowing DNA to spill out and form a nuclear halo (compare to Fig. 2B). DNA in loop domains is solubilized by restriction enzyme digestion resulting in formation of the nuclear matrix (compare to Fig. 2A). In the *endogenous* assay, DNA from the supernatant and DNA remaining with the insoluble matrices are purified, separated by electrophoresis, blotted and probed with a sequence to be tested. In the *exogenous* assay, the capacity for exogenous, end-labeled fragments to bind to purified nuclear matrices is determined. See text for details and Figure 4 for examples.

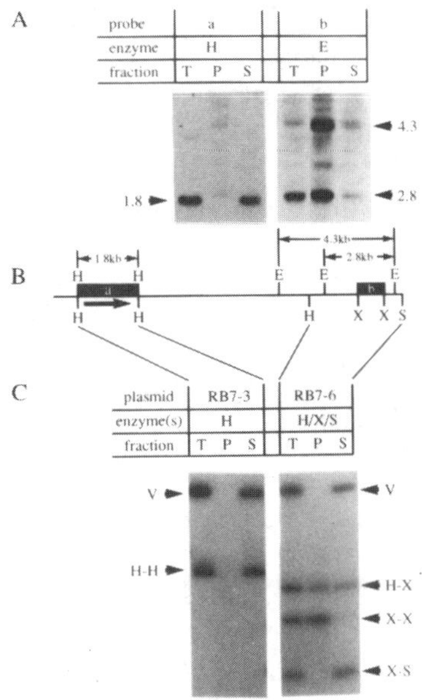

Figure 4. Examples of the endogenous and exogenous nuclear matrix binding assays using a genomic clone from the tobacco root-specific RB7 gene (see text). In panel B is a map of the RB7 gene indicating probes used for the endogenous assay (panel A) and the exogenous assay (panel C). The solid arrow in panel B shows the position and orientation of the RB7 coding region. On the top part of the map, restriction sites and fragment sizes are shown as they pertain to the endogenous assay. *Endogenous assay:* Probe= DNA fragments used to probe Southern blots, as shown in panel B; enzyme= restriction endonucleases used to digest nuclear halos; H= HindIII; E=EcoRI; fraction= the fraction of DNA run on the gel for the Southern blot; T= total purified tobacco (NT1) genomic DNA; P= pellet or matrix-bound fragment; S= supernatant or released fraction. Cutting is inefficient in the middle EcoRI site, resulting in a prominent 4.3 kb fragment. This site may be partially protected due to matrix association. *Exogenous assay:* Plasmid- subclones used to make end-labeled probes (RB7-3 contains a 1.8 kb HindIII/HindIII fragment and RB7-6 contains a 3.4 kb HindIII/SalI fragment); enzymes= restriction enzymes used to make the end-labeled probes; fraction= the fraction of DNA run on the gels; T= total input end-labeled fragment; P= pellet or matrix associated fraction; S- supernatant or unbound fraction. For these assays, equal proportions of each of the fractions are loaded onto a gel and detected by direct autoradiography after electrophoresis. Arrows indicate the bands of interest. V= vector; other fragments are identified by the restriction sites at their ends. (Reproduced form Hall and Spiker, 1994).

Thus, this fragment contains a MAR, and the MAR is strong. DNA containing this strong tobacco MAR was used in our experiments (described below) to test the effects of MARs in transgene expression in plant cells. The HindIII-XbaI (H-X) fragment also contains a MAR. This MAR is much weaker, however, as it partitions about equally between the pellet and supernatant. The yeast MAR that we used in our initial investigations of the effects of MARs on transgene expression in plant cells (Allen et al., 1993) was even weaker than the H-X fragment. There appears to be a correlation between matrix binding activity and effect on gene expression, as will be outlined below.

One might ask if the difference in apparent affinities of the weak yeast MAR and the strong tobacco MAR for the tobacco nuclear matrix might be explained by species specificity. This does not appear to be the case. We have characterized a large number of tobacco MARs (many of which were isolated by cloning DNA bound to isolate nuclear matrices) and found apparent affinities ranging from that of the yeast MAR to that of the strong tobacco MAR (Susan Michalowski, unpublished data).

What then determines if a DNA fragment will bind specifically to isolated nuclear matrices? A number of sequences have been found to be part of MARs. Among the first to be noted were the "A-box" (consensus AATAAAT/CAAA) and the "T-box" (consensus TTA/TTA/TTTA/TTT) (Gasser and Laemmli, 1986). Gasser and Laemmli also noted frequent occurrence of matches to the six bp core of the *Drosophila* topoisomerase II consensus sequence (A/TAT/CATT). Bode and co-workers (Bode et al., 1992) have stressed the importance of DNA sequences that can serve as nucleation sites for DNA unwinding including the sequence AATATATTT. Boulikas (1993) has suggested other consensus sequences. It is probable, however, that searches for consensus sequences are unlikely to tell us what determines if a DNA sequence can bind to the nuclear matrix. Part of the difficulty is that MARs are large. They are usually in the range of 1000 base pairs, although regions of direct contact with the matrix may be smaller (Gasser and Laemmli, 1986B). MAR affinity for the nuclear matrix is somewhat size dependent (around 300 bp is the lower limit for the exogenous MAR assay; see Gasser et al., 1989). Large MARs can often be cut into smaller pieces with each of the smaller pieces retaining matrix binding capacity, albeit often with lower apparent affinity (Hall et al., 1991). Thus is appears as though overall structural features of AT-rich DNA, rather than sequence per se determine matrix affinity of MARs. DNA sequences containing tracts of A/T have a number of unusual properties including a narrow minor groove, reduced capacity to form nucleosomes, tendency of phased runs to form bent DNA structures (Nelson et al., 1987) and tendency toward singe-strandedness (Bode et al., 1992). The idea that overall structure, rather than sequence per se, dictates matrix binding is supported by a correlation between apparent binding strength and numbers of runs of twenty base pairs of at least 90% AT nucleotides (S. Michalowski, unpublished data). We cannot as yet specify structural features of DNA that confer matrix binding. Although high AT-content appears to be involved, it is clear that high AT-content alone is not enough. Several At-rich fragments have been shown to lack matrix-binding capacity (Amati and Gasser, 1988; Bode et al., 1992; Gasser and Laemmli, 1986; Jarmon and Higgs, 1988).

It is obvious that the nuclear matrix is a complex structure containing many different types of proteins (Beven et al., 1991; Capco et al., 1982; Grabher et al., 1992; Hall et al., 1991; Ivanchenko and Avramova, 1992; Ivanchenko et al., 1993; Krachmarov et al., 1991: McNulty and Saunders, 1992). Several approaches have been used to investigate the proteins involve in binding MAR sequences. For example, screening an expression library for clones producing MAR-binding proteins resulted in the isolation of a cDNA for the SATB1 protein (Dickinson et al., 1992). Cross-linking procedures have been used (Dworetzky et al., 1992; Ferraro et al., 1992) as has Southwestern blotting (Romig et al., 1992; Tsutsui et al., 1993; von Kries et al., 1991). If there is a single protein that is responsible for the bulk of the MAR binding, such a protein has not as yet been identified. At present several seemingly unrelated proteins have been shown to have such an activity in vitro. In addition to the proteins identified by the methods mentioned above, other proteins including H1 histones, HMG proteins and topoisomerase II have been shown to interact with MAR sequences in vitro (Adachi et al., 1989; Dickinson et al., 1992; Ivanchenko and Avramova, 1992; Izaurralde et al., 1989; von Kries et al., 1991; Zhao et al., 1993) Of these, H1 histones would not be expected to be bonafide nuclear matrix proteins, although they often appear in nuclear matrix preparations (Capco et al., 1992). HMG proteins also would not be expected to be a part of the nuclear matrix. There is one report indicating that HMG proteins are a part of the nuclear matrix (Ivanchenko and Avramova, 1992), but we find no evidence for NMG proteins in plant nuclear matrix preparations by a Western blotting approach (T. Phelan, unpublished data). Earnshaw and Heck (Earnshaw and Heck, 1995; Earnshaw and Heck, 1988) have demonstrated that topoisomerase II is a major protein associated with metaphase chromosome scaffolds in chicken cells grown in culture. This is an intriguing observation as it evokes an image of topoisomerase II residing at the base of a chromosomal loop domain and directly controlling the topology of the DNA in that domain. The generality of such a role for topoisomerase II must be questioned, however, because this protein does not appear to be a part of the chromosome scaffold in quiescent tissues (Anderson and Roberge, 1992).

EFFECT OF MARS ON TRANSGENE EXPRESSION IN ANIMAL SYSTEMS

Several papers have been published concerning the effect of MAR sequences on expression of stably integrated genes in cells and organisms. In general, the results of these experiments have been interpreted to demonstrate that MARs increase levels of expression and decrease variability of expression. In some cases "position-independent" and "copy number-dependent" expression has been claimed, although it is not always clear precisely what is meant by these terms. Strictly speaking, position independence should mean that all single copy transformants (discounting deletions and rearrangements) would transcribe the transgene at essentially the same rate regardless of location in the genome. In the absence of complicating phenomena, additional copies of the transgene should also be transcribed at the same rate, resulting in a simple linear relationship between copy number and total expression. In practice, however, variability in expression of transgenes flanked by MARs is still considerable. In fact, in absolute terms, variability in expression of transgenes flanked by MARs is usually greater than controls--not surprising in view of the much higher overall levels of expression of MAR-bounded transgenes. Only when the higher mean values of MAR transformants are taken into account by using logarithmic transformations of the data or by using the coefficient of variation (standard deviation divided by the man) as a measure of variability does the decrease in variation become apparent (Allen et al., 1993; Mlynárová et al., 1995; Mlynárová et al., 1994; Phi-Van et al., 1990).

In the first paper specifically designed to test the effects of MARs (Stief et al., 1989) a reporter gene, CAT (Chloramphenicol acetyl transferase) with promoter and enhancer was flanked by chicken lysozyme MARs and used to transform chicken promacrophage cells by transfection. In these experiments, the overall CAT expression form the MAR transformants was about 10-fold greater than the non-MAR controls. Variability of expression in low copy number MAR transformants was slightly less than 10-fold. Non-MAR transformants varied slightly more than 10-fold. Thus, although variability in expression was reduced somewhat, convincing evidence for position-independent gene activity is lacking according to the criteria mentioned above. In the MAR transformants, higher copy number (up to nearly 30 copies per genome) resulted in generally higher levels of expression, although the correspondence between copy number and expression levels was not spectacularly better than in controls. In parallel experiments, similar constructs lacking the enhancer element were used. Again an approximate 10-fold increase in CAT expression was realized in the MAR transformants, but there was no evidence for position independent or coy number dependent expression. Thus, in this work there is some support for MAR-mediated reduction in variability, but the most obvious and striking effect of the MARs is the 10-fold increase in average gene expression. It should be noted that in this paper most of the work was done with measurement of CAT activity, but it was demonstrated that CAT activity correlated with steady-state levels of CAT messenger RNA. These workers also showed that the effect of MARs is dependent upon stable integration into the host genome. In transient expression assays, the MAR sequences had either no effect or a slightly inhibitory effect. That MARs must be stably integrated into the host genome in order to have their effects has been consistently demonstrated (Allen et al., 1993; Klehr et al., 1991).

In a similar set of experiments, Phi-van and co-workers (Phi-Van et al., 1990) used the same reporter gene and the same MAR to transfect rat fibroblasts. Very few low copy number transformants were produced. Thus, the claims for dampening of position effects were based on copy number dependency of gene expression. The evidence here is somewhat more convincing than that of Stief et al.

(1989), and certainly the claim of dampening of position effects is well supported. Nevertheless, the most obvious effect of MARs in these experiments is the approximate 10-fold increase in transgene expression.

In another early paper, Klehr and co-workers (Klehr et al., 1991) used a variety of MARs (including a MAR of plant origin) and found a 20 to 30-fold enhancement of transcription in transfected mouse L cells. No claims for reduction in variability of expression were made in this work.

Determining MAR effects in transgenic animals has been less straightforward than it has been in cells in culture. Bonifer and co-workers (Bonifer et al., 1990) introduced a 21.5 kb fragment carrying the entire chicken lysozyme gene locus into the germ line of mice by injection into fertilized oocytes. In the resulting transgenic animals, a good correspondence between copy number (up to 70) and gene expression was demonstrated. Evaluation of the data is complicated by the fact that only 7 mice were analyzed, and that the DNA introduced contained may sequences in addition to the MARs. Further studies by these same investigators (Bonifer et al., 1994) in which selective regions of the 21.5 kb fragment were deleted, indicated that an interplay between the MAR elements and a locus control region (LCR) affects the level and tissue specificity of transgene expression. LCRs are regions with high densities of DNase I hypersensitive sites that are thought to be involved in determination of chromatin structure in native, developmentally regulated chromatin domains.

McKnight and co-workers (McKnight et al., 1992) used a similar germ-line injection procedure to investigate the effect of the chicken lysozyme MAR on the expression of the mouse whey acidic protein in transgenic mice. In this study, no evidence for MAR-mediated higher overall expression or copy-number dependency were found. Position-independent regulation was claimed, however, based on the effect of MARs on tissue specificity and hormonal regulation.

Based on studies such as those just mentioned, the generalization has arisen that MARs can mediate increased levels of overall gene expression, reduce transformant to transformant variability in transcription, dampen position effects and allow copy number-dependent expression. We have recently published work (Allen et al., 1993) that confirms MAR-mediated increases in gene expression in transformed plant cells. Our data, however, conflicted with published data from animal systems in that copy-number-dependent gene expression was not observed. In fact, high copy number appeared to inhibit transgene expression. We rationalized The differences between our data and the data from animal systems by postulating that homology dependent gene silencing is more prevalent in plant systems than in animal systems (see later).

However, two recently published papers on animal systems have presented data, which appear to correspond more closely to our work with transformed tobacco cells than to previous work with animal cells (Kalos and Fournier, 1995; Poljak et al., 1994). Poljak and co-workers used the CAT reporter gene driven by the SV40 promoter and enhancer. This construct, either alone or flanked by MARs from *Drosophila* histone or heat shock gene, was transfected into HeLa cells or mouse L cells. The MAR sequences did not reduce variability in expression, nor did they result in copy number-dependent expression. In fact, in accordance with our data from tobacco cells, CAT expression per DNA copy was dramatically reduced at higher copy numbers. Overall, MAR-mediated stimulation of transgene activity was about 40-fold. The authors conclude that MARs stimulate transgene expression but do not confer position-independent expression, and invoke a model in which MARs are considered to increase expression by providing entry points, or nucleation sites, for HMGs or related proteins to replace H1 histones. However, they do not explain how their data differ from previously published data on animal cells in culture.

Kalos and Fournier (1995) used both human and rat hepatoma cells for their experiments on human apolipoprotein B (*apoB*) 5' and 3' MARs. The reporter gene was ß-galactosidase driven by the *apoB* promoter. For analysis, these workers divided the transformed cell lines into 1-2 copy lines and multicopy lines. In the 1-2 copy lines, the cells transformed with MAR-containing constructs showed consistently higher expression than control lines (approximately 200-fold). The authors attributed the MAR effect to position-independent transgene expression. In stark contrast to the results of Stief et al., (1989) and Phi-Van et al., (1990), Kalos and Fournier did not observe copy number dependent expression in their experiments. In fact, in their multicopy lines, transgene expression was strongly repressed. The difference between these data and the early data on the effects of MARs on transgene expression remains unexplained.

MAR EFFECTS IN PLANT SYSTEMS

In plant systems, most of the reported experiments have used *Agrobacterium* and T-DNA vectors for transformation. An initial report by Breyne et al. (Breyne et al., 1992) described a slightly negative effect of MARs on reporter gene expression, but there are now several reports of moderate increases, averaging a few fold in magnitude (Table 1) (Mlynárová et al., 1995; Mlynárová et al., 1994; Schöffl et al., 1993; van der Geest et al., 1994). In contrast, we have shown that a yeast MAR contained in the *ARS-1* element can increase average expression of a *35S::GUS* reporter gene more than 20 fold (per transgene copy) in tobacco suspension culture cells stably transformed by microprojectile bombardment (Allen et al., 1993). Thus, it is clear that large MAR effects can be obtained in plant cells under appropriate circumstances, and that MARs from other organisms can work in plants.
Additional experiments with a MAR derived from the tobacco genome (Allen et al., 1995) have produced even larger effects, with the

increase in expression per transgene copy averaging over 130 fold. This particular MAR was isolated from a genomic clone of the *RB7* gene kindly provided by our colleague, Mark Conkling (Conkling et al., 1990) and shown to bind very strongly to tobacco matrix preparations (Hall et al., 1991). It is tempting to speculate that its greater effect on expression in vivo is related to its stronger matrix binding activity in vitro, but further investigation well be required before this issue can be properly resolved.

The lower magnitude of the MAR effects obtained with T-DNA may be a function of the close linkage between the selectable marker and the reporter genes in these vectors. We assume that most, although probably not all, cases of inactivation involve both the selectable marker and the reporter gene, and thus will be eliminated by drug selection. Since many such events are thus excluded from the analysis, any MAR-mediated reduction in their frequency will be underestimated. Viewed from this perspective, it is remarkable that any MAR effect can be observed in such experiments.

Table 1. MAR Effects in Transgenic Plants and Plant Cells.

Plant System	Source of MAR	Promoter-Reporter	DNA Transfer	Effect on Expression Level	Variability	Reference
Tobacco cells	Soybean	*nos-GUS*	T-DNA	small decrease	small decrease	Breyne et al. (1992)
	Human	*nos-GUS*	T-DNA	no effect	small decrease	
Tobacco plants	Soybean	Heat Shock-*GUS*	T-DNA	9 fold increase	no effect	Schöffl et al. (1993)
Tobacco cells	Yeast	*35S-GUS*	Bombardment	12 fold increase	small decrease	Allen et al. (1993)
Tobacco plants	Chicken	*Lhca3-GUS*	T-DNA	4 fold increase	3 fold decrease	Mlynárová et al. (1994)
	Chicken	*Lhca3-GUS*[1]	T-DNA	3 fold increase	7 fold decrease	
Tobacco plants	Bean	Phaseolin-*GUS*	T-DNA	3 fold increase	2 fold decrease	van der Geest et al. (1994)
Tobacco plants	Chicken	*Enh35S-GUS*	T-DNA	2 fold increase	2 fold decrease	Mlynárová et al. (1995)
	Chicken	*Enh35S GUS*[1]	T-DNA	2 fold increase	7 fold decrease	
Tobacco cells	Tobacco	*35S-GUS*	Bombardment	60 fold increase	no effect	Allen et al. (1995)

1. In these constructs, the selectable marker and the reporter gene were between the MARs in a single domain.

In contrast, the microprojectile bombardment experiments employed a co-transformation procedure in which the reporter gene and selectable marker (the *NptII* gene) were delivered on separate plasmids. Our data on NptII protein levels indicate that this gene is unaffected by the MARs on the GUS plasmid. Thus, we have been able to achieve an operational separation between selectable marker and reporter genes in this system, in spite of the known tendency for co-transformed plasmids to integrate at the same genetic locus (Christou et al., 1989; McCabe et al., 1988; Peng et al., 1995; Registar et al., 1994).

Wherever tested, these MAR effects require that the transgene DNA be integrated into the host plant genome. Only small effects are seen in transient expression assays, consistent with the notion that MARs work by altering chromatin structure rather than as classical enhancers.

DO MARS REDUCE POSITION EFFECTS?

It is clear from work cited above that MARs can increase transgene expression in both plants and animals, sometimes by a very large factor. Although it does not necessarily follow from this result that MARs should also reduce position effect variation, there are several reports suggesting that in fact they do. In most instances, these data come from experiments with stably transformed cell lines or transgenic plants. Without MARs, gene expression varies from line to line in a largely random fashion, showing little or no dependence on gene copy number. However, when MARs are placed at the ends of otherwise identical constructs there is an increase in the copy number dependence and/or a decrease in the apparently random variation normally attributed to position effect.

A Elements and Copy Number Dependence

Interestingly, most instance in which MARs have been reported to reduce position effect variation involve the so-called "A element" from the chicken lysozyme gene. This element is the 5' MAR in a chromosomal domain studied by Bonifer et al., (1990). These authors tested a 21.5 Kb fragment from the chicken genome containing the entire lysozyme gene domain, including all known regulatory elements and both 5' and 3' MARs. When the entire domain was introduced into transgenic mice, it showed developmentally appropriate expression that was proportional to transgene copy number, indicating that it functions as a position independent regulatory unit.

The 5' MAR from this domain was also tested in chimaeric gene constructs in which it was placed both 5' and 3' to a reporter gene (Phi-Van et al., 1990; Stief et al., 1989). In both cases, overall expression in the transformant population was increased significantly, and the expression of individual cell lines varied in rough proportion to transgene copy number. These results were interpreted as indicating that MARs could insulate genes from chromosomal position effects.

In plant systems, the best evidence for reduction in position effect variation also comes from experiments with the chicken lysozyme A element. Nap and his colleagues have studied a large number of transgenic tobacco plants prepared with reporter gene constructs with or without flanking A elements (Mlynárová et al., 1995; Mlynárová et al., 1994). Moderate increases in average GUS expression (roughly 2 to 5 fold) were obtained, coupled with a dramatic reduction in the number of plants in which activity was low or undetectable. The reduction in the number of low expressors is consistent with the hypothesis that MARs substantially reduce the impact of gene silencing effects, and that such effects contribute a significant part of the normal variability.

Inconsistencies

In contrast to the data for A elements, there are now several examples in which MARs increase expression but do not produce copy number dependence. The first such observation was the report by Allen et al., (1993) describing the effect of a MAR contained in the yeast ARS-1 element. Tobacco suspension culture cells stably transformed by microprojectile bombardment showed approximately 25 fold higher expression per gene copy when the 35S::GUS reporter construct was flanked by MARs than when it was not. However, the increase in expression was not copy number dependent, and in fact expression was maximal in cell lines with relatively few transgene copies.

Similar results were obtained for a strongly binding MAR isolated from the tobacco genome (Allen et al., 1995). The increase in expression was much greater than with the yeast MAR, in excess of 130 fold on a per gene basis. However, maximal expression was still only observed in lines with low copy numbers.

Poljak et al., (1994) tested "minimal" Drosophila MARs chosen to minimize the chance that they contained any other regulatory information. CAT reporter gene constructs were tested in either HeLa or L cells. Expression of genes bounded by two MARs was stimulated by 20-40 fold on average, but results from an analysis of individual transformed cell lines were quite similar to those of Allen et al. (Allen et al., 1993) in that high variability persisted in the presence of MARs, and expression was not proportional to copy number.

Kalos and Fournier (1995) tested MARs derived from the human *apoB* domain in transient and stable transfection assays with hepatoma cell lines. In these assays, control constructs were efficiently expressed in transient assays but showed low and variable expression in stable transgenic clones. In contrast, single copy transformants containing constructs flanked by the 5' and 3' *apoB* MARs were expressed at consistently higher levels. Thus, it could be argued that MARs reduced position effects on expression of single copy transformants. However, multicopy transformants yielded poor expression with or without MARs, indicating that MARs are unable to overcome certain classes of chromosomal influences. These results were interpreted in terms of gene silencing phenomena similar to those invoked by Allen et al., (1993). This point will be discussed further below.

Since a chicken lysozyme MAR (the A element) figured prominently in early reports of copy number dependence, it is of particular interest that its role has recently been questioned. In transgenic mice, deletion of both boundary elements from transgenes containing the entire lysozyme domain causes ectopic expression of the gene, but position-independent expression is retained as long as all of the several cis-regulatory elements in both enhancer complexes are retained (Bonifer et al., 1994). In the same set of experiments, deleting one of the enhancer complexes reduced position independence, even though both boundary elements remained intact. These data on the natural gene in transgenic mice offer a striking contrast to earlier reports of A element effects in transfection experiments with cultured cells. The mouse data also contrast sharply with recent data suggesting that A elements normalize expression of reporter genes in transgenic plants (Mlynárová et al, 1995; Mlynárová et al., 1994).

The reason(s) for these differences remain unclear. However, it is worth noting that transforming either plants or cultured cells involves integrating foreign DNA into the genomes of rapidly dividing cells where the incoming DNA is likely to insert into transcriptionally active chromatin (Herman et al., 1990; Kertbundit et al., 1991). In contrast, transgenic mice are produced by injecting DNA into fertilized oocytes. Because oocytes are transcriptionally inactive at the time of injection (Latham et al., 1992; Henerey et al., 1995), there may be less contrast in the accessibility of potentially active and inactive portions of the genome, and incoming DNA may not be so efficiently targeted to active chromosomal regions. In mice, therefore, transgene expression assays may measure *de novo* activation of genes from a condensed chromosomal environment, whereas in cell cultures and plants transgene expression may be largely a function of the extent to which transcription and RNA accumulation can persist in the face of precesses that lead to gene silencing. If this hypothesis is correct, one might reasonably conclude that MARs play a more important role in resisting silencing than they do in promoting activation.

DO MARS AFFECT GENE SILENCING?

It is well known that expression of the same transgene construct can vary over several orders of magnitude in individual transgenic plants (Dean et al., 1988; Jones et al., 1985; Peach and Velten, 1991). This variability was once thought to arise entirely from genomic position effects, but recent evidence indicates that one or more 'gene silencing' phenomena are also involved (reviewed by Finnegan and McElroy, 1994; Flavell, 1994; Jorgensen, 1993; Jorgensen, 1995; Kooter and Mol, 1993; Matzke and Matzke, 1993; Matzke et al., 1994; Matzke and Matzke, 1995a; Matzke and Matzke, 1995b). The distinction between position effects and gene silencing phenomena is not always a sharp one. In principle, however, position effects on transgene expression reflect pre-existing features of the insertion site, such as proximity to genomic enhancers and degree of chromatin condensation, while gene silencing usually results from homology-dependent interactions involving the transgene itself (although chromosomal position may influence the severity of these interactions).

Forms of Gene Silencing

As we use the term, gene silencing depends in one way or another on homology between transgenes, between transgenes and endogenous genes, or between flanking genomic sequences and sequences elsewhere in the genome. Thus, it has been termed "homology-dependent gene silencing" (Matzke and Matzke, 1993; Matzke et al., 1994) or "repeat-induced gene silencing" (Assaad et al, 1993). Included under this general heading are *co-suppression* (Napoli et al., 1990; van der Krol et al., 1990; Jorgensen, 1992; Jorgensen, 1993; Jorgensen, 1995), in which both a transgene and a homologous endogenous gene are inactivated, and *trans-inactivation* (Matzke and Matzke, 1993; Matzke et al., 1994; Matzke and Matzke, 1995b; Matzke et al., 1989), in which introduction of a second transgene with homology to a previously introduced transgene leads to silencing of one or both. These two types of silencing usually operate in *trans*, but it is also known that silencing can be induced when multiple transgenes are arranged in *cis*, as in the case of tandem arrays or inverted repeats (Assaad et al., 1993; Hobbs et al., 1993).

The molecular basis for gene silencing remains somewhat mysterious, although evidence is accumulating for at least two different mechanisms. Several recent reports support a post-transcriptional mechanism, especially in cases of co-suppression (e.g., Dehio and Scheell, 1994; Metzlaff et al., Van Blokland et al., 1994), while in other cases silencing seems to occur at the transcriptional level and is frequently correlated with increases in DNA methylation (e.g., Brusslan et al., 1993; Matzke et al., 1989; Matzke et al., 1993; Vaucheret, 1994). Several of the clearest examples of transcriptional silencing seem to be triggered by homology in promoter regions that are not present in transcripts. Thus far, the reported cases of post-transcriptional silencing seem to involve either multicopy transgene insertions or homology between transcribed sequences of a transgene and an endogenous gene (as in co-suppression).

The distinction between transcriptional and post-transcriptional mechanisms is not always clean, however. Even where there is good evidence for a post-transcriptional component, the inactivation process sometimes leads to increases in transgene DNA methylation reminiscent of those associated with transcriptional inactivation (Smith et al., 1994; Ingelbrecht et al., 1994; Wassenegger et al., 1994). One way to resolve this apparent contradiction would be to postulate that transcription can proceed through methylated DNA. However, it is interesting to also consider the hypothesis that the tissues being examined are heterogeneous in their response, with transcriptional silencing occurring in some cells and post-transcriptional silencing in others. There are as yet no data that address the issue of mechanistic heterogeneity, but there are clear indications that somatic gene silencing events occurring during plant development can produce regular or random patterns of activity and inactivity within an organ or tissue (Barnes, 1990; Napoli et al., 1990; Neuhuber et al., 1994; van der Krol et al., 1990; Vaucheret, 1994).

Both transcriptional and post-transcriptional gene silencing mechanisms find precedent in studies with animal cells. Degradation of aberrant or otherwise untranslatable mRNAs has been observed in several laboratories (e.g., Cheng and Maquat, 1993; Pulak and Anderson, 1993), while chromatin condensation and DNA methylation are commonly observed in diverse examples of gene inactivation ranging from X chromosome inactivation and imprinting in mammals to silencing of the HML and HMR mating type genes in yeast (Rivier and Pillus, 1994). In *Drosophila,* position effect variegation (PEV) occurs when chromosomal rearrangements place genes in the vicinity of heterochromatic regions. Such genes frequently become part of the adjacent heterochromatin and are thus inactivated, as described by Henikoff elsewhere in this volume. Of particular interest is the recent observation that mechanisms genetically similar to those that induce heterochromatinization in PEV can also inactivate transgenes at ectopic locations far removed from any large blocks of heterochromatin (Dorer and Henikoff, 1994). Transgenes present in multiple copies are frequently subject to inactivation in this system, especially when arranged as inverted repeats - a result that closely parallels current observations on plant transgenes.

Some informative exceptions to the rule that only multicopy or homologous genes are subject to silencing have been described by Peter Meyer and his colleagues (Meyer and Heidmann, 1994; Meyer et al., 1993; Pröls and Meyer, 1992). The Al gene of maize, with no known sequence homology to any other sequence in Petunia was shown to undergo inactivation even in single copy transformants. However, inactivation may well depend on the homology between sequences near the insertion site and similar sequences elsewhere in the genome, as inactivation was more frequent when the transgene integrated into repetitive DNA.

Each of the several different types of gene silencing can be viewed as one example of a general tendency toward epigenetic inactivation of repeated sequences. It is likely that similar mechanisms account for natural phenomena such as paramutation and imprinting in maize (Brink, 1973; Kermicle, 1978; Patterson et al., 1993), cytokinin habituation in tobacco (Meins, 1989), and the activity phase changes often observed in transposon studies (Federoff, 1989; Martienssen et al., 1990). Similarly, as originally suggested by Waddington (Waddington, 1953), there are a variety of developmental phase transitions that seem to require epigenetic changes in gene activity (Poethig, 1990). Thus, the phenomena revealed by transgene studies, originally viewed as annoying obstacles to the application of molecular biotechnology, may provide keys to understanding previously obscure gene control mechanisms of fundamental importance in plant evolution and development.

MARs and Silencing

The MAR effects we and others have studied are largest in situations that should strongly favor gene silencing interactions. Both in plants and in animals, increases in gene expression are most prominent in direct transformation experiments that produce multicopy inserts. In most cases, multicopy transformants prepared in this way are expected to contain complex arrays of transgenes at single locus, an arrangement that would facilitate either pairing between transgenes or interactions among their transcripts that would lead to transcriptional or post-transcriptional gene silencing, respectively.

Figure 5 shows that the MAR effects we observe are most prominent in low copy number transformants. No comparable data are available for other plant systems, and the early work with animal cell transfection did not show a similar trend, as noted above. However, recent work with animal cell transfection has also provided examples in which the effect is maximal in low copy number transformants (Poljak et al., 1994). All the available data so far are compatible with an hypothesis suggesting that MARs can resist homology-dependent gene silencing, for example by preventing pairing between multiple copies of the transgene or otherwise affecting the properties of transgene chromatin or the localization of transgenes in the nucleus. To account for the decline in expression at high copy number, we assume that as transgene copy numbers increase it becomes increasingly unlikely that MARs will be able to completely suppress all possible pairing interactions, and that pairing leads to chromatin structure changes that silence not only paired genes but other genes in their immediate vicinity.

The transgenic plant data on MAR effects are also consistent with the idea that MARs primarily reduce gene silencing. As noted above, the magnitude of the increase in expression observed in these experiments is somewhat variable but never more than a few fold, in contrast to the much larger effects seen in cell lines. Where large numbers of transformants have been analyzed, however, it can be seen that one of the major effects of MARs is a reduction in the number of transgenic individuals with very low reporter gene expression (Mlynárová et al., 1995; Mlynárová et al., 1994). As noted above, it seems likely that many insertion events are eliminated from

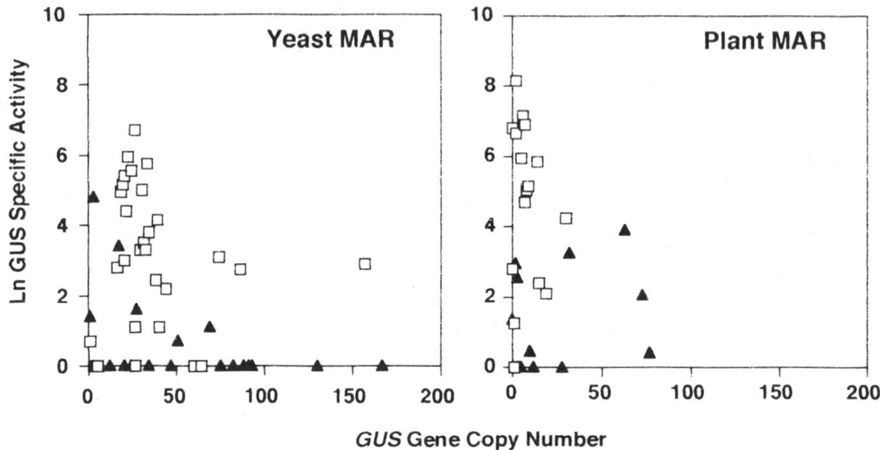

Figure 5. Effects of yeast and plant MARs on *GUS* gene expression following transformation by microprojectile bombardment. Each point represents a cell line established from an independent transformant. Cell lines containing MAR constructs are represented by open squares, and control lines by closed triangles. *GUS* copy numbers were estimated by a combination of quantitative PCR and Southern hybridization analysis. The data are plotted as natural logarithms to more effectively display the variance at low expression levels. For the plant MAR experiment, not all the control data points are visible, even in this log transformation. The full data set contains 16 points for the control and 17 for the MAR construct.

these populations by the selection for a linked drug resistance marker that is inherent in *Agrobacterium* transformation procedures. This selection almost certainly truncates the populations and minimizes any difference in reporter gene activity, such as that attributable to the presence or absence of MARs. However, even after eliminating most of the non-expressors from the starting population in this way, there are still individual that fail to express the reporter gene at high levels. These individuals may represent a subgroup in which the transgenes underwent silencing later in development, after drug selection was removed, or in which the reporter gene has been silenced but the selectable marker is still expressed.

SUMMARY

From the data available thus far, it is reasonable to assume that MARs will soon be included as useful modifiers of gene expression in many of the transgenic plants produced for research and commercial purposes. Their ability to increase gene expression is most dramatically seen in work involving direct DNA transformation, but their ability to reduce the number of low expressors even in populations produced by *Agrobacterium* transformation may still be of considerable importance both in plant breeding and in research applications, such as promoter analysis, in which one must make quantitative comparisons between different gene constructs. It is also possible that MARs will affect generational stability of gene expression, reducing the incidence of transgene silencing in advanced generations or upon crossing into a different genetic background. There are as yet very few data in the public domain that bear on this question (see Finnegan and Mcelroy, 1994; Neuhuber et al., 1994), although significant levels of transgene instability could affect many potential applications.

At present, it seems most logical to think of MARs primarily as increasing gene expression rather than as reducing position effect variation. In addition, it appears likely that their effect on expression most often results from their ability to somehow resist gene silencing. As yet, it is impossible to specify a detailed molecular mechanism for MAR effects, although it is attractive to speculate that in one way or another they may resist chromatin condensation. The latter point might prove especially important for artificial transgene constructs in which the distance between MARs is quite small, as the resulting "mini-domain" might be too small to undergo effective condensation (Allen et al., 1995).

In future studies it will be of obvious interest to characterize more MARs, to continue characterizing the proteins that can bind to them, and to further elucidate the structure of the nuclear matrix. Very little such work has yet been done on plant systems, and it will be of interest to determine the extent to which these basic elements of subnuclear structure differ from those in other organisms.

Future studies will also be required to determine whether MARs can affect both transcriptional and post-transcriptional forms of gene silencing, and to learn more about the molecular mechanisms underlying their effects. In particular, it will be of interest to determine whether or not the MAR effects on gene expression that have been documented in plant and animal systems result from a physical association with to the nuclear matrix. Since we know that exogenous DNA can bind to the matrix during its isolation, even the "endogenous" assay does not permit us to specify whether any particular MAR is bound or not bound *in vivo*. Resolving this question is likely to require a combination of careful structure/function studies for both gene expression and binding, together with structural analyses including transgene and transcript localization studies. Such studies should also contribute to an understanding of structural constraints on gene expression and the interactions that lead to gene silencing.

ACKNOWLEDGMENTS

Our research has been supported by grants from USDA and NSF. We gratefully acknowledge helpful discussions with Arthur Weissinger, Mary Dell-Chilton, and Michael Murray, as well as discussion, assistance, and moral support from members of our laboratories.

REFERENCES

Adachi, Y., Käs, E., and Laemmli, U.K., 1989, Preferential, cooperative binding of DNA topoisomerase II to scaffold-associated regions., *EMBO J.* 8:3997.

Allen, G.C., Hall, G., Michalowski, S., Newman, W., Spiker, S., Weissinger, A.K., and Thompson, W.F., 1995, High level transgene epxression in plant cells: Effects of a strong SAR from tobacco, submitted.

Allen, G.C., Hall, G.E., Childs, L.C., Weissinger, A.K., Spiker, S., and Thompson, W. F., 1993, Scaffold attachment regions increase reporter gene epxression in stably transformed plant cells, *Plant Cell* 5: 603.

Amati, B.B., and Glasser, S.M., 1988, Chromosomal ARS and CEN elements bind specifically to the yeast nuclear scaffold, *Cell* 54:967.

Anderson, H.J., and Roberge, M., 1992, DNA topoisomerase II: A review of its involvement in chromosome structure, DNA replication, transcription and mitosis, *Cell Biol. Int. Rep.* 16:717.

Assaad, F.F., Tucker, K.L., and Signer, E.R., 1993, Epigenetic Repeat-Induced Gene Silencing (RIGS) in Arabidopsis, *Plant Mol. Biol.* 22:1067.

Barnes, W.M., 1990, Variable patterns of expression of luciferase in transgenic tobacco leaves, *Proc. Natl. Acad. Sci. USA* 87:9183.

Bevin, A., Guan, Y.H., Peart, J., Cooper, C., and Shaw, P., 1991, Monoclonal antibodies to plant nuclear matrix reveal intermediate filament-related components within the nucleus, *J. Cell Sci.* 98:293.

Bode, J., Kohwi, Y., Dickinson, L., Joh, T., Klehr, D., Mielke, C., and Kohwi-Shigematsu, T., 1992, Biological significance of unwielding capability of nuclear matrix-associating DNAs, *Science* 255:195.

Bode, J. and Maass, K., 1988, Chromatin domain surrounding the human interferon-a gene as defined by scaffold-attached regions, *Biochemistry* 27:4706.

Bonifer, C., Hecht, A., Saueressig, H., Winter, D.M., and Sippel, A.E., 1991, Dynamic chromatin: The regulatory domain organization of eukaryotic gene loci, *J. Cell. Biochem.* 47:99.

Bonifer, C., Vidal, M., Grosveld, F., and Sippel, A., 1990, Tissue specific and position independent expression of the complete gene domain for chick lysozyme in transgenic mice, *EMBO J.* 9:2843.

Bonifer, C., Yannoutsos, N., Krüger, G., Grosveld, F., and Sippel, A.E., 1994, Dissection of the locus control function located on the chicken lysozyme gene domain in transgenic mice, *Nucleic Acids Res.* 22:4202.

Boulikas, T., 1993, Nature of DNA sequences at the attachment regions of genes to the nuclear matrix., *J. Cell. Biochem.* 52:14.

Breyne, P., Van Montagu, M., Depicker, A., and Gheysen, G., 1992, Characterization of a plant scaffold attachment region in a DNA fragment that normalizes transgene expression in tobacco, *Plant Cell* 4:463.

Brink, R.A., 1973, Paramutation, *Annu. Rev. Genet.* 7: 129-152.

Brusslan, J.A., Karlin-Neuman, G.A., Huang, L., and Tobin, E. M., 1993, An *Arabidopsis* mutant with a reduced level of *cab140* RNA is a result of cosuppression, *Plant Cell* 5:667.

Capco, D.G., Wan, K.M., and Penman, S., 1982, The nuclear matrix: three-dimensional archetecture and protein composition, *Cekk* 29:847.

Cheng, J., and Maquat, L.E., 1993, Nonsense condons can reduce the abundance of nuclear mRNA without affecting the abundance of pre-mRNA or the half life of cytoplasmic mRNA, *Mol. Cell. Biol.* 13:1892.

Christou, P., Swain, W.F., Yang, N.S., and McCabe, D.E., 1989, Inheritance and expression of foreign genes in transgenic soybean plants, *Proc. Natl. Acad. Sci. USA* 86:7500.

Conkling, M.A., Cheng, C.-L, Yamamoto, Y.T., and Goodman, H.M., 1990, Isolation of transcriptionally regulated root-specific genes from tobacco, *Plant Physiol.* 93:1203.

Dean, C., Jones, J., Favreau, M., Dunsmuir, P., and Bedbrook, J., 1988, Influence of flanking sequences on variability in expression levels of an introduced gene in transgenic tobacco plants, *Nucleic Acids Res.* 16:9267.

Dehio, C., and Schell, J., 1994, Identification of plant genetic loci involved in a posttranscriptional mechanism for meiotically reversible gene silencing, *Proc. Natl. Acad. Sci. USA* 91:5538.

Dickinson, L.A., Joh, T., Kohwi, Y., and Kohwi-Shigematsu, T., 1992, A tissue-specific MAR/SAR DNA-binding protein with unusual binding site recognition, *Cell* 70:631.

Dorer, D.R., and Henikoff, S., 1994, Expansions of transgene repeats cause heterochromatin formation and gene silencing in Drosophila, *Cell* 77:993.

Dworetsky, S.I., Wright, K.L., Fey, E.G., Penman, S., Lian, J. B., Stein, J. L., and Stien, G.S., 1992, Sequence-specific DNA binding proteins are components of a nuclear matrix-attachment site, *Proc. Natl. Acad. Sci. USA* 89:4178.

Earnshaw, W.C., and Heck, M.M.S., 1985, Localization of topoisomerase II in mitotic chromosomes, *J. Cell. Biol.* 100:1716.

Earnshaw, W.C., and Heck, M.M.S., 1988, The ultrastructure of the mitotic chromosome scaffold: studies using whole-mount electron microscopy and immunocytological techniques, *in:* "Chromosomes and Chromatin," K. W. Adolph, ed., CRC Press.

Farache, G., Razin, S.V., Rzeszowska-Wolney, J., Moreau, J., Fecillas Targa, F., and Scherrer, K., 1990, Mapping of structural and transcription-related matrix attachment sites in the b-globin gene domain of avian erythroblasts and erythrocytes., *Mol. Cell. Biol. 10*:5349.

Federoff, N.V., 1989, About maize transposable elements and development, *Cell* 56:181.

Ferraro, A., Grandi, P., Eufemi, M., Altieri, F., and Turano, C., Cross-linking of nuclear proteins to DNA by cis-Diamminedichloroplatinum in intact cells-involvement of nuclear matrix proteins, *FEBS Lett.* 307:383.

Finnegan, J., and McElroy, D., 1994, Transgene inactiviation: Plants fight back!, *Biotechnology* 12:883.

Flavell, R.B., 1994, Inactivation of gene expression in plants as a consequence of specific sequence duplication, *Proc. Natl. Acad. Sci. USA* 91:3490.

Gasser, S., and Laemmli, U., 1986, Cohabitation of scaffold binding regions with upstream/enhancer elements of three developmentally regulated genes of D. melanogaster, *Cell* 46:521.

Gasser, S.M., Amati, B.B., Cardenas, M.E., and Hofmann, J.F.-X., 1989, Studies on scaffold attachment sites and their relation to genome function, *Int. Rev. Cytol.* 119:57.

Gasser, S.M., and Laemlli, U.K., 1986b, The organization of chromatin loops: characterization of a scaffold attachment site, *EMBO J.* 5:511.

Goldman, M.A., 1988, The chromatin domain as a unit of gene regulation, *BioEssays* 9:50.

Grabber, A., Eberharter, A., Gstraunthaler, G., and Loidl, P., 1992, Characterization of nuclear matrix proteins of Physarum polycephalum and mammalian cells, *Cell Biol. Int. Rep.* 16:1151.

Hall, G., Allen, G.C., Loer, D.S., Thompson, W.F. and Spiker, S., 1991, Nuclear scaffolds and scaffold attachment regions (SARS) in higher plants, *Proc. Natl. Acad. Sci. USA* 88:9320.

Hall, G., Jr., and Spiker, S., 1994, Isolation and characterization of nuclear scaffolds, *in*: "Plant Molecular Biology Manual," S. B. Gelvin and Schilperoort, R. A., ed., Kluwer Academic Publishers, Dordrecht, The Netherlands.

Henerey, C.C., Miranda, M., Wiekowski, M., Wilmut, I., and DePamphilis, M. L., 1995, Repression of gene expression at the beginning of mouse development, *Devel. Biol.* 169:448.

Herman, L., Jacobs, A., Van Montagu, M., and Depicker, A., 1990, Plant chromosome/marker gene fusion assay for study of normal and truncated T-DNA integration events, *Mol. Gen. Genet.* 224:248.

Hobbs, S.L.A.., Warkentin, T.D., and DeLong, C.M.O., 1993, Transgene copy number can be positively or negatively associated with transgene epxression, *Plant Mol. Biol.* 21:17.

Huber, M., Bosch, F.X., Sippel, A.E., and Bonifer, C., 1994, Chromosomal position effects in chicken lysozyme gene transgenic mice are correlated with suppression of hypersensitive site formation, *Nucleic Acids Res.* 22:4195.

Ingelbrecht, I., Van Houdt, H., Van Montagu, M., and Depicker, A., 1994, Posttranscriptional silencing of reporter transgenes in tobacco correlates with DNA methylation, *Proc. Natl. Acad. Sci. USA* 91:10502.

Ivanchenko, M., and Avramova, Z., 1992, Interaction of MAR-sequences with nuclear matrix proteins, *J. Cell Biochem.* 50:190.

Ivanchenko, M., Tasheva, B., Stoilov, L., Christova, R., and Zlatanova, J., 1993, Characterization of some nuclear matrix proteins in maize, *Plant Sci.* 91:35.

Izaurralde, E., Käs, E., and Laemmli, U. K., 1989, Highly preferential nucleation of histone H1 assembly on scaffold-associated regions, *J. Mol. Biol.* 210:573.

Jack, R.S., and Eggert, H., 1992, The elusive nuclear matrix, *Eur. J. Biochem.* 209:503.

Jarmon, A.P., and Higgs, D. R., 1988, Nuclear scaffold attachment sites in the human globin gene complexes, *EMBO J.* 7:3337.

Jones, J.D.G., Dunsmuir, P., and Bedbrook, J., 1985, High level expression of introduced chimaeric genes in regenerated transformed plants, *EMBO J.* 4:2411.

Jorgensen, R., 1992, Silencing of plant genes by homologous transgenes, *AgBiotech News and Information* 4:265N.

Jorgensen, R., 1993, The germinal inheritance of epigenetic information in plants, *Phil. Trans. Roy. Soc. Lond. B* 339:173.

Jorgensen, R.A., 1995, Cosuppression, flower color patterns, and metastable gene expression states, *Science* 268:686.

Kalos, M., and Fournier, R.E.K., 1995, Position-independent transgene expression mediated by boundary elements from the apolipoprotein B chromatin domain, *Mol. Cell. Biol.* 15:198.

Kermicle, J., 1978, Imprinting of gene action in maize endosperm, *in*: "Maize Breeding and Genetics," D. B. Walton, ed., Wiley-Interscience, NY.

Kertbundit, S., De Greve, H., Deboeck, F., Van Montagu, M., and Hernalsteens, J.-P., 1991, *In vivo* random ß-glucuronidase gene fusions in *Arabidopsis thaliana*, *Proc. Natl. Acad. Sci. USA* 88:5212.

Klehr, D., Maass, K., and Bode, J., 1991, Scaffold-attached regions from the human interferon ß domain can be used to enhance the stable expression of genes under the control of various promoters, *Biochemistry* 30:1264.

Kooter, J.N., and Mol, J. N.M., 1993, Trans-inactivation of gene expression in plants, *Curr. Opinion Biotechnol.* 4:166.

Krachmarov, C., Stoilov, L., and Zlatanova, J., 1991, Nuclear matrices from transcriptionally active and inactive plant cells, *Plant Science* 76:35.

Latham, K.E., Stolter, D., and Schultz, R., 1992, Acquisition of a transcriptionally permissive state during the 1-cell stage of mouse embyrogenesis, *Develop. Biol.* 149:457.

Levy-Wilson, B., and Fortier, C., 1989, The limits of the DNase I sensitive domain of the human apolipoprotein B gene coincide with the locations of chromosomal anchroage loops and define the 5' and 3' boundaries of the gene, *J. Biol. Chem.* 264:21196.

Martienssen, R., Barkan, A., Taylor, W.C., and Freeling, M., 1990, Somatically heritable switches in the DNA modification of *Mu* transposable elements monitored with a suppressible mutant in maize, *Genes and Dev.* 4:331.

Matzke, M., and Matzke, A.J.M., 1993, Genomic imprinting in plants: Parental effects and *trans*-inactivation phenomena, *Ann. Rev. Plant Physiol. Plant Mol. Biol.* 44:53.

Matzke, M., Matzke, A.J.M., and Mittelsten Scheid, O., 1994, Inactivation of repeated genes -- DNA-DNA interaction? *in:* "Homologous Recombination and Gene Silencing in Plants," J. Paszkowski, ed., Kluwer Academic Publishers, Amsterdam.

Matzke, M.A., and Matzke, A.J.M., 1995a, Homology-dependent gene silencing in transgenic plants: What does it really tell us?, *Trends Genet* 11:1.

Matzke, M.A., and Matzke, A.J.M., 1995b, How and why do plants inactivate homologous (trans)genes?, *Plant Physiol.* 107:679.

Matzke, M.A., Neuhuber, F., and Matzke, A.J.M., 1993, A variety of epistatic interactions can occur between partially homologous transgene loci brought together by sexual crossing, *Mol. Genl. Genet.* 236:379.

Matzke, M.A., Primig, M., Trnovsky, J., and Matzke, A.J.M., 1989, Reversible methylation and inactivation of marker genes in sequentially transformed tobacco plants, *EMBO J.* 8:643.

McCabe, D.E., Swain, W.F., Martinell, B.J., and Christou, P., 1988, Stable transformation of soybean (Glycine max) by particle accerlation, *Biotechnology* 6:923.

McKnight, R.A., Shamay, A., Sankaran, L., Wall, R.J., and Hennighausen, L., 1992, Matrix-attachment regions can impart position-independent regulation of a tissue-specific gene in transgenic mice, *Proc. Natl. Acad. Sci. USA* 89:6943.

McNulty, A.K., and Saunders, M.J., 1992, Purification and immunological detection of pea nuclear intermediate filaments - evidence for plant nuclear lamins, *J. Cell Sci.* 103:407.

Meins, F., 1989, Habituation: heritable variation in the requirement of cultured plant cells for hormones, *Ann. Rev. Genet.* 32:395.

Metzlaff, M., O'Dell, M., and Flavell, R.B., 1995, Suppression of chalcone synthase A activities in Petunia by the addition of a transgene encoding chalcone synthase, *in:* "Gene Silencing," D. Grierson, ed., Plenum Press, NY, in press.

Meyer, P., and Heidmann, I., 1994, Epigenetic variants of a transgenic petunia line show hypermethylation in transgene DNA: an indication for specific recognition of foreigh DNA, *Mol. Gen. Genet.* 243:390.

Meyer, P., Heidmann, I., and Niedenhof, I., 1993, Differences in DNA-methylation are associated with a paramutation phenomenon in transgenic petunia, *Plant J.* 4:89.

Mirkovitch, J., Mirault, M.-E., and Laemmli, U.K., 1984, Organization of the higher-order chromatin loop: specific DNA attachment sites on nuclear scaffold, *Cell* 39:223.

Mlynárová, L, Jansen, R.C., Conner, A.J., Stiekema, W.J., and Nap, J.-P., 1995, The MAR-meidated reduction in position effect can be uncoupled from copy number-dependent expression in transgenic plants, *Plant Cell* 7:599.

Mlynárová, L., Loonen, A., Heldens, J., Jansen, R.C., Keizer, P., Stiekema, W.J., and Nap, J.-P., 1994, Reduced position effect in mature transgenic plants conferred by the chick lysozyume matrix-associated region, *Plant Cell* 6:417.

Napoli, C., Lemieux, C., and Jorgensen, R., 1990, Introduction of a chimeric chalcone synthase gene into petunia results in reversible co-suppression of homologous genes *in trans, Plant Cell* 2:279.

Nelson, H.C. M., Finch, J.T., Luisi, B.F., and Klug, A., 1987, The structure of an oligo(dA)_oligo(dT) tract and its biological implications, *Nature* 330:221.

Neuhuber, F., Park, Y.-D., Matzke, A.J.M., and Matzke, M.A., 1994, Susceptibility of transgene loci to homology-dependent gene silencing, *Mol. Gen. Genet.* 244:230.

Paranjape, S.M., Kamakaka, R.T., and Kadonaga, J.T., 1994, Role of chromatin structure in the regulation of transcription by RNA polymerase II, *Ann. Rev. Biochem.* 63:265.

Patterson, G., Thorpe, C.J., and Chandler, V.J., 1993, Paramutation, an allelic interaction, is associated with a stable and heritable reduction of transcription of the maize *b* regulatory gene, *Genetics* 135:881.

Paulson, J.R., and Laemmli, U.K., 1977, The structure of histone-depleted metaphase chromosomes, *Cell* 12: 817.

Peach, C., and Velten, J., 1991, Transgene expression variability (position effect) of CAT and GUS reporter genes driven by linked divergent T-DNA promotrs, *Plant Mol. Biol* 17:49.

Peng, J.Y., Wen, F.J., Lister, R.L., and Hodges, T.K., 1995, Inheritance of *gusA* and *neo* genes in transgenic rice, *Plant Mol. Biol* 27:91.

Phi-van, L., and Strätling, W.H., 1988, The matrix attachment regions of the chick lysozyme gene co-map with the boundaries of the chromatin domain, *EMBO J.* 7:655.

Phi-Van, L., von Kries, J.P., Ostertag, W., and Strätling, W.H., 1990, The chicken lysozyme 5' matrix attachment region increases transcription from a heterologous promoter in heterologous cells and dampens position effects on the expression of transfected genes, *Molec. Cell. Biol.* 10:2302.

Poethig, R.S., 1990, Phase change and the regulation of shoot morphogenesis in plants, *Science* 250:923.

Poljak, L, Seum, C., Mattioni, T., and Laemmli, U.K., 1994, SARs stimulate but do not confer position independent gene expression, *Nucleic Acids Res.* 22:4386.

Pröls, F., and Meyer, P., 1992, The methylation patterns of chromosomal integration regions influence gene activity of transferred DNA in Petunia hybrida, *Plant J.* 2:465.

Pulak, R., and Anderson, P., 1993, mRNA surveillance by the *Caenorhabditis elegans smg* genes, *Genes and Develop.* 7:1885.

Reeves, R., 1984, Transcriptionally active chromatin, *Biochem. Biophys. Acta* 782:343.

Register, J.C., Peterson, D.J., Bell, P.J., Bullock, W.P., Evans, I.J., Frame, B., Greenland, A.J., Higgs, N.S., Jepson, I., Jiao, S., Lewnau, C.J., Sillick, J.M., and Wilson, M.H., 1994, Structure and function of selectable and non-selectable transgenes in maize after introduction by particle bomardment, *Plant Mol. Biol.* 25:951.

Rivier, D.H., and Pillus, L., 1994, Silencing speaks up, *Cell* 76:963.

Romig, H., Fackelmayer, F.O., Renz, A., Ramsperger, U., and Richter, A., 1992, Characterization of SAF-A, a novel nuclear DNA binding protein from HeLa cells with high affinity for nuclear matrix/scaffold attachment DNA elements, *EMBO J.* 11:3431.

Schöffl, F., Schröder, G., Kleim, M., and Rieping, M., 1993, An SAR sequence containing 395 bp DNA fragment mediates enhanced, gene-dosage-correlated expression of a chimaeric heat shock gene in transgenic tobacco plants, *Transgenic Res.* 2:93.

Smith, H.A., Swaney, S.L., Parks, T.D., Wernsman, E.A., and Dougherty, W.G., 1994, Transgenic plant virus resistance mediated by untranslatable sense RNAs: Expression, regulation, and fate of nonessential RNAs, *Plant Cell* 6:1441.

Stalder, J., Larsen, A., Engel, J.D., Dolan, M., Groudine, M., and Weintraub, H., 1980, Tissue-specific DNA cleavages in the globin chromatin domain introduced by DNAaseI, *Cell* 20:451.

Stief, A., Winter, D.M., Strätling, W.H., and Sippel, A.E., 1989, A nuclear DNA attachment element mediates elevated and position-independent gene activity, *Nature* 341:343.

Tsutsui, K., Tsutsui, K., Okada, S., Watarai, S., Seki, S., Yasuda, T., and Shohmori, T., 1993, Identification and characterization of a nuclear scaffold protein that binds the matrix attachment region DNA, *J. Biol. Chem.* 268:12886.

Van Blokland, R., Van der Geest, N., Mol, J.N.M., and Kooter, J.M., 1994, Transgene-mediated suppression of chalcone synthase expression in Petunia hybrida results from an increase in RNA turnover, *Plant J.* 6:861.

van der Geest, A.H.M., Hall, G.E., Spiker, S., and Hall, T.C., 1994, The beta-phaseolin gene is flanked by matrix attachment regions, *Plant J.* 6:413.

van der Krol, A.R., Mur, L.A., Beld, M., Mol, J.N.M., and Stuitje, A.R., 1990, Flavonoid genes in petunia: Addition of a limited number of gene copies may lead to a suppression of gene expression, *Plant Cell* 2:291.

Vaucheret, H., 1994, Promoter-dependent trans-inactivation in transgenic tobacco plants: kinetic aspects of gene silencing and gene reactivation, *C. R. Acad. Sci. Paris* 317:310.

von Kries, J.P., Buhrmester, H., and Strätling, W.H., 1991, A matrix/scaffold attachment region binding protein-identification, purification and mode of binding, *Cell* 64:123.

Waddington, C.H., 1953, Epigenetics and evolution, *Symp. Soc. Exp. Biol.* 7:186.

Wassenegger, M., Heimes, S., Riedel, L., and Sänger, H. L., 1994, RNA-directed de novo methylation of genomic sequences in plants, *Cell* 76:567.

Weintraub, H., and Groudine, M., 1976, Chromosomal subunits in active genes have and altered conformation, *Science* 193:848.

Zhao, K., Käs, E., Gonzalez, E., and Laemmli, U., 1993, SAR-dependent mobilization of histone H1 by HMG-I/Y *in vitro*: HMG-I/Y is enriched in H1-depleted chromatin, *EMBO j.* 12:3237.

THE MERCURIAL GERMLINE GENOME OF HYPOTRICHOUS CILIATES

David M. Prescott and Michelle L. DuBois

Department of Molecular, Cellular and Developmental
Biology, University of Colorado, Boulder, CO 80309-0347

INTRODUCTION

A hypotrichous ciliate contains both a germline nucleus and a
somatic nucleus. The germline nucleus, called the micronucleus,
contains chromosomes with very high molecular weight DNA. It
divides mitotically and undergoes meiosis at cell mating. Its genes are
transcriptionally silent during vegetative growth. The somatic
nucleus, called the macronucleus, contains only gene-sized DNA
molecules ranging in size from ~400 bp to ~15,000 bp with an average
of 2200 bp in *Oxytricha* species. Each different sized molecule is
present in the macronucleus in ~1000 copies. There are ~24,000
different sized molecules, or a total of 24×10^6 molecules per
macronucleus (Lauth et al., 1976). A gene-sized molecule contains a
single transcription unit. The ~66 gene-sized, macronuclear molecules
characterized for hypotrichs so far consist of a 5´ DNA leader, a
transcription unit or open reading frame (ORF) for protein-encoding
molecules, and a 3´ DNA trailer (Hoffman et al., 1995. All the
molecules possess terminal repeats of 3´ dG4T4 with a 14-base or 16-
base 3´ tail of the repeat sequence.

CELL MATING

Hypotrichs can be induced to mate by starvation. Two cells join
and form a transient cytoplasmic channel between them. The
micronuclei undergo meiosis, and the two cells exchange haploid
micronuclei through the channel. In each cell the exchanged

Genomes of Plants and Animals: 21st Stadler Genetics Symposium
Edited by J. Perry Gustafson and R. B. Flavell, Plenum Press, New York, 1996

micronucleus fuses with a resident haploid micronucleus to make a
new diploid micronucleus. The unused haploid micronuclei degenerate
quickly, and the macronuclei degrade completely during the next two
days. The two cells separate, and the new, diploid micronucleus
divides by mitosis, without cell division. One mitotic product remains
as the new micronucleus, and the other develops into a new
macronucleus during ~3 days.

MACRONUCLEAR DEVELOPMENT — POLYTENIZATION AND GENE EXCISION

The first event in macronuclear development is the formation of
polytene chromosomes to ~64 C by ~5 rounds of DNA replication in the
original micronuclear chromosomes. The polytene chromosomes are
subsequently destroyed, ~95% of the unique sequences and all
repetitious sequences are eliminated, and gene-sized molecules to
which telomere sequences are added then emerge. No cis-acting
sequences in the DNA of micronuclear (polytene) chromosomes that
might guide the excision of the gene-sized molecules have been
identified in *Oxytricha,* and how excision is controlled remains a
mystery. The final step in macronuclear development is the
amplification of the gene-sized molecules to an average copy number
of ~1000.

INTERNAL ELIMINATED SEQUENCES IN MICRONUCLEAR GENES

Similarly mysterious is the presence in all 10 micronuclear genes
examined so far in *Oxytricha* and *Stylonychia* of multiple, usually
short, AT-rich sequences that interrupt the genes (Bierbaum et al.,
1991; DuBois and Prescott, 1995; Gray et al., 1991; Greslin et al., 1988;
Greslin et al., 1989; Herrick et al., 1987; Hicke et al., 1990; Klobutcher
et al., 1984; Mansour et al., 1994; Mitcham et al., 1992; Mitcham et al.,
1994; Ribas-Aparicio et al., 1987; Williams and Herrick, 1991; R. C.
Anderson, D. C. Hoffman, K. R. Lindauer, and D. M. Prescott,
unpublished). Internal eliminated sequences (IESs) are spliced out of
genes, and the remaining segments, called macronuclear-destined
sequences (MDSs) are joined to yield the gene-sized molecules of the
macronucleus. Figure 1 shows the structure of the micronuclear gene
encoding the b-telomere binding protein in *O. nova* (Mitcham et al.,
1994). It contains one IES in the first part of its protein-encoding
region and two IESs in what is to become the 3´ DNA trailer in the
eventual macronuclear gene.

A direct repeat of 2 to 7 bp is present at the junctions of an IES
with its two adjacent MDSs in the micronuclear versions of various
genes. These presumably represent part of the information required
for signalling correct IES excision and MDS splicing, but the direct

repeats obviously do not contain sufficient sequence information to specify IES excision. How excision is guided is unknown.

SCRAMBLED MICRONUCLEAR GENES

In some micronuclear genes the MDSs that must be spliced to make a macronuclear gene are in a scrambled order. The micronuclear version of the gene encoding actin I protein is made up of 9 MDSs separated by 8 IESs (Greslin et al., 1989). In the macro-

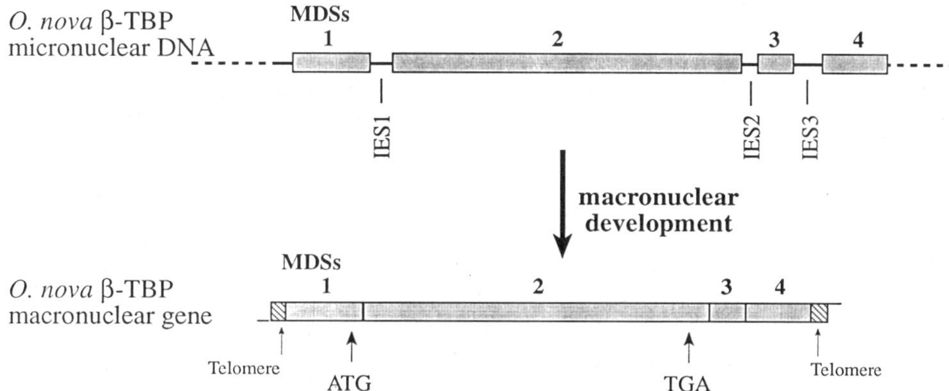

Figure 1. The b-telomere binding protein (b-TBP) gene of *O. nova*. The micronuclear version of the b-TBP gene is separated into four MDSs and three IESs, in a nonscrambled organization. During macronuclear development, the IESs are excised and the MDSs are spliced in order to form the macronuclear b-TBP gene.

nuclear version the MDSs are arranged 1 through 9 consecutively (Figure 2). In the micronuclear version the MDSs are arranged in the order 3-4-6-5-7-9-2-1-8. During macronuclear development the MDSs are reordered and spliced to produce a transcriptionally competent gene encoding actin I. Pairs of repeat sequences 9 to 13 bp long are present at junctions between scrambled MDSs and IESs. For example, the right end of MDS5 has the sequence GCCAGCCCC and the left end of MDS6 has the same sequence; the right end of MDS6 has the sequence CAAAACTCTA and the left end of MDS7 has the same sequence. These pairs of repeat sequence are thought to guide the splicing of MDSs in the correct order by a recombination event between the members of a repeat pair. MDS2 is inverted, and ends in sequences that are repeated in inverted configuration at the appropriate ends of MDS1 and MDS3.

Nothing is known about the molecular machinery that accomplishes removal of IESs and unscrambling during macronuclear development, but it is remarkably precise. How scrambling is imposed in evolution isn't known. The functional significance of IESs and scrambling remain obscure.

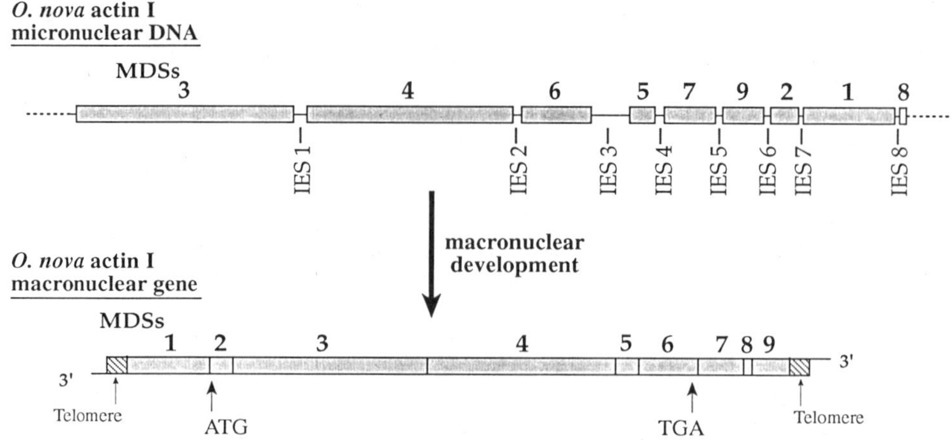

Figure 2. The actin I gene of *O. nova*. The micronuclear version of the actin I gene is scrambled in the pattern MDS3-4-6-5-7-9-2-1-8. The MDSs are unscrambled and IESs are excised during macronuclear development so that in the macronuclear gene, the MDSs are in the orthodox order.

Of the 10 micronuclear genes sequenced so far, 4 are scrambled . These are the actin I gene in *O. nova* (Greslin et al., 1988; Greslin et al., 1989), the actin I gene in *O. trifallax WR* (DuBois and Prescott, 1995), the gene encoding the a-telomere binding protein in *O. nova* (Gray et al., 1991; Mitcham et al., 1992), and the gene encoding the large catalytic subunit of DNA polymerase a in *O. nova* (Mansour et al., 1994). Genes encoding b-telomere-binding protein (Hicke et al., 1990; Mitcham et al., 1994), heat-shock protein 70 (K. R. Lindauer, R. C. Anderson, and D. M. Prescott, unpublished), R1 (Ribas-Aparicio et al., 1987), C2 (Klobutcher et al., 1984), EF1a (Bierbaum et al., 1991), and pMAC81 (Herrick et al., 1987; Williams and Herrick, 1991) contain IESs but are not scrambled.

THE SCRAMBLED STRUCTURE OF THE ACTIN I GENE IN OTHER HYPOTRICH SPECIES

With the expectation of gaining clues about the evolution of IESs and gene scrambling, the actin I genes in two species closely related to *O. nova*, *O. trifallax WR* and *O. sp* (Aspen), have been analyzed. The coding region (ORF) of the macronuclear version of the actin I gene in *O. trifallax WR* is 85% identical in nucleotide sequence to the actin I gene of *O. nova* and 92% identical in amino acid sequence. The AT-rich 5´ leaders and 3´ trailers are very different in the two species. The MDSs in the micronuclear versions of the actin I gene are similarly scrambled in both *O. nova* and *O. trifallax WR* with the exception that the gene in *O. trifallax WR* has an additional, short MDS (MDS10; 17

bp) interposed between MDS9 and MDS2 and an additional, long IES (276 bp) (Figure 3).

The striking similarity in scrambled pattern suggests that the actin I gene became scrambled before the two species diverged from a common ancestor in evolution. The difference in MDS number (9 vs 10) and IES number (8 vs 9) suggests that the scrambled pattern may change after it has been imposed. In addition, the sequences of the AT-rich IESs are totally different in the two species, although coding sequences in MDSs are highly conserved. The ability of the micronuclear version of the actin I gene to change structure is also further borne out by differences in the exact locations of IESs in the two species.

SHIFTING OF JUNCTIONS BETWEEN MDSs AND IESs

Although the overall pattern of scrambling of the micronuclear version of the actin I gene is very similar in *O. nova* and *O. trifallax WR*, there are important differences beyond the presence of an

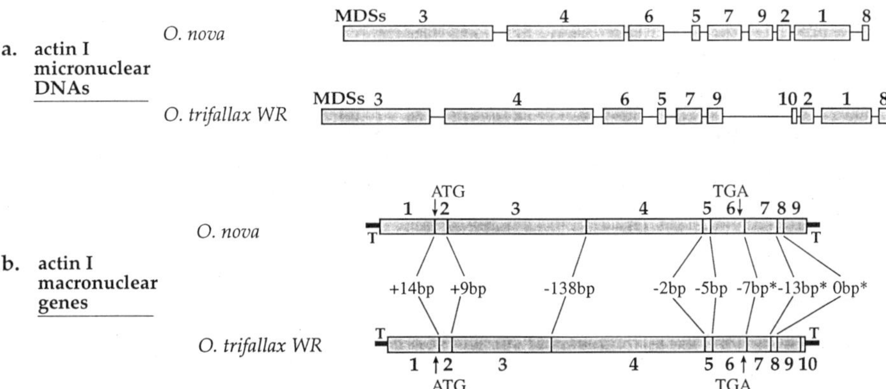

Figure 3. A comparison of the actin I genes of *O. nova* and *O. trifallax WR*. The micronuclear versions of the genes are scrambled in similar patterns, with the exception of an additional MDS (added in a nonscrambled orientation—MDS10) and an additional IES in *O. trifallax WR* (Figure 3a). The MDS-IES junction shifts are shown in the comparisons of the macronuclear actin I genes (Figure 3b). The shifts are presented with respect to the *O. nova* gene; a shift of +9 bp means that the MDS-IES junction of *O. trifallax WR* is 9 bp downstream of the junction of *O. nova*. Asterisks beside the MDS-IES junction shifts indicate that the MDS-IES junctions are located in the 3′ DNA trailer. These shifts are approximated because the noncoding sequences of the trailers are not well conserved between the species.

additional IES and MDS in *O. trifallax WR*. In the *O. nova* micronuclear gene, pairs of repeat sequences occur at the ends of MDSs. The repeat pairs are believed to guide a recombination process in which one copy of a repeat is eliminated when 2 MDSs are spliced in the appropriate order. The pairs of repeats serve to define the beginnings and endings of IESs. In comparison with the *O. nova* actin gene I, in the genes of *O. trifallax WR* and *O. sp* (Aspen), the MDS/IES junctions are shifted. The shifting is accomplished by moving 2 to 138 bp from the 3´ end of one MDS to the 5´ end of the numerically next MDS or vice versa, as illustrated in Figure 4 for the segment of the actin I gene containing MDS4 — IES2 — MDS6 — IES3 — MDS5. For example, in *O. trifallax WR* the repeat sequence at the right end of MDS4 begins and ends two bases earlier than the corresponding repeat in *O. nova*. For this to happen two bases (AG) have been recruited into the upstream (internal) end of the twopeat and two bases (AT) are missing from the downstream end that adjoins IES2. The two missing bases are now present in *O. trifallax WR* immediately following the downstream

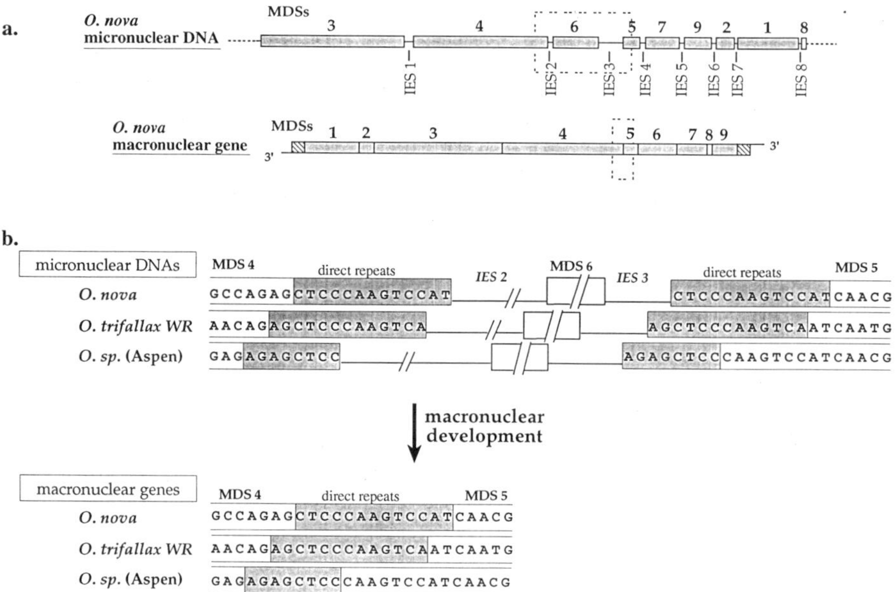

Figure 4. MDS-IES junction shifts of MDS4-MDS5 of the actin I gene in three species, *O. nova*, *O. trifallax WR* and *O. sp* (Aspen). MDSs are boxed and IESs are represented as horizontal lines. Direct repeats are boxed and shaded. A diagram of the micronuclear and macronuclear versions of the actin I genes is in Figure 4a. The dashed boxes outline the regions in the micronuclear and macronuclear genes that are described in Figure 4b. The MDS-IES junctions between MDS4-MDS5 are shown in Figure 4b. One copy of the repeat is located at the 3´ end of MDS4, and the other is located at the 5´ end of MDS5. Because of the scrambled organization of the genes, the repeats are separated by IES2, MDS6, and IES3. During macronuclear development, MDS4 is spliced to MDS5, and one copy of the repeat is retained in the macronuclear sequence. IES2 and IES3 are excised and destroyed, and MDS6 is reordered to the 3´ end of MDS5.

(internal) end of the repeat on the 5´ end of MDS6, and are no longer part of the repeat. In the actin I gene of *O. sp* (Aspen) a total of four bases (AGAG) have been recruited (compared to the *O. nova* gene) into the repeat sequence and a total of 9 bases (CAAGTCCAT) are missing (compared to the *O. nova* gene) from the end that adjoins IES2. The missing 9 bases are present just inside the inner end of the repeat that is at the 5´ end of MDS6 in *O. sp* (Aspen), and are not part of the repeat.

The overall effect is the shifting of bases (really bp) from the 3´ end of one MDS to just inside the repeat at the 5´ end of the next numerical MDS. This shifting is accompanied by a change in the sequence of the repeat, and in the case of *O. sp* (Aspen) a shortening of

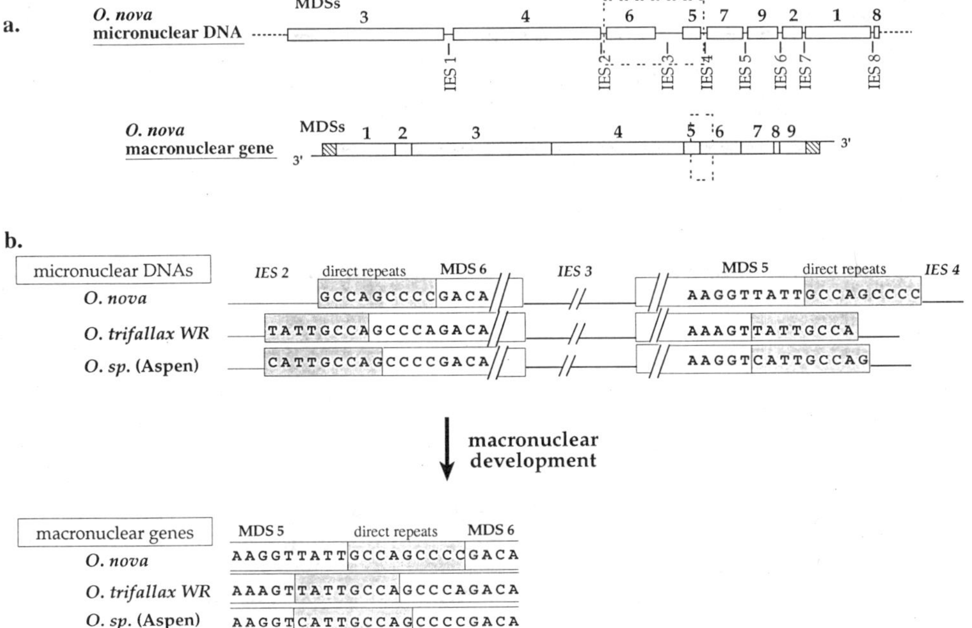

Figure 5. MDS-IES junction shifts of MDS5-MDS6 of the actin I gene in three species, *O. nova*, *O. trifallax WR* and *O. sp* (Aspen). MDSs are boxed and IESs are represented as horizontal lines. Direct repeats are boxed and shaded. A diagram of the micronuclear and macronuclear versions of the actin I genes is in Figure 5a. The dashed boxes outline the regions in the micronuclear and macronuclear genes that are described in Figure 5b. The MDS-IES junctions between MDS5-MDS6 are shown in Figure 5b. Due to scrambling, MDS6 is upstream of MDS5, but one copy of the repeat still is found at the 3´ end of MDS5 and one at the 5´ end of MDS6. When the MDSs are rearranged during macronuclear development, IES3 is excised, MDS5 and MDS6 are spliced, and one copy of the repeat is retained in the macronuclear sequence.

one of the repeat pairs by 5 bp. It is remarkable that when the segment, IES2 — MDS6 — IES3, is removed from interposition between MDS4 and MDS5 during macronuclear development and MDS4 and MDS5 are spliced (presumably by recombination between the repeats with elimination of one copy of the repeat), the exact same macronuclear sequence is generated in the joining region in all three *Oxytricha* species.

A second example of MDS/IES junctions shifting in the actin I gene is shown for the segment --- IES2— MDS6— IES3— MDS5— IES4 (Figure 5). In this case in *O. nova*, the 9-bp repeat on the left end of MDS6 is repeated on the right end of MDS5. In comparison with *O. nova*, in *O. trifallax WR* 5 bp have been shifted from the repeat on MDS6 into the body of MDS6. In *O. sp* (Aspen) 4 bp have been shifted in the same manner. The splicing of MDS6 and MDS5 produces the same macronuclear sequence in the joining region for the three species, except for three separate nucleotide changes among the species.

The shifting of MDS/IES junctions adds to the list of the unusual evolutionary malleability (introduction of IESs and MDS scrambling) and developmental processing (cutting, splicing, elimination, and amplification) of the germline DNA in hypotrichs. This most recent discovery of junction shifting means that the evolution of structural changes in micronuclear genes continues after IESs have been introduced and MDSs have been scrambled.

ACKNOWLEDGEMENTS

This work was supported by NSF Research Grant MCB-9506287 to D.M.P.

REFERENCES

Bierbaum, P., Donhoff, T., and Klein, A., 1991, Macronuclear and micronuclear configurations of a gene encoding the protein synthesis elongation factor EF1a in *Stylonychia lemnae, Mol. Microbiol.* 5:1567.

DuBois, M., and Prescott, D.M., 1995, Scrambling of the actin I gene in two *Oxytricha* species, *Proc. Natl. Acad. Sci. U S A* 92:3888.

Gray, J.T., Celander, D.W., Price, C.M., and Cech, T.R., 1991, Cloning and expression of genes for the *Oxytricha* telomere-binding protein: specific subunit interactions in the telomeric complex, *Cell* 67:807.

Greslin, A.F., Loukin, S.H., Oka, Y., and Prescott, D.M., 1988, An analysis of the macronuclear actin genes of *Oxytricha, DNA* 7: 529.

Greslin, A.F., Prescott, D.M., Oka, Y., Loukin, S.H., and Chappell, J.C., 1989, Reordering of nine exons is necessary to form a functional actin gene in *Oxytricha nova, Proc. Natl. Acad. Sci. U S A* 86:6264.

Herrick, G., Cartinhour, S.W., Williams, K.R., and Kotter, K.P., 1987, Multiple sequence versions of the *Oxytricha fallax* 81-MAC alternate processing family, *J. Protozool.* 34:429.

Hicke, B.J., Celander, D.W., MacDonald, G.H., Price, C.M., and Cech, T.R., 1990, Two versions of the gene encoding the 41-kilodalton subunit of the telomere binding protein of *Oxytricha nova, Proc. Natl. Acad. Sci. U S A* 87:1481.

Hoffman, D.C., Anderson, R.C., DuBois, M.L., and Prescott, D.M., 1995, Macronuclear gene-sized molecules of hypotrichs, *Nucleic Acids Res.* 23:1279.

Klobutcher, L.A., Jahn, C.L., and Prescott, D.M., 1984, Internal sequences are eliminated from genes during macronuclear development in the ciliated protozoan *Oxytricha nova, Cell* 36:1045.

Lauth, M.R., Spear, B.B., Heumann, J., and Prescott, D.M., 1976, DNA of ciliated protozoa: DNA sequence diminution during macronuclear development of *Oxytricha, Cell* 7:67.

Mansour, S.J., Hoffman, D.C., and Prescott, D.M., 1994, A gene-sized DNA molecule encoding the catalytic subunit of DNA polymerase alpha in the macronucleus of *Oxytricha nova, Gene* 144:155.

Mitcham, J.L., Lynn, A.J., and Prescott, D.M., 1992, Analysis of a scrambled gene: the gene encoding alpha-telomere binding protein in *Oxytricha nova, Genes Dev.* 6:788.

Mitcham, J.L., Prescott, D.M., and Miller, M.K., 1994, The micronuclear gene encoding b-telomere binding protein in *Oxytricha nova. J. Euk.. Microbiol,* 41:478.

Ribas-Aparicio, R.M., Sparkowski, J.J., Proulx, A.E., Mitchell, J.D., and Klobutcher, L.A., 1987, Nucleic acid splicing events occur frequently during macronuclear development in the protozoan *Oxytricha nova* and involve the elimination of unique DNA, *Genes Dev.* 1:323.

Williams, K.R., and Herrick, G., 1991, Expression of the gene encoded by a family of macronuclear chromosomes generated by alternative DNA processing in *Oxytricha fallax, Nucleic Acids Res.* 19:4717.

Mathematical Meiotic Models of Genome Analysis: Comparison With Molecular Approaches

J. Sybenga

Department of Genetics, Wageningen Agricultural University
Dreijenlaan 2
6703 HA Wageningen, The Netherlands

SUMMARY

The analysis of the evolutionary distance between genomes is one of the justifications of genome analysis based on molecular analysis and high resolution mapping. It is also the main objective of estimating affinity differences between genomes by mathematical meiotic models, based on the relative frequencies of diakinesis/metaphase I configurations in polyploid hybrids. Polyploid hybrids are preferred over diploid hybrids because they combine several genomes in the same meiotic cellular environment. The first stage of meiotic pairing, although DNA dependent, is indirect and less DNA specific than the second phase. This involves accurate DNA homology search, often leading to crossing-over and chiasmata. The combination of the two stages results in configurations that contain detailed information on evolutionary divergence at both the DNA level and in a more general biological sense. In polyploids these configurations include multivalents of various shapes. In order to interpret this information in terms of evolutionary distance, different mathematical models have been developed. These correspond in some essential respects, but differ in others. Their relative merits are discussed and it is indicated that some frequently adopted simplifications are not acceptable. Certain autopolyploids suggest considerable spurious divergence between identical genomes as a result of the properties of their meiotic pairing and chiasma system. It is advisable, therefore, not to draw conclusions from hybrids alone.

When this and other complicating factors, including variation in pairing and chiasma patterns, are taken into account, diakinesis and metaphase I configurations can give good information on genome differentiation. This includes differences in DNA sequence and gross structural homology as well as other forms of evolutionary divergence.

INTRODUCTION

The first applications to genome analysis of meiotic diakinesis and metaphase I observations were meant to determine the (haploid) genomes that together form the composite genome of an allopolyploid plant species. Already in this century Kihara in Japan (Kihara, 1930) and Goodspeed and Clausen (1928) in the U.S. made a start to elucidate the genomic composition of bread wheat (*Triticum aestivum* L., 2n=6x=42) and tobacco (*Nicotiana tabaccum*, 2n= 4x=48), respectively. The observations had quantitative aspects, but the interpretations were essentially qualitative. The number of allopolyploids of which the contributing diploids have been identified increased continuously until, at present, there is hardly any widely distributed allopolyploid that has not been analyzed. When uncertainties still exist as to which diploid species are involved, there may be a number of reasons: the original diploid is extinct, or too rare to be found, or it has changed since the origin of the allopolyploid, or a special ecotype was involved. It is also possible that the allopolyploid has mutated, or that one or more of the diploid genomes in the allopolyploid have a hybrid origin.

Later, meiotic studies were extended to more types of polyploid hybrids, often in the context of attempts to transfer genes from one species to another. It appeared that the information obtained could be used for estimating relative affinities between the genomes involved. This is also one of the aims of molecular, and high resolution gene mapping forms of genome analysis, and it is interesting to compare the two, quite different, approaches to fundamentally the same problem.

For a quantitative meiotic analysis the different genomes are combined in a hybrid, which, in practice, is only possible in plants. Essentially, the affinity estimates made are estimates of the relative rates of meiotic pairing between the chromosomes of the different genomes in the hybrid. Although pairing in a diploid hybrid can give some indication of the relation between two genomes (Wang, 1989), there are complications that make this unsatisfactory (Jauhar, 1990). The main advantage of polyploid hybrids is that the pairing of more than two genomes can be compared in the same genetic and physiological environment. For a number of reasons, the observations are primarily made at diakinesis or metaphase I of meiosis, and their interpretaion in terms of evolutionary affinity requires interpretative mathematical models. These are in part based on models made in the sixties and seventies for determining the pairing pattern of autopolyploids on the basis of configurations observed at metaphase I.

In some plant species, pairing can be studied directly in synaptonemal complexes at pachytene, but this has disadvantages. Collecting a sufficient amount of observations for a quantitative analysis is problematic in most species and, at least as important, much of (late) pachytene pairing appears to be non-homologous "saturation" pairing rather than hom(oe)ologous pairing and thus is not of interest for the study of affinity relations. On the contrary, it can be quite misleading.

Diakinesis/metaphase I configurations are much easier to obtain and study in large numbers, but they require chiasmata to be maintained. Therefore, models used to transform meiotic configuration frequencies into pairing relations should contain chiasma parameters in addition to pairing parameters. Occasionally, the chromosomes of the different species in the hybrid can be distinguished at metaphase I, for instance by C-banding patterns or by (genomic) *in situ* hybridization. Then the interpretation of the observations is direct, and simple models are sufficient for transforming observations into pairing parameters. However, in most materials this is still not well possible. Then, only the frequencies of specific types of configurations, including univalents, bivalents and multivalents, are available.

In order to evaluate the models in respect of their ability to relate pairing affinities to evolutionary molecular genome composition, three critical questions must be answered:

i. To what extent does meiotic pairing reflect DNA-sequence correspondence, and

ii. To what extent do metaphase I configurations reflect pairing?

iii. How realistic are the explicit and implicit assumptions underlying the different models?

MEIOTIC PAIRING AND DNA COMPOSITION

There is no reason to doubt that meiotic pairing depends on and thus reflects DNA correspondence, and that the stronger the deviations from homology, the more pairing is inhibited.

There are two stages in meiotic pairing:

i. Primary, long distance attraction.

ii. Short distance attraction, under normal conditions leading to the formation of a synaptonemal complex (SC), and involving homology search that may regularly lead to genetic exchange. Both steps are critical, but very little, even in a descriptive sense, is known of the first stage.

Long distance attraction

Primary, long distance attraction must necessarily preceed intimate pairing, whether or not this leads to an SC or any form of genetic exchange. There are good reasons to assume that primary attraction is not brought about directly by DNA:

i. The distance to be bridged is too great in higher eukaryotes, and the homologous segments are too few in number, even when arranged in long identical strands. Or, the concentration of unique homologous DNA segments is too low. The situation is in no way comparable to single strand DNA association in the test tube.

ii. Mutual attraction between homologous segments of double stranded DNA, if it exists at all, is insufficient.

iii. Large segments of single strand DNA, without complex protection (which would greatly restrict finding the homologue), would primarily return to their original homologue rather than finding a far-away complementary strand, and would be too vulnerable to survive for any length of time.

iv. Not strictly localized repetitive DNA would have to be systematically exluded from pairing to prevent an inextricable knot to be formed.

v. Pairing in hybrids with large differences in DNA content can be completed efficiently, and then lead to correct genetic exchange (for instance *Gossypium* and *Lolium*, see Sybenga, 1992).

vi. In contrast, pairing in hybrids of species with only minor differences in quantity and quality of their DNA may show very restricted pairing, also when synaptic genes are not involved.

vii. In many species synaptic initiation is restricted to specific chromosomal segments, often distally located (many grasses), occasionally proximally (several *Allium* species), independent of DNA composition. Simple genetic differences may regulate this pattern, as appears from hybrids between species with different patterns. Preferential pairing in allopolyploids is often regulated by relatively simple genetic systems.

For these reasons the concept of the "zygomere" has been proposed (Sybenga, 1966) for an as yet not clearly definable unit realizing long distance attraction between hom(oe)ologous chromosome segments. It could be a DNA-specific proteinaceous substance that takes care of the actual attraction, locally as well as cellularly regulated. It has been proposed that zygomeres are arranged in series, more or less closely associated (Figure 1), and independently controlled, although with a certain coordination (Sybenga, 1975, 1988, 1992). The concept of the zygomere is purely hypothetic, but is is inevitable that a strictly regulated and DNA specific, but not directly DNA mediated long distance attraction system must exist. This implies that the first stage in meiotic pairing only indirectly reflects DNA correspondence, and that an evolutionary factor apart from local DNA divergence is involved. Such a factor is of interest in itself, but detracts from the possibility to study DNA homologies purely on the basis of preferential pairing between specific genomes.

Role of rearrangements

Chromosomal rearrangements changing the gross order of genes in a chromosome affect pairing preferences in the experiment in otherwise strict autopolyploids (Sybenga, 1973), without changing DNA sequences as such. They are certain to occur in nature and must

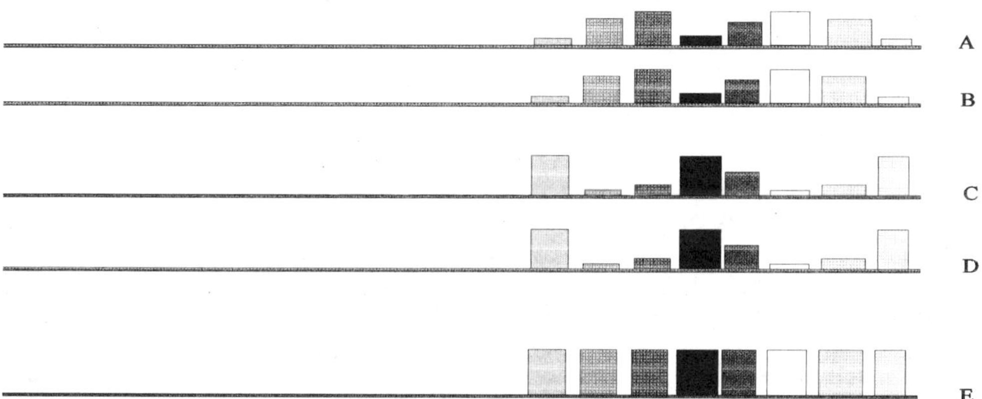

Figure 1. Graphical representation of "zygomere" specificity and activity. Horizontal bars: chromosome segments. Vertical columns: series of chromosome pairing initiations sites (zygomeres). Patterns denote DNA dependent specificity, height denotes activity. The specificites are the same, but the activation patterns differ. A and B are identical as are C and D. A and C can pair, as can B and D, but when all four are present together, there will be strong preferential pairing between A and B and between C and D. E has all zygomers active and will pair equally well with A and B as with C and D. In the triploid A, C, E, trivalents will be formed, with E predominantly in the middle.

be a factor to be accounted for in preferential pairing studies. Large rearrangements tend to show up by forming characteristic configurations, but smaller rearrangements, even of microscopically demonstrable size, may not do so and may affect pairing without being recognized as rearrangements. On the other hand, rearrangements outside regions where pairing initiation and progression are concentrated, may go undetected, but they also do not induce pairing preferences (Sybenga, 1973). The role of gross rearrangements, therefore, is not uniform. This is clearly a matter of disturbance of long-distance attraction, and has little relevance for purely DNA homology studies.

Slight nonstructural or substructural differences have been reported to reduce the effectiviness of pairing and crossing-over between otherwise homologous chromosomes in hybrids between varieties of the same species. Crossway and Dvorak (1984) studied chiasma formation between telocentrics in hybrids of different varieties of wheat. Chiasmata were much less frequent between marked chromosomes (telocentrics) in hybrids than between the same marked chromosomes of the same variety. It appeared that after one generation, when chromosomes passed on to the next generation may be expected to have had a chiasma, and consequently were recombined, pairing between one original chromosome and a recombined chromosome was still reduced.

This is certainly not always the case. Lack of preferential pairing between genetically differentiated chromosomes was reported in rye (*Secale cereale* L.) trisomics by Sybenga (1976). In rye tetraploids, Orellana and Santos (1985), Benavente and Orellana (1989) and later publications, reported that in some instances homologous chromosomes differentiated by C-bands paired preferentially over identical chromosomes, in other instances identical chromosomes over differentiated chromosomes. Only in particular cases would identical chromosomes systematically pair preferentially.

Genetic regulation of homoeologous pairing

Genetic restriction of homoeologous pairing in natural allopolyploids has been studied most extensively in wheat and relatives, but similar systems are found in several other allopolyploids. In the presence of specific genes, pairing differentiation between homoeologous genomes is stronger than when the system is not operative. In bread wheat and durum wheat a major gene on the long arm of chromosome 5B is responsible. Absence of 5B or a mutation at the locus concerned (Sears, 1984) results in considerable homoeologous pairing, which is absent when the gene is functional. In addition to this gene, several with less strong effects have been reported in wheat. Genes affecting homoelogous pairing in amphidiploids have been detected in diploid species, but their role in nature is unclear. It is not certain at which stage genetic pairing regulation is effective, but it may well be during long distance pairing. Even premeiotic somatic association has been suggested as the cause (Feldman et al., 1966). The subject has attracted considerable attention (Sybenga, 1992), but for the present purpose the main conclusion is that systems exist that seriously complicate the analysis of meiotic pairing in polyploid hybrids.

Short distance pairing

Short distance pairing follows primary attraction after this has brought the chromosomes close enough together for DNA interactions. The most important initial activity is homology search, which is directly DNA dependent, and consequently a good measure of DNA correspondence. Repetitive DNA is almost systematically excluded, but the mechanism involved is not clear. Homology search involves processes that can lead directly to genetic exchange (crossing-over or conversion), and it has often been argued, starting with Maguire (1968) that the first initimate contact consistently results in (or from?)

commitment to crossing-over. Effective homology search involves only a very small fraction of the total DNA. It may require specialized DNA sequences (minisatellites?), but this has not been confirmed. Steinmetz et al. (1987) report recombinational hot-spots in the major histocompatibility complex of mice and suggest that special target substances may be involved. Ashly et al. (1993) found that transfer of a special telomeric sequence from the chromosomal end to an interstitial position, would greatly enhance chiasma formation in a segment slightly removed from the insertion point. There are more reports of the same effect.

Effective contact at the DNA level normally results in the formation of a synaptonemal complex (SC), but in the SC only minute amounts of DNA are actually included. Later stages of SC formation do not involve any homology at all. They are just "saturation" pairing, bringing together as many unpaired chromosome segments as possible, independent of homology. Mutants not capable of SC formation have been shown to form very close local associations, and to be able to complete reciprocal exchange as well as conversion quite efficiently in the absence of an SC. Only the distributional pattern is disturbed: there is no interference (yeast: Sym and Roeder, 1994; potato: Jongedijk et al., 1991, for example, older literature in the latter). The fact that interference occurs only in the presence of an SC implies that a crossing-over event as the initiation of SC pairing must be relatively infrequent in normal material, and that most of the later crossing-over occurs inside the SC. This also implies that, when the level of crossing-over in asynaptic mutants nevertheless is almost the same as in normal organisms, attempts to pairing initiation resulting in a cross-over must be much more frequent in mutants. Probably, the period in which this takes place is extended, because if not, numerous pairing initiation points and absence of interference would be expected in normal material.

The normal SC is formed by linear association of proteinaceous globules, which together form a scaffold, from which the bulk of the chromatin extends in the form of large loops. It is probable that there are specific attachment sites on the DNA for the globules forming the SC, but there seems to be much freedom for the sites to be or not to be involved in SC formation, as exchange is not restricted to a limited number of sites on the chromosome. This does not exclude the existence of hotspots of recombination (Steinmetz et al., 1987). The SC scaffold is different from the scaffold of condensed mitotic and later meiotic stages, and from the scaffold of the nuclear reticulum. Exchange takes place inside the SC in special, electron-microscopically visible structures, recombination nodules (RNs). It is very probable that in the mutants just mentioned, similar structures are involved. They have not been reported in the yeast mutant referred to, which does not show any recognizable trace of SCs. They have been seen in partially asynaptic mutants of the tomato (Havekes et al., 1993), where RNs are formed and where, similarly, no interference is observed as a result of a failing SC.

The conclusion is, that the essentials of long distance pairing and even the first actions leading to exchange, are not a function of the DNA directly, but mediated by specific structures, associated with the DNA. Although there is no doubt about a close relation between DNA homology and initial pairing, there are too many separate controlling elements involved to use pairing as a perfect measure of DNA match. Yet the parallel is sufficiently interesting to make use of the pairing pattern for studying global correspondence and differentiation. Part of this differentiation is a result of the co-evolution (or parallel differentiation) of DNA and pairing pattern, and this is in itself an important subject in the context of genome analysis.

Short distance pairing involves homology search which is directly dependent on DNA homology. Although per cell the number of sites sampled is small, each cell is a new test, and when a large number of cells is analyzed, DNA homology sampling by crossing-over is a good way to analyze DNA correspondence. This is one reason why observations made at diakinesis and metaphase I of meiosis are in principle a good source of information on DNA divergence between species. The limitations will be discussed below.

SYNAPTONEMAL COMPLEX ANALYSIS

Since synaptonemal complexes (SCs) are the direct result of pairing, it would seem that SCs offer the best possibilities for studying the evolutionary relationships between the genomes in polyploid hybrids. However, here too, complications appear.

i. In most materials it is difficult to collect sufficient quantitative data, and few details of the chromosomes, often even not the centromere, can be recognized.

ii. Multivalent frequency is one of the most fundamental criteria of homology in polyploids. It appears that it is not only determined by preferential pairing, but also by the frequency of pairing partner switch (PPS). With only one site of pairing initiation, there is no PPS, and only bivalents can be formed, irrespective of the number of chromosomes that can pair (Figure 2a). With three points of pairing initiation, two PPSs are possible, but how many are formed in each particular case, depends on chance in addition to presence or absence of preferential pairing between specific chromosomes. Loidl (1986) concludes that the number of autonomous points of pairing initiation equals 1.5 times the maximum number of PPS + 1. A Poisson distribution of PPSs would suggest a large number of points, distributed evenly about the chromosomes, each with a small probability of initiating pairing. A binomial distribution would point to a smaller number, each with a greater probability. Loidl and coworkers (Loidl, 1986; Loidl and Länger, 1993), Callow and coworkers (Callow and Gladwell, 1984; Hamey et al., 1988) and Jones and coworkers (Jones, 1994; Jones and Vincent, 1994; Vincent and Jones,

1993) have analyzed the relation between PPS and points of pairing initiation.

iii. The distribution of PPSs over the chromosomes is expected to be random only when they are formed in a sequential order from the end onwards, or when their number is small. With random order, two PPSs at distal locations can not be followed by a single one type. If randomness is nevertheless found, either the number in excess of two is too small to significantly affect the analysis of distribution, or the distribution is in fact biassed.in the middle. There must be an even number, and of a reciprocal

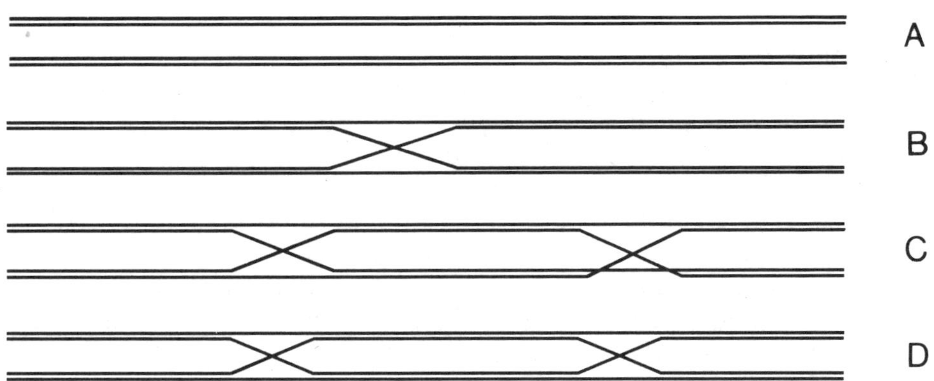

Figure 2. Pairing partner switch (PPS) between four homologous or homoeologous chromosomes. A: no PPS, there is one point of pairing initiation, or more, but then pairing initiation is between the same chromosomes in all cases. B: one PPS. In the two end segments pairing has started between different chromosomes. C: Two PPSs: in addition to pairing initiation in both end segments, there is one in between<. D. Two PPSs, involving different chromosomes than in C. Without a chiasma in the middle segment, the pairing initiation point in the middle goes undetected at MI.

iv. Even in pure autopolyploids of the same species, triploids have a higher PPS frequency than tetraploids, although, theoretically they should have the same number. The number of trivalents is also higher than the number of quadrivalents in tetraploids (Jones, 1994). The reason is not known. The consequence is that estimates of pairing affinity with the same basic materials will be different when made at the triploid compared to the tetraploid level. This is important not only when studying SCs, because at metaphase I too, trivalents are more frequent in autotriploids than quadrivalents in autotetraploids.

v. When the number of PPS is at its maximum, it is late pachytene, the stage when "saturation" pairing has been active for an unknown period of time, involving non-homologous as well as homologous and homoeologous pairing, but not leading to genetic exchange. Some of the PPSs have their origin in this period, and do not reflect switches of effective pairing between homologous partners. As a matter of fact, most pairing at late pachytene does not reflect homology at all. This is the most important drawback of using SC studies as such for analyzing evolutionary relations between genomes.

Only when recombination nodules (RNs) can be analyzed, is the SC a good source of information on homoeology relations. RNs are the sites of genetic exchange, and equivalent to chiasmata at later stages. They must reflect close homology at the DNA level. In only few species, however, is detailed analysis of a sufficient number of SCs with RNs possible, and no results have been reported for polyploid hybrids.

DIAKINESIS — METAPHASE I CONFIGURATIONS: THE ROLE OF CHIASMATA.

Diakinesis - metaphase I configurations depend on chiasmata for their formation. This adds to the sensitivity of homology testing, because chiasma formation requires good DNA correspondence. However, segments with good DNA correspondence, and paired at zygotene - pachytene, are not recognized as homologously paired at metaphase I when no chiasmata are formed. Chiasma formation is subject to different regulation systems and is sensitive to different internal and external factors. This must be considered when interpreting meiotic configuration frequencies in terms of homology. Some chiasma factors have been mentioned above. Others will be discussed below.

One question concerns the extent to which chiasmata represent homologous pairing. Maguire (1968) proposed that each initial pairing event would lead to a chiasma: pairing results in a commitment to exchange. She based this on the strict correlation found between pachytene pairing of an inversion segment and an interchange trisomic of maize, and the frequency a chiasma appeared at later stages. Later, especially by yeast geneticists, the order of events was even reversed: commitment to crossing-over comes first, and pairing follows. This is an exaggeration, as discussed above: most probably, chiasmata are also formed in the SC after initial pairing.

Very little is known about long distance attraction, resulting in a first choice between alternative partners in a polyploid hybrid. The molecular processes involved in recombinational exchange resulting in observable chiasmata at meiotic metaphase I, are better understood. However, this is not true for the first steps of recombinational exchange, probably a prerequisite for the initiation of intimate pairing, and leading directly to exchange between DNA strands.

For (indirectly) studying genetic, including DNA sequence correspondence through meiotic analysis, the chiasmata are important. The fact that at metaphase the relative association frequencies between specific chromosomes can be studied only through chiasmata, therefore, need not be a disadvantage. Possible complicating factors in chiasma formation need consideration.

Chiasma localization

It has long been known that chiasmata tend to be localized in specific chromosomal segments, possibly, but not necessarily as a result of general lack of long distance attraction in other segments. In hybrids between species of *Allium* with very different localization patterns, one strictly proximal, the other more distal, there is no clear localization. In later generations different patterns segregate, including types not present in the parents, and chiasma frequencies vary greatly (Darlington, 1965). In most Triticinae chiasmata are localized in the distal segments of the chromosomes, and rarely occur in proximal regions. In hybrids between subspecies of rye (*S. cereale* L.) both with distal chiasma localizaton, chiasma formation was randomly distributed, and in later generations different patterns were observed (Jones, 1967, 1984; Sybenga, 1992). Localization is obviously genetically determined and does not mean that regions free from chiasmata can never pair homologously, and even less that they would have little or no DNA sequence homology. Information on evolutionary relations is simply not available.

Chiasma suppression, physiological and genetic imbalance

Pairing partner switch (PPS) suppresses chiasma formation in its surrounding as a result of disturbed pairing. Distal PPSs, and to some extent also interstitial PPSs may even prevent chiasmata to be formed in the entire distal segment where normally chiasmata might be concentrated. Such segments, as a consequence, appears unpaired. Multivalent formation as a result of PPS will then not persist till metaphase, and this affects the analysis (Sybenga, 1972, 1992). It is even possible that as a result of lack of interference chiasmata are formed more proximally, resulting in ring bivalents that in most models are considered to point to bivalent instead of multivalent pairing.

Chiasma frequencies may be severely reduced by unfavorable external conditions or genetic factors causing asynapsis or desynapsis. In species hybrids similar effects may be found when the genetic balance is disturbed. This may not be interpreted to be lack of homology.

Low chiasma counts

In a few cases of diploid and allopolyploid species the genetic map has been found to be larger than what corresponds with chiasma counts. This is especially so since restriction fragment length polymorphism (RFLP) markers have come into use for high resolution mapping (Nilsson et al., 1993). Some of the descrepancy is due to classification errors in RFLP analysis, and the phenomenon of spurious "map inflation" is presently a subject of study. It occurs especially when large numbers of molecular markers from various data sources are combined in computer programs based on simple mapping functions. The possible role of conversion is probably low, because in most organisms where it can be studied, its frequency is well below that of reciprocal exchange. In addition, in many plant species it is difficult to establish exact numbers of chiasmata, and consequently chiasma frequencies tend to be underestimated. It is not certain yet that the discrepancy is real and affects the interpretation of meiotic configurations based on chiasma formation.

There is another complication with chiasma counts. In several cases, when genetically very different combinations of chromosomes or even alien substitutions or additions are studied, it appears that the number of exchanges, especially double exchanges close together involving the same chromatids, are formed without observable chiasma formation (Gill et al., 1994). This might be a consequence of reduced pairing resulting in the abolition of interference which, occasionallly can lead to two exchange events close together. Subsequently, the small segment between them is incapable of maintaining chromatid cohesion, necessary for chiasma maintenance (compare Maguire, 1980, 1985). There is double exchange, but no chiasma is seen at MI. Always the same two chromatids must be involved and the three remaining combinations of chromatids in two cross-overs should never occur. This is improbable. More likely, meiotic exchange is not involved, but a somatic event somewhere in the generative phase. Univalents at metaphase often lag at anaphase and subsequently misdivide or desintegrate. Misdivision resulting in chromosomes broken at the centromere can reunite with other broken centromeres and form centromere translocations. However, chromosomes associated in normal bivalents and not broken during anaphase, can also get involved (Lukaszewski and Gustafson, 1983). This is apparently induced by repair processes at the broken end. The exact stage when this occurs is uncertain, but may well be immediately after meiosis. Similarly, desintegration of lagging chromosomes activates repair processes, and some of the resulting fragments can invade intact chromosomes. In somatic tissue, the phenomenon of spontaneous introgression of chromosome segments after anaphase lagging and subsequent chromosome loss and desintegration, has been observed on several occasions. It is especially clear when alien chromosomes have a deviant mitotic cycle, lag and desintegrate. During subsequent repair, segments are incorporated into a host chromosome (Sybenga, 1992). Thus, in the case of apparent double "meiotic" exchange involving the same two chromatids, a chiasma is probably not involved. Both cases of too low chiasma counts are, at present, no reason for concern, the more so because in most models instead of single chiasmata, whole arm associations are considered.

Phenomena like chiasma localization, reduced chiasma formation resulting from genetic imbalance and suppression by pairing partner switch, should be considered when interpreting metaphase configuration data in terms of relative homoeologies between the genomes of different species in polyploid hybrids.

IMPLICIT AND EXPLICIT ASSUMPTIONS OF MATHEMATICAL MEIOTIC MODELS

The number of different basic configurations from which estimates of pairing preferences and chiasma frequencies in polyploids must be made is limited. In normal triploids there are four: chain trivalents, ring bivalents with univalent, open bivalent with univalent, three univalents. There are three degrees of freedom, and that there only three parameters can be estimated. In normal tetraploids there are six different configurations: ring quadrivalents, chain quadrivalents, trivalents with univalent, ring bivalents, open bivalents and univalent pairs. More complicated configurations are observed, but can not satisfactorily be dealt with by the models. They are assigned to one of the more simple categories. If it would be possible to recognize individual sets of four chromosomes, several more configurations derived from one set of four chromosomes could be distinguished. For instance, a combination of two ring bivalents may be formed by one set of four chromosomes, or, alternatively, one has its origin in one set, the second in another. Where the models have most reason to be used, different sets of hom(oe)ologues can not be distinguished, and it is not clear from which set the configuations are derived. As a consequence, not more than these six different configuations can be distinguished. The number of degrees of freedom and thus the number of parameters that can be estimated equals five. These parameters can not be derived directly from the data, and interpretative mathematical models are required.

The models can best be explained by an example. A tetraploid is chosen, and in this particular case it is a combination of two sets of two fully homologous genomes (CCDD, Figure 3), whereas between the sets there exists a certain, unknown evolutionary differentiation, for which a quantitative estimate must be derived. Several other combinations are possible, for instance a combination of four different, equally distant genomes(CCCC or CDEF), or three homologous genomes and one that is different (CCCD). Etc. The models applied to these different situations are based on the same basic assumptions, but they have a different construction. In all models it is assumed that there are two arms in each chromosome. In some models these are considered to be equal in length and in chiasma frequency. In other models they are considered potentially different.

The basic assumption is that the relative frequency of pairing between the different homologous and homoeologous genomes and the frequency of chiasmata formed, together reflect the evolutionary affinity between the genomes.

In order to restrict the number of parameters to what can be handled in terms of the number of degrees of freedom and the possibilities of model construction, several limiting assumptions have to be made. One is that all sets of hom(oe)ologous chromosomes may be pooled. In respect to chiasma frequency differences this has been shown to be a reasonable assumption (Sybenga, 1995a).

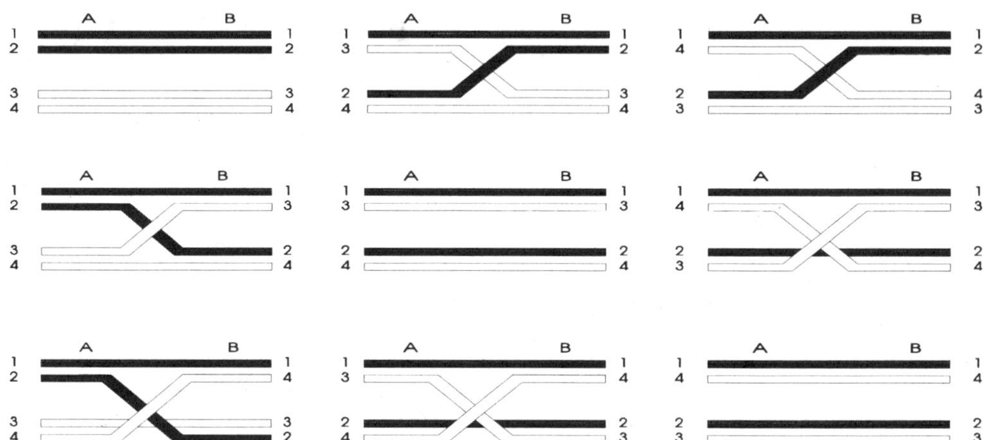

Figure 3. The nine pairing combinations of the four chromosomes (numbered 1, 2, 3 and 4) of an amphidiploid. Three combinations are bivalent pairs, one entirely homologous, the other two entirely homoeologous. Six combinatons are quadrivalents with homologous pairing at one side of the PPS, homoeologous pairing at the other side. Modified from Sybenga (1995a), with permission of CRC Press.

Another limiting assumption is that there are (only) two regions of pairing initiation, located distally. When the chromosomes associated in one arm are not the same as those in the other, there is a pairing partner switch (PPS, Figure 3), resulting in a quadrivalent. This assumption is not made by Jackson and coworkers (Jackson and Casey, 1982; Jackson and Hauber, 1982), who recognize the possibility of several PPS. However, they do not consider the distribution of chiasmata between the PPSs, nor do they consider the effect of more than one PPS on multivalent frequency when only distal chiasmata are formed (Sybenga, 1995a). Callow and coworkers (Callow and Gladwell, 1984; Hamey et al., 1986) have worked out the effects of multiple PPS, but not related to actual diakinesis/metaphase I configuration frequencies with variable chiasma formation.

In view of the often large numbers of PPS observed in synaptonemal complexes (SCs), assuming that only one is formed may seem an oversimplificaton. However, configurations at metaphase I in seem an oversimplificaton. However, configurations at metaphase I in polyploids are almost never of a type ("bird's cage", Darlington, 1965) requiring more than one PPS even in autopolyploids with large numbers of chiasmata (Sybenga, 1995a; Wallace and Callow, 1993). There may be different reasons:

i. Most PPSs exceeding one result from late pairing that is not effective for chiasma formation. This may be the reason why PPSs in excess of two, which increase multivalent frequency at MI even without chiasmata between them (Sybenga, 1995a) do not clearly appear to have this effect.

ii. Chiasma formation is localized, and multiple PPSs tend to occur in regions where chiasmata are rare. This is not expected when effective pairing initiation coincides with commitment to form a chiasma.

iii. PPSs locally suppress pairing and chiasma formation.

Concluding, the assumption that maximally one tends to be effective is an acceptable simplification.

With maximally one PPS, there are nine possible combinations (Figure 3). If pairing is random, all are equally probable. Six have a PPS and result in a quadrivalent. Three lack a PPS, and a set of two bivalents results. For a triploid, a similar diagram can be drawn, but higher polyploids are more complicated (Sybenga, 1995b). In the tetraploid, two of the three bivalents result from fully homoeologous pairing and only one from fully homologous pairing. The quadrivalents all have one arm paired homologously, and one paired homoeologously. The essence of the models is that with affinity differences the two homoeologous bivalents decrease in frequency in favor of more quadrivalents, and the four quadrivalents decrease in frequency in favor of homologous bivalents. There is a net increase of (homologous) bivalent pairing, but not as direct as might intuitively be thought.

The models have two functions:

i. To translate multivalent frequency into relative affinity

ii. To take care of the cases where not all essential segments have a chiasma, and thus their pairing is not recognized at metaphase I.

In respect of treating affinity parameters, different research groups have different starting points. Crane and Sleper (1989a, 1989b), introduce separate affinity parameters for all possible combinations of genomes. Together with the large number of different possibilities of chiasma formation, the number of parameters to be estimated far exceeds the number of degrees of freedom: the models are highly overparameterized. Rigid optimization is necessary to obtain estimates, and merely ranges can be given for the chiasma and pairing parameters. In contrast, Alonso and Kimber (1981), Kimber and Alonso (1981) and many more publications of Kimber and associates in a series including several different authors, directly estimate the chiasma parameter c, (see above) and use the models for estimating a single pairing parameter, x. These models are underparameterized and also require optimization (or a maximum likelihood approach) for obtaining the best single estimate of the pairing parameter. Only one specific situation can be accommodated by each model. If it is not known which situation prevails (is, for instance a triploid could be CCC or CDD or CDE, where different letters represent different genomes), all models, each specific for one combination of genomes, are applied. The model giving the best fit between the observed configuration frequencies and those reconstructed on the basis of the derived c and x, is assumed to represent the actual situation. Later, Chapman and Kimber (1992a, b) introduced refinements, but the basis was not essentially changed (Sybenga, 1995a).

For the triploid Sybenga (1988) considers two situations:

i. The relations between the three genomes are not a priori known, and three pairing parameters must be estimated. This is more than the number of degrees of freedom available, and only ranges can be given.

ii. There are two identical genomes and one is different. The single pairing parameter is estimated by an analytical approach, in addition to two chiasma parameters, one for each arm. For the tetraploid Sybenga (1994) assumes that the specific situation is known at the start. He estimates one pairing parameter, and four chiasma parameters: one for each of the two chromosome arms, and for bivalents as well as for multivalents, a total of five parameters. This number equals the number of degrees of freedom. Analytical solutions are possible, unless the assumptions on which the models are based do not correspond with the observations (Sybenga, 1988, 1994, 1995a).

Whereas preferential pairing and affinity estimates represent the relative relations between the genomes, the chiasma parameters represent, at least in part, the level of the correspondence. This is particularly stressed by Kimber and associates, and it is one of the reasons why they prefer to estimate a single chiasma parameter: the level is assumed to be the same for all arms, long or short. Sybenga (1988, 1994) agrees that chiasmata represent levels, but argues that there are other factors affecting chiasma formation: genetic and environmental factors, chiasma localization, hybrid dysgenesis, etc. Level of affinity and variation in chiasma formation are confounded. In addition, arm length does have an effect, even when only initial pairing is considered. Further, in amphidiploids, quadrivalents have half of their arms homologously paired, half homoeologously. With low levels of pairing differentiation, bivalents are either fully homologous or fully homoeologous. With high levels of differentiation, homoeologous bivalents are rare, and bivalent chiasma frequencies are essentially those between homologous chromosomes. Then only quadrivalent chiasma frequencies are (in part) homoeologous frequencies and of interest for affinity levels, and it is important to distinguish between quadrivalent and bivalent pairing when estimating chiasma frequencies. Finally, the arm length difference is determined solely by the location of the centromere in bivalents, but by centromere location plus the location of the PPS in quadrivalents.

As indicated above, affinity estimates may differ between triploids and tetraploids even when the same genomes are combined, because, for as yet not entirely known reasons, multivalent frequencies are systematically higher in triploids. For more obvious reasons, differences between affinity estimates from different combinations of the same genomes in tetraploids may be found. In an amphidiploid CCDD, all chromosomes can find a homologous pairing partner, and homoeologous pairing resulting in quadrivalents is a matter of simple competition. In the CCCD tetraploid, however, one of the C genomes is consistently free, either as a complete chromosome or as segments, together constituting the equivalent of a complete chromosome. These segments are available for pairing with D without competition from a homologous partner. The result is a slightly higher quadrivalent frequency. A similar effect, but less strong, acts in the CCD triploid: any segment of the CC chromosomes failing to pair, will be available for D, which lacks a homologous partner.

MODEL CONSTRUCTION

Often, the first step in model construction is the formulation of the expected numbers of different meiotic configurations for the autotetraploid with all critical segments having at least one chiasma. This is the situation of Figure 3, but without differences between the genomes. All nine pairing combinations have the same expected frequency and there is either no or one PPS. This basic model is discussed by Sybenga (1975, 1992). Many models agree on this point, although Jackson and Casey (1982) and Jackson and Hauber (1982) recognize the possibility of more than one PPS. However, their treatment of chiasma formation does not satisfactorily deal with it, and they do not actually estimate pairing affinity parameters. Callow and coworkers (Callow and Gladwell, 1984), give pairing models for cases with more than one PPS, but their consideration of the consequences for metaphase configurations is incomplete and not suitable for affinity estimates. The basic autotetraploid model is discussed by Sybenga (1975, 1992, 1995a).

The next step is an attempt to formulate the effect of less than the saturating number of chiasmata. Here differences start appearing between the different models, originating from differences in views on basic chiasma distributions. Driscoll et al. (1979) on which all models of Kimber and associates are based, starting with Alonso and Kimber (1981) and Kimber and Alonso (1981), assume a different chiasma distribution than Sybenga (1965, 1972, 1975, 1992, 1995a), Jackson and Casey (1982), Jackson and Hauber (1982), and Crane and Sleper (1989a, 1989b). Kimber c.s. do not start with a pairing model, but assign the available chiasmate associations directly to different arm pair combinations. The frequency of different arms to associate is determined by affinity relations. The difference with the first mentioned models has effects only for the category of two chiasmata per set of four chromosomes in tetraploids. For higher ploidy levels the consequences of the difference have not sufficiently been analyzed. The problem is discussed in detail by Sybenga (1992, 1995a), and arguments are presented in favor of the first approach.

There is also an essential difference between different models in distinction between chromosome arms. Alonso and Kimber (1981), Kimber and Alonso (1981) and all following papers of Kimber and associates follow Driscoll et al. (1979) in assuming that all chromosome arms are equal in chiasma formation, as do Jackson and Casey (1982), Jackson and Hauber (1982) and following publications. Although Chapman and Kimber (1992a, b) treat the two arms separately, they do not assign different values to them. Sybenga (1965, 1975, 1988, 1994, 1995a, 1995b) distinguishes between the two arms of the chromosomes in respect of chiasma formation, for triploids and higher ploidy levels, both after bivalent and after multivalent pairing. Crane and Sleper (1989a,b) and subsequent papers not only distinguish between arms, but also between different pairing combinations of the same arm, but not between bivalent and multivalent pairing. In the examples of Table 1 it can be seen that assigning different chiasma frequencies to the two arms results in a very much better fit between the observations and the reconstruction made on the basis of the models, with the pairing and chiasma parameters as estimated. It is difficult to judge the effect of distinguishing different chiasma frequencies after different pairing combinations (Crane and Sleper, 1989a, 1989b), but theoretically this should be important. However, it leads to great over-parameterization, and an analytical approach to such estimates is impossible.

In Sybenga's 1975 model, an estimate is made of the frequency (f) of quadrivalent pairing from metaphase I configuration frequencies by correcting for chiasma failure in critical segments of the quadrivalent. From the three multivalent frequencies (ring quadrivalents, chain quadrivalents, trivalents with univalent) and as the fourth class the remaining configurations combined, f can be derived:

$$f = \frac{(triv + 2cq + 4rq)2}{16rq}$$

The arm chiasmate association frequencies after quadrivalent pairing $a(qu)$ and $b(qu)$ can be estimated along similar lines. The bivalent arm association frequencies ($a(biv)$ and $b(biv)$) are derived separately.

From f, the preferential pairing parameter p can be derived: p = 1/3 (4-6f) (Sybenga, 1994), where p ranges between 0 (no preferential pairing, all genomes are identical) and 2/3. The basic frequency of any association of two chromosomes of the four with random pairing equals 1/3 for the amphidiploid of Figure 3. The maximal preferential pairing thus equals p=2/3, when pairing is completely preferential.

If the chiasmate arm association frequencies for bivalent and quadrivalent pairing are identical, f as expressed above should have the same value as when derived from the ratio of the ring quadrivalent frequency (rq) and the square of the ring bivalent frequency ($rbiv^2$). This is usually not the case.

An example of a formula expressing the frequency of a specific configuration on the basis of p and the arm association frequencies is that for chain quadrivalents:

$$cq = 2(a(qu)^2.b(qu) + a(qu).b(qu)^2 - 2a(qu)^2.b(qu)^2).(2/3 - 1.5p^2)$$

where cq is the chain quadrivalent frequency, $a(qu)$ and $b(qu)$ the two chiasmate arm association frequencies after quadrivalent pairing and p the preferential pairing factor. Comparable formulae can be given for the other five configurations, and from these formulae equations are derived from which the five pairing and chiasma parameters, as referred to above, are estimated. The equations for deriving the pairs of arm association frequencies are quadratic equations, and occasionally the discriminant is negative, which points to negative interference (Sybenga, 1975). Then, the difference between the roots, determining the differences between the arm association frequencies, can not be found, and for convenience it is assumed that the two are identical.

The frequency of chain quadrivalents according to Kimber and Alonso (1981) is:

$$cq = 8c^3(1-c).(x^2y + xy^2 + y^3)/(x^3 + 2x^2y + 2xy^2 + 4y^3)$$

where c is the average arm association frequency for all configurations, x is the affinity factor for the most closely related genomes, ranging between 0.5 and 1, and $y = (1-x)$, the affinity between the least related genomes. Comparable formulae are provided for the other five configurations, and from these x is solved by optimization; c can be derived directly.

The formulae given by Crane and Sleper (1989a, 1989b) will not be shown here.

In Table 1, three examples are given: one amphidiploid and two autotetraploids. When all roots can be solved, the reconstruction of the configuration frequencies using the formulae of Sybenga (1994) and the derived parameters, gives an exact fit. When the roots can not be solved, the fit is less exact, as expected. It is seen that the fit of the Kimber and Alonso (1981) reconstructions is poor. This is mainly due to their failure to distinguish between arms. For the *Lycopersicum* amphidiploid the fit is reasonable, and the estimates both for p and x look realistic. For autotetraploid *Tradescantia virginiana* the p value is very low, which is expected. The x estimate, however, is far too large,

Table 1. Observed and reconstructed numbers of configurations in an amphidiploid and two autotetraploids.

Amphidiploid (somatic hybrid) *Lycopersicon esculentum* + *L. peruvianum*: 4x=48; 60 cells; f=0.497; a(qu)=b(qu)=0.796; a(biv)=1; b(biv)=0.521; c=0.778 (Sybenga, 1994)

	rq	*cq*	*triv*	*rbiv*	*obiv*	*univ*	
A.	145	145	41	371	402	5	
B.	143.5	147.3	37.8	377.7	394.3	10.9	p=0.336
C.	112.6	163.5	64.8	461.4	247.5	49.4	x=0.701

Tradescantia autotetraploid cultivar: 4x=24; 50 cells; f=0.658; a(qu)=0.950; b(qu)=0.604; a(biv)=0.935; b(biv)=0.0.836; c=0.814 (Sybenga, 1975, 1994).

	rq	*cq*	*triv*	*rbiv*	*obiv*	*univ*	
A.	65	92	9	161	101	6	
B.	65	92	9	161	101	6	p = 0.077
C.	72.1	72.9	22.1	174.1	78.6	13.2	x = 0.643

Hyoscyamus niger second generation artificial autotetraploid: 4x=68; 206 cells; f=0.236; a(qu)= 0.909; b(qu)=0.909; a(biv)=0.979; b(biv)=0.302; c=0.704; (Lavania et al., 1991, Sybenga, 1994 and unpublished).

	rq	*cq*	*triv*	*rbiv*	*obiv*	*univ*	
A.	595	170	85	1598	3604	102	
B.	564.5	225.8	22.6	1580.5	3718.6	79.2	p = 0.536
C.	221.3	625.2	393.6	2272.3	1769.1	482.5	x = 0.775

A. observed numbers of configurations; B. expected, arm association factors (for both arms after quadrivalent and bivalent pairing); preferential pairing factor p (Sybenga, 1994). C. expected, with affinity factor x and average arm association frequency c estimated according to Kimber and Alonso (1981).
rq: ring quadrivalents; *cq*: chain quadrivalents; *triv*: trivalents with univalent; *rbiv*: ring bivalents; *obiv*: open bivalents; *univ*: pairs of univalents. Extracted from Sybenga (1994).

suggesting that shortcomings of the model not only affect the fit between observed and reconstructed configuration frequencies but also the estimate of affinity.

The completely aberrant behaviour of the induced second generation *Hyoscyamus* autotetraploid is important. The p and x values suggest strong preferential pairing. Yet is is a complete autotetraploid. Possible reasons have been discussed by Lavania et al.

(1991) and Sybenga (1992). The same problem occurs with all autotetraploids with low quadrivalent frequencies at metaphase I. There is no reason to expect that this does not occur in some amphidiploids, where it would be interpreted as true preferential pairing. It is necessary, therefore, for reliable interpretation of *p* and *x* estimates, to analyse the two parents of an amphidiploid at the autotetraploid level, and to compare their behavior with that of the amphidiploid. The latter is made readily as the hybrid between the two autotetraploids. Although it must be admitted that parents and hybrids may well behave differently, the value of preferential pairing and affinity estimates remains uncertain if this precaution is not taken. For other polyploid hybrids, where the genomic composition is different, including triploids (CCD, CDE, CCCD, CCDE, CDEF etc.) a comparison with the autotetraploid parents is less straightforward, but may nevertheless give an indication of the risk involved.

REFERENCES

Alonso, L.C., and Kimber, G., 1981, The analysis of meiosis in hybrids II. Triploid hybrids, *Can.J.Genet.Cytol.* 23:221.

Ashley, T., Cacheiro, N.L.A., Russell, L.B., and Ward, D.C., 1993, Molecular characterization of a pericentric inversion in mouse chromosome 8 implicates telomeres as promoters of meiotic recombination, *Chromosoma* 102:112.

Benavente, E., and Orellana, J., 1989, Pairing competition between identical and homologous chromosomes in autotetraploid rye heterozygous for interstitial C-bands, *Chromosoma* 98:225.

Callow, R.S., and Gladwell, I., 1984, A general treatment of chromosome synapsis in even-numbered polyploids, *J.Theor.Biol.* 106:455.

Chapman, C.G.D., and Kimber, G., 1992a, Developments in the meiotic analysis of hybrids. II. Amended models for tetraploids, *Heredity* 68:105.

Chapman, C.G.D., and Kimber, G., 1992b, Developments in the meiotic analysis of hybrids. V. Second order models for tetraploids and pentaploids, *Heredity* 68:205.

Crane, C.F., and Sleper,D.A., 1989a, A model of meiotic chromosome association in triploids, *Genome* 32:82.

Crane, C.F., and Sleper, D.A., 1989b, A model of meiotic chromosome association in tetraploids, *Genome* 32;691.

Crossway, A., and Dvorak, J., 1984, Distribution of nonstructural variation along three chromosome arms between wheat cultivars Chinese Spring and Cheyenne, *Genetics* 106:309.

Darlington, C.D., 1965, "Cytology". J.& A. Churchill Ltd., London.

Driscoll, C.J., Bielig, L.M., and Darvey, N.L., 1979, An analysis of frequencies of chromosome configurations in wheat and wheat hybrids, *Genetics* 91:755.

Feldman, M., Mello-Sampayo, T. and Sears, E.R., 1966, Somatic association in *Triticum aestivum*, *Proc. Nat. Acad. Sci.* USA 56:1192.

Gill, K.S., Gill, B.S., Endo, T.R., and Friebe, B., 1994, Lack of correspondence between chiasmata and crossovers in wheat, *IV Kew Chromosome Conf.* p. 71.

Goodspeed, T.H., and Clausen, R.E., 1928, Interspecific hybridization in *Nicotiana* VIII. The *sylvestris-tomentosa-tabacum* hybrid triangle and its bearing on the origin of *tabacum*, *Univ. Cal. Publ. Bot.* 11:245.

Hamey, Y., Abberton, M.T., Wallace, A.J., and Callow, R.S., 1988, Pairing autonomy and chromosome size, *in*: "III Kew Chromosome Conference," P.E. Brandham, ed., HMSO London:241.

Havekes, F., Jong, J.H. de, and Heijting, C., 1993, Synapsis and chiasma formation in tomato, Abstracts XVII *Internat. Congr. Genetics* Birmingham p. 132.

Jackson, R.C., and Casey, J., 1982, Cytogenetic analyses of autopolyploids: models and methods for triploids to octoploids, *Am. J. Bot.* 69:487.

Jackson, R.C., and Hauber, D.P., 1982, Autotriploid and autotetraploid analyses: correction coefficients for proposed binomial models, *Am. J. Bot.* 69:644.

Jauhar, P.P., 1990, Multidisciplinary approach to genome analysis in the diploid species, *Thinopyrum bessarabicum* and *Th. elongatum* (*Lophopyrum elongatum*), of the Triticeae, *Theor. Appl. Genet.* 80:523.

Jones, G.H., 1967, The control of chiasma distribution in rye, *Chromosoma* 22:69.

Jones, G.H., 1984, The control of chiasma distribution, *in*: "Controlling events in meiosis," C.W.Evans, and H.G.Dickinson, eds., 38th Symp. Soc. Exp. Biologists; Comp. Biologists Cambridge p. 293.

Jones, G.H., 1994, Meiosis in autopolyploid *Crepis capillaris.* III. Comparison of triploids and tetraploids; evidence for non-independence of autonomous pairing sites, *Heredity* 73:215.

Jones, G.H., and Vincent, J.E., 1994, Meiosis in autopolyploid *Crepis capillaris.* II Autotetraploids, *Genome* 37:497.

Jongedijk, E., Ramanna, M.S., Sawor, Z., and Hermsen, J.G.T., 1991, Formation of first division restitution (FDR) 2n-gametes through pseudo-homotypic division in ds-1 (desynapsis) mutants of diploid potato: routine production of tetraploid progeny from 2xFDR-2xFDR crosses, *Theor. Appl. Genet.* 82:645.

Kihara, H., 1930, Genomanalyse bei Triticum und Aegilops, *Cytologia* 1:263.

Kimber, G., and Alonso, L.C. 1981, The analysis of meiosis in hybrids. III. Tetraploid hybrids, *Can.J.Genet.Cytol.* 23:235.

Lavania, U.C., Srivastava, S., and Sybenga, J. 1991, Cytogenetics of fertility improvement in artificial autotetraploids of *Hyoscyamus niger* L., *Genome* 34:190.

Loidl, J., 1986, Synaptonemal complex spreading in *Allium.* II. Tetraploid *Allium vineale, Genome* 28:754.

Loidl, J., and Länger, H., 1993, Evaluation of models of homologue search with respect to their efficiency on meiotic pairing, *Heredity* 71:342.

Lukaszewski, A. J., and Gustafson, J. P., 1983, Translocations and modifications of chromosomes in triticale x wheat hybrids, *Theor. Appl. Genet.* 64:299.

Maguire, M.P., 1968, Chromosome pairing in altered constitutions and models of synapsis and crossing-over, *Genet. Res.* 12:21.

Maguire, M.P., 1980, Adaptive advantage for chiasma interference: a novel suggestion, *Heredity* 45:127.

Maguire, M.P., 1985, Evidence on the nature and complexity of the mechanism of chiasma maintenance in maize, *Genet.Res.* 45:37.

Nilsson, N.-O., Säll, T., and Bengtson, B.O., 1993, Chiasma and recombination data in plants: are they compatible? *Trends in Genet.* 9:344.

Orellana, J., and Santos, J.L., 1985, Pairing competition between identical and homologous chromosomes in autotetraploid rye. I. Submetacentric chromosomes, *Genetics* 111:933.

Sears, E.R., 1984, Mutations in wheat that raise the level of meiotic chromosome pairing, *in*: "Genetic manipulation in plant improvement," J.P. Gustafson, ed., 16th Stadler Genetics Symp. Columbia MO, pp. 295.

Steinmetz, M., Uematsu, Y., and Lindahl, K.F., 1987, Hotspots of homologous recombination in mammalian genomes, *Trends In Genet.* (January):7.

Sybenga, J., 1965, The quantitative analysis of chromosome pairing and chiasma formation based on the relative frequencies of MI configurations. II. Primary trisomics, *Genetica* 36:339.

Sybenga, J., 1966, The zygomere as hypothetical unit of chromosome pairing initiation, *Genetica* 37:186.

Sybenga, J., 1972, "General Cytogenetics," North Holland/American Elsevier, Amsterdam, London, New York.

Sybenga, J., 1973, Allopolyploidization of autopolyploids. 2. Manipulation of the chromosome pairing system, *Euphytica* 22:433.

Sybenga, J., 1975, "Meiotic Configurations," Springer-Verlag, Berlin Heidelberg New York.

Sybenga, J., 1976, Quantitative variation in chromosome pairing affinities within a species, Secale cereale, *in*: "Current Chromosome Research", K. Jones, and P.E. Brandham, eds., Elsevier/North Holland Biomedical Press p. 143.

Sybenga, J., 1988, Mathematical models for estimating preferential pairing and recombination in triploid hybrids, *Genome* 30:745.

Sybenga, J., 1992, "Cytogenetics in Plant Breeding," Springer-Verlag, Berlin, Heidelberg, New York.

Sybenga, J., 1994, Preferential pairing estimates from multivalent frequencies in tetraploids, *Genome* 37:1045.

Sybenga, J., 1995a, Limitations and pitfalls in the use of quantitative polyploid meiotic models for genome analysis, *in*: "Methods of Genome Analysis in Plants: their Merits and Pitfals," P.P. Jauhar, ed., CRC Press, Boca Raton. in press.

Sybenga, J., 1995b, Meiotic pairing in autohexaploid *Lathyrus*: a mathematical model, *Heredity*, in press.

Sym, M., and Roeder, G.S., 1994, Crossover interference is abolished in the absence of a synaptonemal complex protein, *Cell* 79:283.

Vincent, J.E., and Jones, G.H., 1993, Meiosis in autopolyploid Crepis capillaris. I Triploids and trisomics; implication for models of chromosome pairing, *Chromosoma* 102:195.

Wallace, A.J., and Callow, R.S., 1993, Synaptic responses to concerted genomic evolution in Lathyrus. II. Intragenomic effects, *Heredity* 70:92.

Wang, R.R.-C., 1989, An assessment of genome analysis based on chromosome pairing in hybrids of perennial Triticeae, *Genome* 32:179.

GENOME RESEARCH — IMPLICATIONS FOR SCIENCE AND SOCIETY

R.B. Flavell

John Innes Centre, Norwich Research Park,
Colney, Norwich, Norfolk, NR4 7UH
United Kingdom

Bringing a meeting to a close is very difficult if one is not able to summarise it in a concise intellectually sound way. So I decided to address another issue. I often ask myself "are our genome studies driving the world's societies to be much more overtly based on genetic manipulation and genetic selection?" Of course, all the organisms in existence today are the result of mutation, recombination and selection, because there would be no evolution without these processes, but our forefathers were blessed with not knowing about genetics. It is Darwin, Mendel, and the geneticists of this century who have revealed to us the origins and consequences of variation in organisms. Their hypotheses have induced much controversy. There has been the debate about Lysenkoism and we still have the debate over evolution versus creationism.

We are now obviously in the era of genome projects, genetic analysis and genetic modification on a new unprecedented scale. There are vigorous debates over genetic engineering and the release of transgenic organisms into natural environments. We have many new laws to cover these issues and these laws are becoming pervasive around the world, and are affecting researchers, industries, growers, consumers and parents. We have politicians responding to the fact that people are watching how they vote on gene technology issues. The evidence is abundant, therefore, that <u>firstly</u> our science is creating remarkably more extensive options for changing organisms and the populations that are propagated on the planet to serve us, and <u>secondly</u>, that societies are becoming aware of these options and are expressing concerns. This sets the scene for a renaissance of innovation, for controversial debate, and social chaos.

I believe one of the special things about the revelations of recent molecular genetics is the recognition of harmony across the kingdoms. This week it has been a delight for me a plant scientist to see the technical, biological, and infrastructural systems being developed in human genome studies; to see the conserved synteny bringing together cattle, cat, and human genomes; and to see the conserved synteny across the grass genomes. The sharing of information across the kingdoms is contributing enormously to the power base of genetics and genetic manipulation for the future. The genetic information and technology train is accelerating rapidly and I do not see how it can be stopped. I think it is hard to dispute that we are driving gene technology into the world's societies.

Alleles which predispose human individuals to diseases are being discovered at a rapid rate through the exploitation of molecular markers in family and population studies. Origins of cancer are being understood. Specific alleles are being associated with predictable consequences - for diseases and for social behaviour disorders. These advances bring options for learning our predisposition to degenerative processes during our lives. Do we want this sort of information? Can we as individuals cope with knowing predicted personal shortcomings? These advances also bring options for prenatal diagnosis and the need for decisions on whether to abort fetuses or not. Abortion in most societies is a very sensitive issue. The debate is connected with the right or not to take a life. Assuming abortion is allowed, on what grounds should it be permitted? Should all parents have their genomes fingerprinted to look for suspect alleles? Should all fetuses be screened in rich societies because they can afford it and probably not in the poor societies, because they cannot afford it? What should be done with the information? Who should decide what is a an unwanted allele - the State, the Church, the mother alone or the mother and father. In all societies I think there will be major problems whoever decides, but the mother is perhaps the best placed to decide, within a framework of options defined by the State. Mothers or at least individuals are probably better judges of the value of unborn individuals than governments or the Church which are heterogeneous in beliefs and do not have universal or even majority support. The state needs to provide guidelines but if the State decides to take over the interpretation of genetic fingerprints and the definition of unwanted alleles, and uses this to order abortions, we would start ethnic cleansing and such abhorances.

In the UK, a few months ago, parliament gave the police go-ahead to keep databases of DNA fingerprints. They can be collected for anyone thought to be involved in suspect activity and stored for ever for future reference. The State has used DNA fingerprints to resolve paternity suits, familial relationships of individuals wanting to settle in the UK from other countries and to convict rapists and murderers. Such fingerprints have also, of course, been used to eliminate many innocent people from police investigations. So, genome scientists have clearly added very significant tools for detection of criminals. Will the tools be abused? Will any genetic predisposition to crime be recognised by the courts? Will insurance companies be allowed access to such files and refuse to insure individuals with suspect alleles? These well-known concerns that I have described, and many more, are concerns because knowledge of the genomes of individual citizens can be exploited to undermine freedom of the individual. Furthermore, they could be used to direct breeding programmes on a scientifically rational basis and provide a means for eliminating individual sets of genes from societies. It has been done before, it is happening now in parts of the world, and it will happen again.

Yet for plants, yeasts, bacteria, and domesticated animals - organisms bred by man to help serve his needs, the situation is very different. Breeders of these species for decades, if not hundreds of years, have selected vigorously, thrown away most genotypes and propagated elite forms. Very few people hold ethical concerns about strain selection in plant or microbial breeding. The concerns here are about how we use the organisms. Do the strains carry any forms of toxins for us or other organisms in the environment? Will unhelpful alleles be spread from selected strains into related strains that contribute to valuable ecosystems and so undermine these ecosystems? Will use of the new strains provoke desecration of previous natural environmental systems, or demand the use of unfriendly chemicals? Will use of new strains mean we lose the old ones and in consequence loss of valuable germplasm?

It is usually governments who define systems of when and how organisms are to be used. They must do benefit/risk analyses which are often very complex. The outcome of benefit/risk analyses depend on who you are and where you are living. You and I are perhaps unlikely to get a benefit directly from a transgenic plant that will really change the quality of our life in any substantial way. We are well fed. We do not need to take risks. But people living in poor circumstances in rural economies on subsistence agriculture could benefit. They might be willing to take more risks associated with something beneficial than we whose need is much less. Therefore decisions on what product is worth having need to be made locally and by governments. Yet such local decisions are related to large scale social, economic, and environmental issues - things that affect all societies in the long term and are also connected with global management of trade and the environment. Politicians, indeed all of us, find it increasingly hard to deal with such long term issues of benefits and risks but the Rio Conference of 1992 made it clear that such issues must be on the global agenda.

Genetic diversity, sustainable food supplies in changing environments, healthier food, and a cleaner environment will surely become even bigger issues for societies as more countries become industrialised and the population doubles in the poorer continents. I am convinced that genetic manipulation based on tomorrow's genome research will offer many more options to help manage these problems than the genetics of yesterday and today. Yet the new genetics will continue to frighten some and excite others.

With these examples I am attempting to illustrate that on the one hand individuals, and especially mothers, will need to make big decisions on issues specific to them, provoked by information from genome research, and, on the other hand, governments will need to make big decisions on whether to allow hundreds, thousands, or tens of thousands of new genes to be put into farm animals and crops that will be grown all over the world and be the principal food source of huge populations.

When I ponder the next few decades, I certainly expect there to be a renaissance to innovation, great debate and some chaos coming from the exploitation of our genome research. But I fear the pace of application will be too slow because of fear and uncertainty, and the fear and uncertainty will be there in part because genetics is not understood, and decision makers and scientists will not be trusted. This leaves me with the conclusion that there are things I ought to do, we have to do, with increasing vigour.

1) Keep telling others of what the possibilities/options will be in the future - to provoke early consideration and debate.

2) Teach our students to teach others.

3) Inform school teachers, regularly and diligently. They are so influential that if we had a goal to create genetically literate societies in Europe and the US within 30 years we would have to do it through them.

4) Keep informing politicians and serve as consultants to any opinion - forming bodies to help them find the courage to make the decisions they feel are right, based on sound scientific knowledge and principles.

5) Encourage the establishment of genetic counsellors in societies so that the public, doctors and other opinion formers have professional sources of help to understand the issues behind the decisions they face.

You will have your own list. After seeing the stunning progress reported at the Stadler Genetics Symposium I would hate to see the powerful train of genome research causing unnecessary difficulties and confrontations and indeed tragedies in the lives of individuals and societies because we failed to contribute to making our societies literate in genetics and genome research.

I also think that if the politicians really understood how dependent the planet is on the genetic success of its organisms, they would fund the subject and its public understanding more preferentially.

My conclusion is that we need to change the <u>minds</u> of people, and the <u>genotypes</u> of plants and animals in order to enable man to survive well on this planet over the next century and beyond. Hopefully, changing the minds of people by genetic manipulation or selective breeding will never be entertained. In consequence the genome projects of plants and animals will turn out to be more important to societies around the world than the human genome project, which brings benefit predominatly to families rather than societies in general. This may be the bias of a plant scientist but deserves more debate.

Genome research is on a roll. Listening to it these past few days has convinced me even more that we are driving an unstoppable train. We have got a lot to discover but also a lot to teach and to a huge number of people.

SEG program, 180
selection for compactness, 179
sense Chs transgenote, 161
sense suppression, 159, 164,
    174, 175
        phenotypes, 160
        threshold, 161
sequence
        -based prediction, 185
        comparison methods, 194
        conservation, 206
        conservation analysis,
            180
        diversification, 140
        similarity, 178, 196
Sertoli cell degeneration, 119
sexually heritable, 168
short distance pairing, 286
short interspersed elements
    (SINES), 20
Siberian
        function, 40
        indigenous populations,
            32
simple sequence DNA, 225
Single Strand Conformational
    Polymorphism tests (SSCPs),
    81
sle loci, 142
Sorghum bicolor, 104
sorghum genome organization,
    109
species-specific repetitive
    element, 108
specific combining ability, 95
spontaneous introgression of
    chromosome segments, 292
spreading, 211, 213
spurious
        map inflation, 292
        alignments, 181
        divergence, 281
statistical radiation hybrid
    map, 54
stem-loop structures, 153
STS (sequence tagged site), 60

    marker, 66
Stylonychia, 272
Su(var)s, 227
superclusters, 206
suppression by pairing partner
    switch, 293
synaptic initiation, 284
synaptonemal complex (SC),
    287
synteny
        analysis, 61, 130
        conservation, 129
        map, 61, 129
        mapped markers, 132
synthesized tandem repeats
    (STR's), 3

—T—

tagging genes, 59
Taraxacum officinale, 9
TBLASTN, 181, 193
T-box, 250
T-Chs41, 162
T-DNA vectors, 254
tetraploidization, 105, 145
tobacco MAR, 250
topoisomerase II, 251
tourist elements, 153
TPR, 196
TPR motif conservation, 197
Tradescantia virginiana, 299
trans-acting factors, 243
trans-acting PEV modifiers, 227
transcript map, 53
transcription, 243
transcription factors, 160
transcriptional silencing, 259
transgene manipulations, 222
transgenes, 159
transgenic organisms, 245
transgenotes, 167
trans-inactivation, 218, 259
translocations, 108, 144
transposable elements, 221
Triticum aestivum, 282
Type 1 insertion sequences,
    224

Type I cognate sequence, 225
Type I markers, 131
Type II markers, 131

—U—

undermethylated, 154
under-representation of DNA
  sequences, 232
universal PCR primers, 121
universal Type I (coding gene),
  115
uridine monophosphate
  synthase deficiency (UMPS),
  134
Ursus americanus, 20
Ursus arctos, 20

—V—

V. radiata, 140
V. unguiculata, 140
variable number tandem
  repeat (VNTR), 1,15, 31, 45
viral carcinogenesis, 117

ZOO-FISH, 133
zygomere, 284

—W—

weaver disease, 134
wheat, 61
white gene, 217
white pine blister rust, 96
whole genome sequencing, 67

—X—

X chromosome inactivation, 260

—Y—

YAC, 64, 156
*YAC and cosmid ordered
  libraries*, 64
YACs, 75
yeast MAR, 250
yeast proteins, 179

—Z—

Zea mays, 104